ERRATUM

Green Manure/Cover Crop Systems of Smallholder Farms

Experiences from Tropical and Subtropical Regions

M. Eilittä, J. Mureithi and R. Derpsch (eds.)

ISBN 1-4020-2045-7

The caption of the bottom photo on p. xii was partly printed in the photo and is therefore illegible. It should read:

Mr. Harrison Chongwe of Chipata, Eastern Zambia, shows off a maize cob from a field planted after an improved fallow with *Gliricidia sepium*. In seasons subsequent to the initial 2 years of fallowing, farmers intercrop *G. sepium* with maize and cut it back several times a year. *By* Rosa Katanga, ICRAF, Zambia.

Kluwer Academic Publishers • Dordrecht / Boston / London

Green Manure/Cover Crop Systems of Smallholder Farmers

Experiences from Tropical and Subtropical Regions

Edited by

Marjatta Eilittä

Joseph Mureithi

and

Rolf Derpsch

KLUWER ACADEMIC PUBLISHERS
DORDRECHT / BOSTON / LONDON

A C.I.P. Catalogue record for this book is available from the Library of Congress.

ISBN 1-4020-2045-7 (HB)
ISBN 1-4020-2051-1 (e-book)

Published by Kluwer Academic Publishers,
P.O. Box 17, 3300 AA Dordrecht, The Netherlands.

Sold and distributed in North, Central and South America
by Kluwer Academic Publishers,
101 Philip Drive, Norwell, MA 02061, U.S.A.

In all other countries, sold and distributed
by Kluwer Academic Publishers,
P.O. Box 322, 3300 AH Dordrecht, The Netherlands.

Printed on acid-free paper

Cover photo: Mr. Doroteo Chima, Anselmo Pascual (at rear), Adolfo Ramirez, Pedro Lopez, Saúl Melo and Juan Ramírez harvesting maize in December in the field of Pedro Lopez in the community of Santa Rosa Sintepec, municipality of Huyepan Ocampo, Veracruz, Mexico. Maize has been planted in June and *Mucuna* has been relay-intercropped between maize rows in August. *By:* Marjatta Eilittä.

Printed in the Netherlands.

TABLE OF CONTENTS

Preface

R. Puentes

The Rockefeller Foundation

On 12 April 1997, over 100 researchers and development practitioners from 20 countries in South and Central America, Mexico, East and West Africa and South-East Asia gathered in Florianópolis in the state of Santa Catarina, Brazil. From this state capital situated on an island just off the mainland, they travelled west by bus to Chapecó, near the border with Paraguay, a 400-km trip. On the way, and divided into three groups, they visited farmers in small towns and villages to learn about the ways in which they were practicing systems with conservation tillage and green manure/cover crops (GMCCs). Arriving in Chapecó, they familiarized themselves with the Empresa de Pesquisa Agropecuária e Difusão de Tecnologia de Santa Catarina (EPAGRI), the local research institute, before embarking on the last 4 days of the workshop, Cover Crops and Green Manure Systems for Small-scale Farmers of Tropical and Subtropical Regions.

The location of the event was by no means coincidental. For almost two decades, research and development efforts in Santa Catarina had focused on GMCCs and conservation tillage (see Chapter 1). Following the pioneer work of a small group of progressive farmers and the findings of a team of committed researchers, new GMCC adaptations emerged and rapidly disseminated throughout the state. Individuals and organizations from other countries and regions were eager to find out how the Brazilian farmers had incorporated these technologies in their systems and the results of their innovations.

The workshop was noteworthy for the diversity of its participants, from development specialists working in small non-governmental organizations in the Mexican states of Oaxaca, Chiapas and Veracruz, to researchers from national systems in Kenya, to experts from international centres in Indonesia. The event included multiple field visits, case study presentations from different regions, small working group discussions and exhibitions of farm equipment. These formal activities captured the attention of the visitors but equally important became the informal exchanges of information and experiences. Discussions were rich and visitors left the meeting with renewed enthusiasm and with new knowledge. Many also left with more concrete souvenirs: the various demonstrations of tools and machines developed by the Brazilian farmers and local craftsmen enthused several of the participants so that they boarded their airport buses carrying a *matraca*—a locally invented, affordable and highly useful hand-held seeder.

The exchange of experiences and information in Chapecó was the seed for a new project: the Green Manure/Cover Crop Exploration. Funded by the Rockefeller Foundation, and comprising both researchers and development specialists working on GMCC systems around the world, the programme supported processes of documentation, information sharing and identification of research and development priorities on the topic. The focus was on resource-poor farmers in tropical and subtropical regions and the specific components of the programme emerged from the discussions at Chapecó. One of the outcomes of the programme was the publication of the Catalogue of Green Manure/Cover Crop Systems by the International Centre for Information on Cover Crops (CIDICCO, the Spanish acronym) that is reviewed in Chapter 12 of this volume. Another decision was to launch a project to review and further develop the potential of *Mucuna pruriens* (L.) DC. as a food and feed crop. This project produced two volumes reporting the work of multiple research teams and the discussions at two workshops held in Honduras and Kenya. Another output of the GMCC Exploration Project is this book of case studies analysing diverse GMCC systems of smallholder farmers in tropical and subtropical regions.

From the outset, the purpose of these case studies was to describe a range of GMCC species, cropping systems, regions where they are practiced and, most important, to analyse and better understand the performance of these systems in farmers' fields. Systems' sustainability and productivity are key issues and it had already become clear by the late 1990s that farmers did not always respond enthusiastically to GMCC systems. In some cases, adoption has been slow; in other situations, abandonment has also occurred. But Chapecó showed that the GMCC approach could have a highly significant and positive impact on the livelihoods of smallholder farmers. The cases and the concluding chapters in this book try hard to explore reasons for farmer interest or disinterest and to integrate aspects of, and lessons from, both development and research.

Hopefully, the case studies in this book will benefit the work of a wide group of researchers and development practitioners in tropical and subtropical regions, including those that focus their work on agriculture, livestock, agroforestry and natural resource management. The idea for the publication emerged at Chapecó. Five years later, those who worked hard to bring it to reality, authors and editors, expect that the book will be a different type of *matraca*—one that will disseminate knowledge and experiences and that will hopefully promote further developments and adaptations around the GMCC concept for the benefit of resource-poor farmers around the world.

Acknowledgements

The Editors

This volume of case studies is an outcome of the Green Manure/Cover Crop (GMCC) Exploration Project, which brought together an international group of researchers and development workers with experience in GMCC systems to analyse past work and to deliberate on future directions. The project's members initiated the collection of these case studies on GMCC systems, created the guidelines to help document them and reviewed and commented on them. We would like to gratefully acknowledge the contributions of the Project's members to this book. The members of the team were: Leandro do Prado Wildner (Empresa de Pesquisa Agropecuária e Difusão de Tecnologia de Santa Catarina [EPAGRI], Santa Catarina, Brazil), Albert Etèka (Clearinghouse for Cover Crops Information and Seed Exchange in Africa [CIEPCA], Benin), Phrek Gypmantasiri (Chiang Mai University, Thailand), Milton Flores (International Centre for Information on Cover Crops [CIDICCO], Honduras), Robert Gilbert (Post-doctoral Fellow funded by the Rockefeller Foundation, Malawi), Joseph Mureithi (Legume Research Network Project [LRNP], Kenya Agricultural Research Institute [KARI], Kenya), Ruben Puentes (Rockefeller Foundation-Mexico), Ken Schlather (Cornell University, NY, USA), Sheryl Swinkel (Cornell University, NY, USA), Bernard Triomphe (Post-doctoral Fellow funded by the Rockefeller Foundation, Mexico) and Marjatta Eilittä (Post-doctoral Fellow funded by the Rockefeller Foundation, FL, USA).

The case study authors faced the challenge of integrating materials and experiences from research and development into a coherent study describing the on-farm performance of the GMCC systems. We gratefully acknowledge their perseverance and patience throughout the process of writing and revisions, and recognize their commitment to improving the productivity and sustainability of smallholder farming in the tropical and subtropical regions.

Farmers throughout these regions have developed and experimented with GMCC systems, often leading and always being the most important actors in the efforts described here. Most researchers, including the editors, have gained great inspiration and determination from the example of strong, community-minded farmers who strive to improve not only their own situations but also the future of their communities. The editors acknowledge their gratitude to the farmers whose systems we here describe and analyse.

This volume of case studies would not have been possible without the funding of the Rockefeller Foundation and particularly the efforts, constant encouragement and support of Dr. Ruben Puentes. The Foundation funded the GMCC Exploration Project, supported many of the research/development projects described in this volume, made possible a team residency for the GMCC Exploration Team in Bellagio, Italy, provided a post-doctoral fellowship for the first editor and helped fund the publishing of this book. The constant and kind assistance of Ms. Pilar Palacia (Rockefeller Foundation-Mexico) with arrangements for meetings and many other practicalities is gratefully acknowledged.

Skilfully and cheerfully, Annie Jones carried out the language editing and carefully formatted the camera-ready version of this book. Drs. Robert Carsky and Robert Gilbert gave detailed comments on the chapter Learning from the Case Studies. The geographic information systems (GIS) laboratory of the Centro Internacional de Agricultura Tropical (CIAT) in Colombia generated the maps depicting the case study locations. Various authors provided photographs and the photographers are acknowledged in the photo credits. The editors also fondly acknowledge the patience and support of their families during the preparation of this volume of case studies.

Legume scientific names were checked through the International Legume Database and Information Service (ILDIS) database (www.ildis.org) and non-legume names through the VAST (VAScular Tropicos) database of the Missouri Botanical Gardens (http://mobot.mobot.org/W3T/ Search/ vast.html).

Workers planting maize in EPAGRI training center in Chapecó, Santa Catarina, Brazil, as a part of training in minimum tillage of maize. In the black oat residues, openings are made where the planting is done with a *matraca*, a locally developed hand-held seeder for large-seeded crops. *By* Claudino Monegat, EPAGRI, Brazil.

The late Mr. Francisco Montoya in the community of San Judas, municipality of El Corpus, department of Choluteca, Honduras, demonstrating the mixture of seed types he plants in a *tapado* field and enthusiastically explaining that despite the large biomass of the cover and the regrowth of the natural vegetation, it is possible to get a good harvest. *By* Milton Flores Barahona, CIDICCO, Honduras.

Mr. Daniel Romero from Soroguara, Fransico Morazán, Honduras, demonstrating to the visitors the thick *Mucuna* mulch in his maize field in November, just prior to slashing time. Mr. Romero has been cultivating *Mucuna* in the field for 7 years. *By* Milton Flores, CIDICCO, Honduras.

[illegible]rison Changwe of C[illegible] Zambia, show[illegible] maize [illegible] field planted after an improved fallow with *Gliricidia sepium*. In seasons subsequent to the initial 2 years of fallowing, farmers intercrop *G.sepium* with maize and cut it back several times a year. *By* Rosa Katanga, ICRAF, Zambia.

Farmers in southern Malawi participating in nationwide on-farm cropping system trial comparing unfertilized hybrid maize (background) with CG7 groundnut (foreground). *By* Robert Gilbert, the Rockefeller Foundation, Malawi.

Farmers in Kikambala, Kilifi district, coastal Kenya, slashing in the Napier-*Gliricidia* alley field. Napier and *Gliricidia* fodder are high-quality feed for dairy cows and were introduced to the region's farmers as substitutes to the expensive commercial dairy concentrates. *By* Hemedi Mkuzi Saha, KARI, Kenya.

Selecting and cracking of *Mucuna* seeds during a demonstration of ways to utilize them in foods in the community of Zouzouvou, Mono province, southern Benin. In this region with high land pressure, efforts were also undertaken to develop the food utilization of *Mucuna* to increase the demand for the planted fallow technology. *By* Mark Versteeg, IITA, Benin.

Mr. Utai Chansai from San Sai district, Chiang Mai Province, Thailand, using a two-wheel tractor with fitted seat to incorporate 35-day old *Sesbania rostrata* prior to planting rainy season rice in early August. Mr. Chansai has shortened the *Sesbania* growth period so that he can plant rice at the optimal date; substantial rice yield increases have been reported even with sub-optimal growth of *Sesbania*. *By* Budsara Limnirankul, Chiang Mai University, Thailand.

Introduction

The Editors

OVERVIEW

The world's population, estimated at over 6 billion, is expected to reach 8 billion by 2020 (United Nations 2000), with almost all of the growth taking place in developing countries. Much of this increase is occurring in sub-Saharan Africa and South Asia, two regions with high rates of poverty, where in 1998, 40-48% of the population lived on less than US$1 per day (World Bank 2001). The higher population necessitates greater food production. This, however, will be increasingly difficult because of a number of factors, including declining trends in yield gains in recent years, decreased cropping land per capita and a host of other challenges of biophysical, economic, and policy-related nature.

Future ability to provide for and improve food security and livelihoods of the poorest is also threatened by the quickly declining soil fertility in many developing countries. It has been estimated that worldwide almost 40% of the crop land is degraded to some degree (Oldeman et al. 1991) and significant impacts on productivity are taking place on about 16% of land (Scherr 1997). The reasons for such degradation are complex and include farming practices that promote soil degradation, such as continuous or almost continuous cropping and insufficient or no fertilizer use. This decline in soil fertility most affects the poorest, because they are especially dependent on agriculture, on common property resources (which often are more degraded), on annual crops (which tend to degrade soil more), and they are often unable to make the investments needed to improve land (Scherr and Yadav 1997).

During the 1980s and 1990s, farmers, development organizations and researchers looking for solutions to the problems of productivity and soil fertility decline on smallholder farms turned increasingly to green manure/cover crops (GMCCs) throughout the tropics. Many GMCC species were screened for their soil-improving characteristics and the impacts of diverse GMCC systems on main crop yields were studied in on-station and on-farm conditions. These systems were often researched and diffused as one of the technologies in the Integrated Nutrient Management or Integrated Soil Fertility Management approach, which combines the use of inorganic and organic sources of nutrients, among other technologies. At times, GMCCs were diffused as one of the technologies in low-input, sustainable, or organic production, which often shunned the use of inorganic fertilizers.

In the second half of the 1990s, numerous efforts were also made to collect and synthesize the quickly accumulating information and knowledge on GMCC systems. One such was the Rockefeller Foundation's GMCC Exploration Project (1998-99), whose aim was to document new experiences on these systems – whether from research or development efforts – and make them available to those working in different regions. An equally important objective was to learn lessons for the benefit of future research and extension efforts.

GMCCs are by no means a new technology; on the contrary, they have been utilized for thousands of years and were a common technology in the twentieth century, both in many industrial and developing countries. Today's conditions offer both opportunities and challenges for GMCCs in tropical and subtropical smallholder farms that need to be understood and considered before the impact of GMCCs on farm can be substantial.

This volume is a direct outcome of the GMCC Exploration Project of the Rockefeller Foundation. As a part of the project, different authors researched and documented 12 diverse cases of field experience with GMCC systems. They are published in this book along with a brief chapter on lessons learnt. Since these experiences and their analysis took place in the context of GMCC work in the 1990s, this context and the case studies are briefly discussed below.

GREEN MANURE/COVER CROPS AND RECENT EFFORTS

Promise of GMCCs and Today's Challenges

The concept of GMCCs is broad, to the point of being vague, with different individuals emphasizing different aspects – or considering that different aspects of the concept are essential to the definition. For some, GMCCs should be leguminous crops because of the central role and common deficiency of N in smallholder tropical fields. For others, GMCCs only have significance if combined with minimum tillage.

For the purpose of this volume, we have adopted the following, relatively wide, definition for GMCCs[1]:

'The use of leguminous or non-leguminous plants (e.g. natural or improved fallow with several species) as ground cover and canopy, in various temporal and spatial configurations used in crop or animal production systems. The purpose of using these species is to improve one or more of the following: erosion control, availability and recycling of N, P, K, Ca and other nutrients, soil moisture and water infiltration, and weed and pest control. Improvement of human and/or animal diet and income may be additional goals.'

The definition accommodates the diverse systems included in this book. It emphasizes the fact that, while any kind of plant may be used as a green manure/cover crop, a key reason for utilizing that particular plant often is for soil improvement or weed suppression.

GMCCs are a well-tested and proven technology. Both the ancient Chinese and the early Romans cultivated crops that were ploughed under to benefit soil fertility. Indigenous green manuring or mulching systems were developed also in tropical regions, such as the utilization of a *Mucuna pruriens* (L.) DC. species for soil fertility improvement in Java, Bali and Sumatra in the seventeenth century (Buckles 1995) and the slash-and-mulch, *frijol tapado* system, discussed in Chapter 2, which is a pre-Hispanic system on the hillsides of Central America. In England, ley farming, the rotation of forages with annual crops, was common in the nineteenth century for the double purpose of providing forage for livestock and maintaining soil fertility; later the practice was introduced to Australia. In the early twentieth century, in the USA and many tropical regions, GMCCs were introduced from other places, screened and cultivated by farmers; as with the ley system, a goal additional to soil fertility was the development of nutritious livestock feeds (Buckles 1995, Eilittä and Sollenberger 2001). Utilization of GMCCs quickly declined after the Second World War as conditions to their continued cultivation became less favourable; particularly influential were the good availability and low price of inorganic fertilizers.

Today's conditions in developing regions present both particular opportunities and challenges for a beneficial impact and increased adoption of GMCCs. Decreasing soil fertility offers a motivation and impetus for their adoption. According to various estimates, 5 to 7 million hectares of farmland are lost every year because of soil degradation (Scherr and Yadav 1997). Examples of areas with severe soil degradation abound; for example, in Central Paraguay, a former granary of the country, has now such degraded soils that many annual crops cannot be economically produced (Florentin et al. 2001). Recent International Food Policy Research Institute (IFPRI) and World Resources Institute (WRI) analysis suggests that almost 75% of agricultural land in Central America, 20% in Africa and 11% in Asia is seriously degraded (Wood et al. 2001). The quickly deteriorating soil fertility would seem to offer an impetus for the adoption and cultivation of GMCCs. In sub-Saharan Africa alone, it was estimated in 1990 that cumulative per hectare losses of nutrients over the past 30 years had been 660 kg N, 75 kg P and 450 kg K (Stoorvogel and Smaling, 1990). In Africa, over 8 million tons of nutrients, equal to $US1.5 billion, are annually lost (Henao and Banante 1999).

In addition, the potential of GMCCs is greater now because the high price of inorganic fertilizers puts them, in sufficient quantities to sustain crop production, beyond the reach of most smallholder farmers. In the 1990s, fertilizer use stagnated at extremely low levels in sub-Saharan Africa, with only about 1.1 to 1.3 million metric tons of N, P_20 and K_20 consumed annually between 1988-89 and 1998-99 (IFDC 2003). In contrast, consumption in North America was over twentyfold and in

Western Europe somewhat less. Such low levels of fertilizer use represent a serious threat to sustainability because the crops 'mine' the soil of its nutrients unless they are replaced with plant residues, manures or fertilizers.

Challenges to the adoption and continued use of GMCCs are also evident. The availability of arable land is quickly decreasing, making it difficult to insert GMCCs in the cropping systems. The additional labour needed for the management of GMCCs may not be available. The effects of HIV-AIDS (Human Immunodeficiency Virus-Acquired Immuno-deficiency Syndrome) may become devastating, especially in many African countries. FAO (2001) estimates that, between 1985 and 2020, 23 million agricultural workers will have died from AIDS in the 27 most affected African countries and that, in the 10 most affected countries, the agricultural labour force will decrease by 10-26%. In other locations, as in Mexico and Central America, rural-urban migration is reducing the available labour force, making difficult the practice of resource-conserving technologies (Garcia-Barrios and Garcia-Barrios 1990). Numerous other factors are challenging or limiting the cultivation of GMCCs by smallholder farmers, from low producer prices of agricultural goods, to lack of seed, to the short span of projects that work on these crops.

Exchange and Synthesis of Information on Green Manure/ Cover Crops in the 1990s

In the 1990s, as extensive efforts on GMCC systems were undertaken[2], it became evident that avenues for information exchange were often lacking and that improved regional- or country-level coordination would be beneficial. Diverse media were established to accomplish this, from listservers, to databases, to networks and information clearinghouses. Because they provided much of the framework for the exchange of information on, and experiences with, GMCC systems in the 1990s, they are briefly reviewed here.

A number of listservers focusing on GMCCs were initiated in the 1990s, for example, Mulch-L (in English), Coberagri-L (in Spanish), and EVECS-L (in French)[3], and discussion on them was active throughout the decade. Additionally, some relevant databases were constructed. Examples are LEXSYS (developed by the International Institute of Tropical Agriculture [IITA]), Organic Resources Database (by the Tropical Soils Biology and Fertility Programme [TSBF], Wye College, University of London and Kenya Agricultural Research Institute [KARI]-Muguga Soil Chemistry Lab), and those by the Natural Resources Institute (NRI) (Kiff et al. 1996) and University of California-Davis.[4]

Several regional networks/clearinghouses established during the 1990s had GMCC systems as an important component of their work. In **southern Africa**, since 1994, the Soil Fertility Network has conducted studies from farming systems diagnostics to long-term experiments on productivity trends and more process-oriented research; in recent efforts, multi-disciplinarity and scaling up have been emphasized. In **eastern Africa**, researchers from the KARI and certain non-governmental organizations (NGOs) have been working together since 1994 in the Legume Research Network Project (LRNP).[5] The early emphasis on identifying suitable 'best bet' legume species has evolved into research on the management of GMCCs within the country's cropping systems. In addition to research in 11 sites across Kenya, LRNP distributes information, conducts seed bulking and publishes a biannual newsletter.

In **West Africa**, the Centre for Cover Crops Information and Seed Exchange in Africa (CIEPCA, the French acronym) at IITA was established in 1997 to distribute information on cover crops and to multiply seed.[6] CIEPCA operated through country coordinators, published a newsletter (in English and French) and led an inter-institutional project on the food and feed utilization of *Mucuna*. In **Central America**, the International Centre for Information on Cover Crops (CIDICCO, the Spanish acronym), founded in 1990 in Tegucigalpa, Honduras, strives to identify, document, disseminate and promote research on GMCCs for smallholder farmers through various publications, including a newsletter and an informal information network. Finally, in **South-East Asia**, the Interim Information Support for the South-East Asian Regional Network on Soil Fertility and Improved Fallow Management provides an avenue for exchange of information on soil fertility, fallow management and shifting cultivation in the South-East Asian tropical upland areas. It is implemented by the University of the Philippines Los Baños Foundations Inc., in collaboration with the World Agroforestry Centre (ICRAF) – Regional Research Programme and the International Fund for Agricultural Development (IFAD). Worldwide, the Management of Organic Inputs in Soils of the Tropics (MOIST) Group at **Cornell University** investigates and exchanges information on cover crops, green manures, managed fallows and mulches in tropical farming systems.[7]

Taking Stock of Experiences with Green Manure/Cover Crops

By the late 1990s, experiences with GMCCs had accumulated worldwide. Clearly, some potential was evident. Because GMCC systems require few external inputs (often only seed in the first year of use), they seemed to be a technology well suited to the smallholder farmers. The integration of GMCCs appeared to reinforce synergies within mixed cropping systems, such as intercropping, and importantly, the GMCC systems addressed the problem of declining soil fertility that is worsening rapidly on many smallholder farms. While diverse experiences were available, little synthesis of information had been done.

Increasing experience led to attempts to synthesize information with a view to improving future research and development efforts. Regional syntheses were undertaken in West Africa, where a workshop was held in

October 1996, and in South-East Asia, where a workshop was held in June 1997 (Cairn 1997, Buckles et al. 1998), in south-eastern Mexico, where an inter-institutional assessment on GMCC systems of smallholder farmers was conducted in 1996-97 (Arteaga et al. 1997) and, in the form of overview articles, in Central America (Flores et al. 1997) and southern Brazil (Calegari et al. 1997). A global workshop, Green Manure-Cover Crop Systems for Smallholders in Tropical and Subtropical Regions, was held in Chapecó, Santa Catarina, Brazil, 6-12 April 1997.

Another global effort was the Green Manure/Cover Crop Exploration Project of the Rockefeller Foundation-Mexico (1998-99), a direct precedent to the work described here. One of several projects of the Agricultural Sciences Division (now Food Security) to explore initiatives for future programs, it brought together researchers and development specialists from diverse institutions working on GMCC systems.[8] While the ultimate purpose of the GMCC Exploration was to contribute to the strategic discussion within the Foundation, it also aimed at improving the general understanding of GMCCs. Its main objectives were to describe the extent of GMCC cultivation, to approximate the potential adoption and impacts of GMCC systems, to investigate and identify factors that affected their adoption and impact, and to examine how research could contribute to their improvement. The project primarily worked through e-mail communication among the participants but met three times over the course of the project; the final meeting was held in Bellagio, Italy, in June 1999.

An important specific goal of this project was the collection and synthesis of information and experiences that had been gained on GMCCs worldwide. To that end, a survey of GMCC systems of smallholder farmers was conducted (see Flores and Janssen, this volume). Another such effort was the preparation of in-depth case studies. These cases describe diverse experiences with smallholder GMCC systems in different agro-ecosystems and socio-economic settings in the tropical and subtropical regions.

THE CASE STUDIES

Process and Guidelines

To document these experiences, individuals who were closely familiar with a particular experience with GMCC systems were contacted and asked to review and document it in the form of a case study.

To ensure that important areas of interest would be covered, the GMCC Exploration Project members created guidelines for documenting the case studies. The six main areas in the guidelines were:

(1) Structure and functioning of the GMCC system: crops, products and productivity, management and its variations, costs and benefits, and associated problems;

(2) Use and adoption by farmers: characterization of adopters, trends in adoption, perceived benefits/drawbacks, diffusion strategies, other GMCC systems in the location;
(3) Critical factors affecting number of users and performance of the system;
(4) Seed availability and exchange: types available, ways to maintain seed supply, problems with seed availability, farmer sales/exchanges and seed markets;
(5) Research activities and results: characterization of projects, research needed; and
(6) Institutional actors: main institutions and their activities, inter-institutional collaboration.

Typically, no additional field research was conducted before writing up the cases, since most of the main authors had either initiated the work on the system(s) or had worked with the system(s) for several years. However, in three instances (Chapters 7, 9 and 10), where the first authors had no close prior familiarity with the systems, on-site interviews and data collection were conducted. Most of the cases were researched and/or written in 1999-2000 but more recent information often has been included.

Because the aim of the case studies is to describe these particular experiences of research and/or diffusion from a broad, multi-disciplinary perspective, they typically do not utilize only published research results, nor do they necessarily include an exhaustive list of publications on the topic. Instead, they contain descriptions of diffusion efforts and impressions. Those emphasizing a research viewpoint on the system may therefore find information on research aspects deficient; those looking mainly for development experiences may find that the development context has been neglected at the expense of research. Hopefully, the combination of the two better reflects the real-life experiences with these systems.

An Overview of the Cases

The cases in this volume come from 11 countries in Latin America, Africa and Asia (Figure 1, Table 1). They are a diverse collection of experiences, including those developed by farmer innovation and through research; those that span a few years and those that span hundreds of years; and those that include mainly research, mainly diffusion, or both. The cases are briefly reviewed below.

Four cases come from Latin America. The first, Use of Green Manure/Cover Crops and Conservation Tillage in Santa Catarina, Brazil, by L. do Prado Wildner, V. Hercilio de Freitas and M. McGuire (Chapter 1), brings together a considerable amount of literature on, and experiences with, diverse GMCCs and conservation tillage in the state of Santa Catarina in southern Brazil. This hilly state, with high and well-distributed rainfall but soils that are susceptible to erosion, produces 13% of the commercial agricultural output in Brazil, although it accounts for only 1%

of the national territory. Since the late 1970s, efforts in the state have focused on GMCCs and conservation tillage and have involved various research and diffusion organizations. Diverse ways of utilizing a large number of both leguminous and non-leguminous GMCCs in different seasons and cropping arrangements have been developed for the major crops. Machinery adapted to GMCC management and conservation tillage also has been locally developed. The area in GMCCs in the state has greatly increased in past decades, from 5% of the arable cropland in 1987 to 44% in 1997, and adoption continues to increase.

Chapter 2, Slash-and-Mulch System: *Fríjol Tapado* in Costa Rica, by G. Meléndez describes the functioning of this fallow system that has been practiced in Central America since pre-Hispanic times. Relying on short, typically 1- to 2-year fallows in between 2-3 years of crop production, the system is practiced in hillsides of humid regions, mainly for bean (*Phaseolus vulgaris* L.) production. After site selection, management consists of only slashing paths through the vegetation, seeding and slashing of vegetation over the bean seeds; the field is then left untouched until harvest. In 1980, one half of Costa Rica's beans was produced under this system but, by the late 1990s, the share had reduced to one quarter. The system has been relatively thoroughly researched in the 1990s and, consequently, a substantial amount of information is available on its sustainability and productivity as well as the effect of various interventions on them.

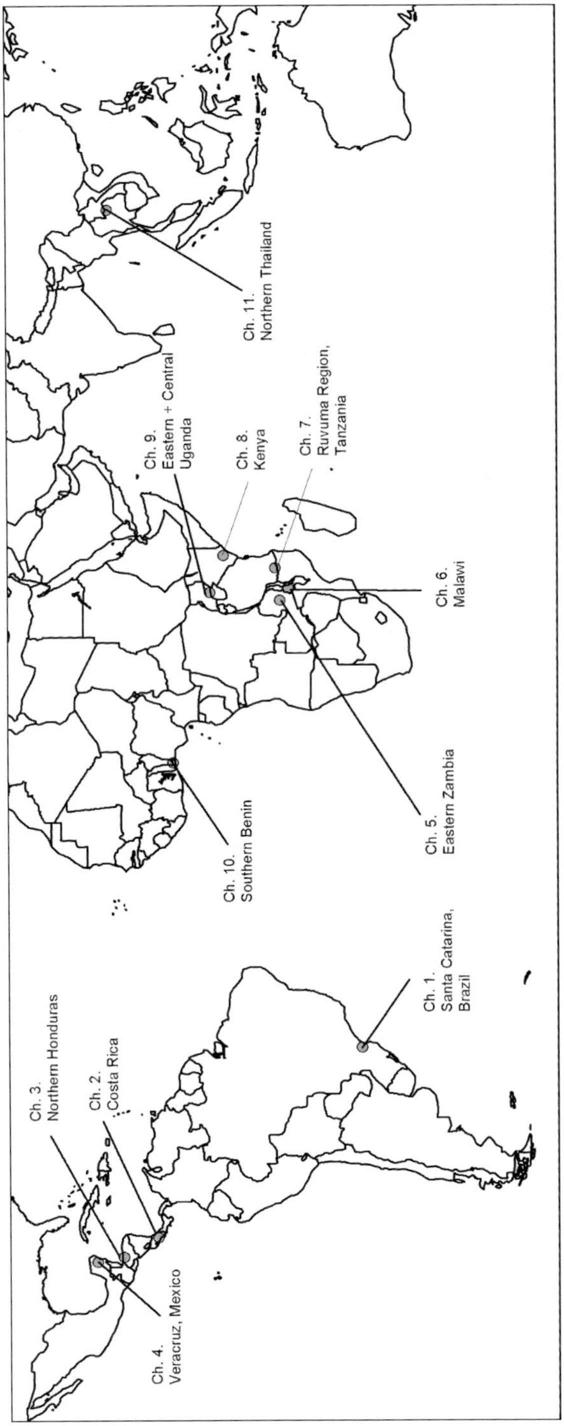

Figure 1. Locations of the 11 case studies in this volume.

Table 1. Green manure/cover crop (GMCC) systems described in the case studies and the environments in which they are practiced.

Chapter	Main crop species	GMCC species	Annual rainfall (mm)	Period of rainfall (months)	Rain-fed cropping seasons (no.)	Farm size	Tillage and burning in cropping system	Level of mechanization
1. GMCCs-Cons. Till. Brazil	Maize, bean, tobacco, onion, etc.	Diverse legume and non-legume species	1200-2370	Year round	Year round	36% <5 ha; 51% 5-10 ha	Previously tilled; increased use of conservation tillage with GMCCs	25% own tractor or mini-tractor; many implements of small type
2. *Tapado* Costa Rica	Bean	Spontaneous vegetation	800-3500	April-Nov, some areas, year round	Two, some areas, year round	Great majority <10 ha	None	None
3. Maize-*Mucuna* Honduras	Maize	*Mucuna*	2000-3000	June-Nov, Dec-May	Two	70% 2.0-7.7 ha	No tillage; burning before first season, no burning before second season	None
4. Maize-*Mucuna* Mexico	Maize	*Mucuna*	1200-2500	June-Dec	Two	Variable; typically 2-7 ha	No tillage except in some low-lying areas; burning before first season	In most areas none
5. Improved fallows Zambia	Mainly maize, also cotton, groundnut, sunflower	*Sesbania*, *Gliricidia*, *Tephrosia*, pigeon pea	600-2000	Nov-April	One	Avg. 2-3 ha	Ridging	Some ridge with oxen, others with hoe

Continued.

Table 1. (Continued)

Chapter	Main crop species	GMCC species	Annual rainfall (mm)	Period of rainfall (months)	Rain-fed cropping seasons (no.)	Farm size	Tillage and burning in cropping system	Level of mechanization
6. Best-bet legumes Malawi	Maize	Soybean, pigeon pea, groundnut, *Tephrosia*, *Mucuna*	960	Nov-April	One	Avg. 1 ha	Common with hoe or with oxen	None
7. *Marejea* Tanzania	Maize	*Marejea*	1180	Nov-May	One	Typically 1-2 ha cultivated	Burning of stover, ridges with hoe	Typically none
8. Forages Kenya	Napier grass	*Leucaena*, *Gliricidia*, *Clitoria*	900-1200	April-June, Oct-Dec	Two	2 ha typical; 75% of farms < 6 ha.	Common, with hoe	Typically none, some ploughing with oxen
9. GMCCs Uganda	Maize, beans, banana	*Lablab*, *Calliandra*, *Mucuna*, *Crotalaria*, *Canavalia*, *Tephrosia*	1255	Feb-June, Aug-Jan	Two	Avg. 2.1 ha	Common, with hoe	Typically none
10. Maize-*Mucuna* Benin	Maize	*Mucuna*	1000-1300	April-July, Sep-Nov	Two	Avg. < 1 ha	Rarely done, with hoe	Typically none
11. Rice-*Sesbania* Thailand	Rice	*Sesbania*	1100-1300	May-Oct	One	Avg. 0.8 ha	Common	Common, of small type

Note: Data from case studies except for farm-size data for Malawi (National Economic Council 2000, p. 61).

Chapter 3, *Mucuna* Use by Hillside Farmers of Northern Honduras, by B. Triomphe and G. Sain, chronicles the spontaneous adoption of the *abonera* system of maize (*Zea mays* L.)/*Mucuna* by hillside farmers of northern Honduras. The area, at below 600 masl, has a rainfall of 2000-3000 mm, allowing for two maize crops. *Mucuna* in the fields usually reseeds itself in the second-season maize, continues growing in the field alone during the first season and is slashed prior to second-season maize planting. Since its introduction to northern Honduras several decades ago, rapid adoption in the 1980s had by the early 1990s resulted in adoption by 65% of the hillside farmers. During the 1990s, disadoption was equally rapid and in 1997, only 40% of farmers were estimated to be practicing the system. As for *fríjol tapado*, a number of researchers worked on this system in the 1990s and quite a lot of information has been available for the case study.

The last case study from the Latin American region, Cultivating Maize with *Mucuna* in the Los Tuxtlas Region of South-eastern Veracruz, Mexico, by M. Eilittä, F. L. Arteaga, M. Diaz, C. Guerrero, B. Herrera, G. Narvaez, L. Paré, G. C. A. Robles and B. L. Triomphe (Chapter 4) recounts a local experience of two NGOs introducing *Mucuna* to smallholder maize cultivation in the 1990s. The region is hilly, with a typical annual rainfall of 1200-2500 mm. Working through farmer extension workers, the NGOs distributed *Mucuna* seed and trained farmers on *Mucuna* cultivation. Farmers mainly chose to intercrop *Mucuna* with summer maize and then either slashed it prior to cultivation of winter maize, or omitted winter maize cultivation by fallowing the field with *Mucuna*. No formal adoption studies have been conducted but at peak years of extension efforts (1993-94), hundreds of farmers were planting *Mucuna*. In the second half of the 1990s, cultivation has waxed and waned but currently *Mucuna* is little grown in most of the communities where the efforts took place. Several individuals have researched the system but certain aspects, such as adoption or socio-economic factors, have been little studied.

Altogether, six case studies come from Africa. Of these, two are from southern Africa. Improved Fallows in Eastern Zambia, by J. Peterson, F. R. Kwesiga, S. Franzel, D. Phiri and C. Gladwin (Chapter 5) recounts the large-scale research and diffusion efforts of the International Centre for Research in Agroforestry and the Zambian Ministry of Agriculture, Food and Fisheries that were initiated through on-farm studies in 1992. On-station studies had started 5 years earlier. Eastern Zambia is a hilly region, with an average rainfall of 960 mm. It has a low population density but continuous cultivation, typically without inorganic fertilizer, is the norm. The efforts have focused on 2-year fallows with four species, *Sesbania sesban* (L.) Merr., *Tephrosia vogelii* Hook. f., *Gliricidia sepium* (Jacq.) Walp. and pigeon pea (*Cajanus cajan* [L.] Millsp.). Because of the number of research institutions involved, a considerable amount of information is available on the system. The approach has involved working with farmers from the beginning, and a network of farmers,

NGOs, research organizations and government extension personnel has been involved in diffusion efforts. It has been estimated that 20 000 farmers have tested the technology and, in a survey of 2000-01, 70% of those who had planted improved fallows in 1996-97 or earlier had adopted the practice.

The second case study from southern Africa, Best-Bet Legumes for Smallholder Maize-Based Cropping Systems of Malawi, by R. Gilbert (Chapter 6), describes research efforts to incorporate GMCCs in the maize-based systems of Malawi. A small country with varied terrain, altitude (0-2000 masl) and rainfall (600-2000 mm), it relies greatly on maize as a staple food, with over 90% of the total cultivated area in maize. Three leguminous systems are discussed: grain legume-maize rotations with groundnut (*Arachis hypogaea* L.) or promiscuous soybean (*Glycine max* [L.] Merr.), legume/maize intercrops with pigeon pea or *Tephrosia*, and *Mucuna*-maize rotations. These systems are compared to continuous maize, the common cropping system in Malawi, applying food security, economic and soil fertility criteria. During the 1990s, GMCCs were a much-studied component of soil fertility technologies in the country, thus a reasonable amount of information is available to compare these systems.

Three of the cases come from East Africa. Promotion of *Marejea* Cultivation in the Ruvuma Region of Tanzania: Experiences of the Catholic Missionaries at Peramiho Mission Centre, by C. K. K. Gachene (Chapter 7) gives an account of efforts to diffuse *marejea* (*Crotalaria ochroleuca* G. Don) cultivation and use of inorganic fertilizer. The Ruvuma region of south-western Tanzania, at an altitude of 500-2000 masl and with a unimodal rainfall of about 1200 mm, is one of the main producers of maize in Tanzania; other crops are grown in varied associations. In the area, the Catholic missionaries of Peramiho initiated efforts in the 1970s to promote practices favouring soil fertility improvement by providing *marejea* seed to farmers and allowing them to exchange the produced seed with inorganic fertilizer. Although no adoption studies have been conducted, the data on seed receipts at the Mission Centre tell of large-scale *marejea* cultivation. In the 1980s, a number of other organizations, as well as some prominent politicians, became involved in the effort but by the early 1990s, the incentives for *marejea* cultivation were withdrawn and the project lost importance within the mission. Consequently, its cultivation decreased and, currently, very little *marejea* is cultivated in the area. Only a small amount of published information is available on the experience and interviews were used to research the case.

The second case from East Africa, Forage Production Systems for Dairy Production in the Coastal Lowlands of Kenya, by D. M. G. Njarui and J. G. Mureithi (Chapter 8) is an account of efforts to improve dairy cow nutrition in the coastal lowlands of Kenya through an alley farming system involving *Leucaena leucocephala* (Lam.) De Wit, *Gliricidia* and *Clitoria ternatea* L., together with Napier grass (*Pennisetum purpureum* Schumach. This region is relatively humid, with an average rainfall of 900 to 1200 mm in bimodal distribution, and mixed cropping systems involving various food and cash crops. Although risky, dairying is a profitable enterprise because of the high milk prices. Early efforts had focused on the screening of forage species but in 1988, KARI and the International Livestock Centre for Africa (ILCA) initiated, together with the Ministry of Livestock Development and Marketing, an on-farm dairy project oriented towards research and extension. Four production systems were developed and introduced to the region's farmers. Adoption of the systems, however, has been relatively low. The information available on the system is quite extensive but data on adoption and on-farm performance are lacking, as for many other case studies.

The final case study from East Africa, Green Manure/Cover Crop Technology in Eastern and Central Uganda: Development and Dissemination, by C. K. K. Gachene and C. S. Wortmann (Chapter 9), recounts the efforts of the Ugandan National Agricultural Research Organization (NARO) and the International Centre for Tropical Agriculture (CIAT, the Spanish acronym) in the Ikulwe District of eastern Uganda. This area, representing the traditional banana (*Musa* sp. L.)-coffee (*Coffea* L.)-based systems of the Lake Victoria Crescent agro-ecological zone, is at an altitude of 1200 masl and has an annual average rainfall of 1300 mm. Utilizing participatory methodologies, the effort included several common GMCC species in various management options. The on-farm trials, initiated in 1992, documented the productivity of the various GMCC options, while farmer experimentation generated new possibilities for integrating GMCCs in the local systems. Adoption of GMCCs in the area has been judged to be slow, but certain systems, such as the cultivation of *Tephrosia* for mole rat control and production of *Mucuna* fodder in banana groves, are better adopted and promising. Many publications are available on this experience but no detailed adoption studies have been conducted.

The one case from West Africa, Integrating *Mucuna* in the Maize-Based Systems of Southern Benin, by M. Eilittä, A. Etèka and R. J. Carsky (Chapter 10), gives an account of on-farm research that led into large-scale extension of *Mucuna* to the smallholder farmers in southern Benin. The area has a bimodal rainfall distribution averaging 1000 to 1300 mm, and high to very high population density. Efforts focused on two systems of maize-*Mucuna* cultivation: sole-cropped *Mucuna* fallow for areas with severe *Imperata cylindrica* (L.) Raeusch. infestation and a more popular system where *Mucuna* was relay-cropped in first-season maize and left to grow alone for the second season. The on-farm research was initiated in the late 1980s by the Beninoise government's Applied Research On-farm (RAMR, the French acronym) project on *Mucuna* cultivation for soil fertility maintenance and *Imperata* suppression. Extension efforts of the government's Regional Action Centre for Development (CARDER, the French acronym) and Sasakawa-Global 2000 led to many farmers testing *Mucuna*, approximating 3000 in 1993 and 10 000 in 1996. These efforts took place in the context of another line of activities, the promotion of an improved maize-fertilizer-pesticide package that included credit. Adoption levels have been estimated at as high as 7% of the region's farmers but current adoption levels are low. Although quite a few written materials are available on the experience, certain aspects, such as on-farm performance, have been poorly documented. Interviews complemented literature research in this case study.

Just one case study describes an experience with GMCCs in Asia. *Sesbania rostrata* in Rice-Based Farming Systems of Northern Thailand, by P. Gypmantasiri, B. Limnirankul and C. Phothachareon (Chapter 11) details the efforts of the Multiple Cropping Centre of Chiang Mai University to introduce *Sesbania rostrata* Bremek. & Oberm. in the rice (*Oryza sativa* L.) systems of the upper north of Thailand. This culturally rich mountainous region, receiving an annual rainfall of 1100-1300 mm, has glutinous rice as a subsistence rainy season crop and, if irrigation is available, one or two cash crops in the dry season. Rainfall commences in April but rice planting takes place in July; rainfall in May and June is usually sufficient to support growth of *Sesbania*. The research has taken place both on station (from 1993) and on farm (from 1997). Efforts on farm are continuing and focus beyond *Sesbania*, on the intensification of the rice system. The efforts are of small scale, with the researchers working in three districts of the Chiang Mai Province. Consequently, the experience and amount of information available is limited.

The final case study, Survey of Green Manure/Cover Crop Systems of Smallholder Farmers in the Tropics, by M. Flores Barahona and M. Janssen (Chapter 12), differs in that it reports on surveying work that was conducted as a part of the GMCC Exploration Project. In all, surveying included 72 cases collected from Africa, Latin America and Asia that will be available through an interactive Web site developed by CIDICCO. The systems surveyed include a total of 27 main crops and 36 GMCCs in various cropping arrangements. Although most of these systems have been adopted recently by farmers, a large share (over 40%) is traditional. Issues emerging from the surveyed systems and future efforts are also discussed.

Brief Characterization of the Cases

The first 11 cases, which describe field-level systems of smallholder farmers, share a number of similarities. Most focus on subsistence farming by smallholder farmers, with cash cropping a secondary goal. An exception to this is the experience in Santa Catarina, Brazil (Chapter 1), where GMCCs have been incorporated in mainly cash crop systems, such as tobacco (*Tabacum nicotianum* Berchtold & Opiz.), onion (*Allium cepa* L.) and maize. Agro-ecologically, the systems described in the case studies are primarily found in sub-humid and humid areas, with rainfall typically ranging between 900 and 1300 mm. In some areas where these systems are practiced, such as in southern Brazil, Costa Rica, Honduras and Mexico (Chapters 1-4), higher rainfall can occur; in contrast, in parts of Malawi (Chapter 6), lower rainfall is common.

The cases span a range from well-known to relatively unknown GMCC systems. Well-known ones include the diverse GMCCs and conservation tillage efforts in the state of Santa Catarina, Brazil (Chapter 1), the pre-Hispanic slash-and-mulch *fríjol tapado* system in Costa Rica (Chapter 2), spontaneous adoption of maize-*Mucuna* system by hillside farmers in northern Honduras (Chapter 3), improved fallows with *Sesbania*, *Gliricidia* and pigeon pea in the Eastern Province of Zambia (Chapter 5), and the maize-*Mucuna* system in Benin (Chapter 10). These well-known cases have also involved extensive farmer utilization of the systems and large-scale research/diffusion efforts. For example, in the case of Honduras, over 60% of the hillside farmers cultivated *Mucuna* in the late 1980s-early 1990s, while in the state of Santa Catarina in southern Brazil over 40% of the land was estimated to be cropped to GMCCs in the late 1990s. Similarly, in Benin, 10 000 farmers were reported to be testing *Mucuna* in 1996 and, in Zambia, 20 000 farmers are reported to have tested improved fallows. In addition, a relatively large amount of information, both from research and from diffusion literature, is available on these well-known cases. In contrast, other systems and experiences described in this book are less known and have less information available. Examples are those researched by faculty at the Multiple Cropping Centre of Chiang Mai University in northern Thailand (Chapter 11), where a few dozen farmers have tested the technology, or in coastal Kenya (Chapter 8), where improved fodder production was the entry point for the efforts.

The origin of the system also varies. Of the 11 cases, two describe systems that evolved without intervention from development or research organizations. One, *fríjol tapado* in Costa Rica (Chapter 2), a short-term fallow system most commonly involving beans, is pre-Columbian in origin, while another, the maize-*Mucuna* system in Honduras (Chapter 3), was developed in the region only in recent decades. Most cases therefore describe experiences that were initiated by either research or development efforts or by the two types of organizations working together. Most commonly, although efforts were initiated by research institutions, development organizations quickly became involved and the two types of organizations collaborated, with a wide range of efforts seen from those that mainly involve research to those that mainly involve diffusion.

Most efforts described involve a number of main crops and GMCCs, such as in Santa Catarina (Chapter 1), or the efforts of farmers and researchers to integrate numerous GMCC species in the banana-based mixed cropping systems of the Lake Victoria basin (Chapter 9). However, six of the cases focus on systems with one main crop, typically maize (Honduras-Chapter 3, Mexico-Chapter 4, Malawi-Chapter 6, Tanzania-Chapter 7 and Benin-Chapter 10; but *fríjol tapado* from Costa Rica involves beans, Chapter 2). Five cases have only one GMCC, most commonly *Mucuna* (in Honduras-Chapter 3, Mexico-Chapter 4 and Benin-Chapter 10; other species include *Sesbania* in Thailand-Chapter 11 and *Crotalaria* in Tanzania-Chapter 7). About one half of the cases therefore describe experiences with a number of species, typically three to five. The diversity of GMCC species has been greatest by far in Santa Catarina, Brazil (Chapter 1), where dozens of species have been utilized as GMCCs. A special case in terms of species diversity is the *fríjol tapado* system, where natural vegetation forming in previously cropped lands is used as the mulch for the following crop.

Although conservation tillage is often considered an important component of the GMCC systems, many of the efforts described have focused on integrating the GMCCs in the existing tillage systems, whether or not they involve tillage. *Fríjol tapado* in Costa Rica (Chapter 2) and maize-*Mucuna* systems in Honduras (Chapter 3) are cropping systems where tillage is not used with the GMCCs, while in other locations, tillage is either occasional (such as in Benin-Chapter 10, where mounding is sometimes practiced) or is used in some fields or communities (as in southern Veracruz-Chapter 4). In the cases from East (Chapters 7-9) and southern (Chapters 5 and 6) Africa, as well as from Thailand (Chapter 11), the species have been incorporated in the existing systems that utilize tillage.

Finally, only five of the cases in this volume document use of the GMCCs for other purposes than soil fertility improvement and weed suppression. This is partly caused by the several cases with *Mucuna* and its limited uses as a food and feed.

These case studies shed light on the performance of GMCC systems in the varied conditions of smallholder farmers in developing regions. They describe both the opportunities and the limitations for these systems by highlighting the context in which smallholders are practicing or experimenting with them. Despite the limited number and varied nature of these cases, the concluding chapters attempt to draw lessons from them and to suggest future work to improve the productivity and sustainability of GMCC systems.

NOTES

1. This definition differs to some extent from that adopted by the GMCC Exploration Project (see Chapter 12).
2. No attempt is made here to list the principal organizations involved; individuals from many of them have participated in the efforts to exchange and synthesize knowledge that are described.
3. For Mulch-L (in English) contact lhf2@ cornell.edu; for Coberagri-L contact milton.flores@cidicco.hn and for EVECS-L contact r.carsky@cgiar.org.
4. LEXSYS has recently been updated and converted to a more user-friendly format and is available online at ftp://ftp.bangor.ac.uk/pub/departments/af/LEXSYS/. Information on the Organic Resources Database is available at http/www.wye.ac.uk/BioSciences/soil/ or at tsbfinfo@tsbf.unon.org. The University of California-Davis database is available at www.sarep.ucdavis.edu/sarep/ccrop/search_ccrop.html.

5. Prior to 1998, Legume Screening Network.
6. At the time of writing, CIEPCA is without funds.
7. MOIST: http://ppathw3.cals.cornell.edu/mba_project/moist/home2. html. CIDICCO and CIEPCA have collaborated with Ft. Myers, Florida-based Educational Concerns for Hunger Organizations in the Tropical Soil Cover and Organic Resources Exchange (TROPSCORE) effort by Cornell University, which maintains the Worldwide Portal to Information on Soil Health (http://mulch.mannlib.cornell.edu).
8. GMCC Exploration Project members were: Leandro do Prado Wildner (Empresa de Pesquisa Agropecuária e difusão de tecnologia de Santa Catarina [EPAGRI], Chapecó, Brazil), Albert Etèka (CIEPCA), Phrek Gypmantasiri (Chiang Mai University, Chiang Mai, Thailand), Milton Flores (International Cover Crops Clearinghouse, Tegucigalpa, Honduras), Robert Gilbert (Post-doctoral Fellow funded by the Rockefeller Foundation, Lilongwe, Malawi), Joseph Mureithi (LRNP, Nairobi, KARI, Kenya), Ruben Puentes (Rockefeller Foundation-Mexico, Mexico City, project leader), Ken Schlather (Cornell University, Ithaca, NY, USA), Sheryl Swinkel (Cornell University, Ithaca, NY, USA), Bernard Triomphe (Post-doctoral Fellow funded by the Rockefeller Foundation, Mexico City, Mexico) and Marjatta Eilittä (Post-doctoral Fellow funded by the Rockefeller Foundation, Gainesville, FL, USA).

REFERENCES

Arteaga L, Carranza T, Eilittä M, González M, Guerrero C, Guevara F, Herrera B, López A, Martinez F, Mendoza A, Narváez G, Puentes R, Reyes H, Robles C, Sohn I, Triomphe B (1997) El uso de sistemas de cultivo con plantas de cobertura en algunas comunidades del Sureste Mexicano: Contexto, resultados y lecciones aprendidas. Paper presented in a workshop on Green Manure/Cover Crop Systems for Smallholders in Tropical and Subtropical Regions, 6-12 April 1997, Chapeco, Brazil.

Buckles D (1995) Velvetbean: A 'new' plant with a history. Econ. Bot. 49(1):13-25.

Buckles D, Etèka A, Osiname O, Galiba M, Galiano G (Eds.) 1998. Cover crops in West Africa: Contributing to sustainable agriculture. International Development Research Centre (IDRC)-International Institute of Tropical Agriculture (IITA)- Sasakawa-Global 2000 (SG2000), Canada. 318 p.

Cairns M (Ed.) 1997 Indigenous strategies for intensification of shifting cultivation in Southeast Asia. http://www.idrc.ca/cbnrm/documents/abstract_main.cfm.

Calegari A, Wildner L do P, de Freitas V (1997) Adubaçâo verde e sistemas de cobertura do solo na regiâo sul do brazil. Paper presented at the International Workshop on Green Manure Cover Crop Systems for Smallholders in Tropical and Subtropical Regions, 6–12 Apr 1997, Chapeco, Santa Catarina, Brazil. Rural Extension and Agricultural Research Institute of Santa Catarina, Santa Catarina, Brazil.

Eilittä M, Sollenberger L E (2001) The many uses of *Mucuna*: Velvetbean in the southern United States in the early 20th century. *In*: Flores B, Eilittä M, Myhrman R, Carew L B, Carsky R J (Eds.) Food and feed from *Mucuna*: Current uses and the way forward. Centro Internacional de Información sobre Cultivos de Cobertura (CIDICCO), Center for Cover Crops Information and Seed Exchange in Africa (CIEPCA) and World Hunger Research Center, Judson College, Tegucigalpa, Honduras. p. 73-100.

FAO – Food and Agricultural Organization (2001) The impact of HIV/AIDS on food security. Committee on world food security, 27th session, 28 May-1 June 2001, Rome, Italy. http://www.fao.org/docrep/meeting/003/Y0310E.htm#P101_12054.

Florentín M A, Peñalva M, Calegari A, Derpsch R (2001) Abonos verdes y rotación de cultivos en siembras directa. Proyecto conservación de suelos, Ministerio de Agricultura y Ganadería-German Agency for Technical Cooperation (MAG-GTZ), Dirección de Investigación/ Dirección de Extensión Agraria (DIA/DEAG), San Lorenzo, Paraguay. 84 p.

Flores M, Paredes M, Alemán R, Sagastume N, Sherwood S, Schlachter K (1997) El uso de cultivos de cobertura en Centroamerica: Más allá del entusiasmo. Centro Internacional de Información sobre Cultivos de Cobertura (CIDICCO), Tegucigalpa, Honduras.

García-Barrios R, García-Barrios L (1990) Environmental and technological degradation in peasant agriculture: A consequence of development in Mexico. World Dev. 18(11):1569-1585.

Henao J, Banante C (1999) Nutrient depletion in the agricultural soils of Africa. 2020 Vision Brief No. 62, International Food Policy Research Institute (IFPRI), Washington, USA.

IFDC - International Fertilizer Development Council (2003) Fertilizer statistics, derived from Food and Agriculture Organisation (FAO) data. http://www.ifdc.org/ Free_ Fertilizer_Statistics/index.html.

Kiff L, Pound B, Holdsworth R (1996) Cover crops: A revision and database for field users. Natural Resources Institute (NRI), Chatham, UK. 180 p.

National Economic Council (2000) Profile of poverty in Malawi, 1998. Lilongwe, Malawi. 110 p.

Oldeman R, Hakkeling R, Sombroeck W (1991) World map of the status of human induced soil degradation: An explanatory note. International Soil Reference and Information Centre (ISRIC), Wageningen, the Netherlands, and United Nations Environment Program (UNEP), Nairobi, Kenya.

Scherr S (1997) Soil degradation: Athreat to developing country food security by 2020? Food, Agriculture and the Environment Discussion Paper no. 27. International Food Policy Research Institute (IFPRI), WA, USA. 63 p.

Scherr S J, Yadav S (1997) Land degradation in the developing world: Issues and policy options for 2020. 2020 Vision Brief 44, International Food Policy Research Institute (IFPRI), Washington, USA.

Stoorvogel J J and Smaling E M A 1990. Assessment of nutrient depletion in sub-Saharan Africa: 1983 - 2000. Report 28, The Winand Staring Centre, Wageningen, the Netherlands.

United Nations (2000) World population prospects: The 2000 revision – highlights. Population Division, Department of Economic and Social Development. Un.org/esa/population/unpop.htm.

Wood S, Sebastian K, Scherr S J (2001) Pilot analysis of global agroecosystems. International Food Policy Research Institute (IFPRI) and World Resources Institute (WRI), Washington, USA. 110 p.

World Bank (2001) Global economic prospects and the developing Ccuntries 2001. International Bank for Reconstruction and Development/The World Bank. Washington, DC, USA.

Chapter 1

Use of Green Manure/Cover Crops and Conservation Tillage in Santa Catarina, Brazil

L. do Prado Wildner[1], V. Hercilio de Freitas[1] and M. McGuire[2]

[1] *Empresa de Pesquisa Agropecuária e difusão de tecnologia de Santa Catarina (EPAGRI), Brazil*
[2] *University of the Virgin Islands, St. Croix, West Indies*

SUMMARY

The shallow, heavy soils, hilly topography and high rainfall of Santa Catarina contribute to intense erosion when management of soil cover is inadequate. Increasingly, hillside farmers practicing annual cropping leave plant residues on the soil surface to minimize erosion and to reduce fluctuations of soil water and temperature. Over time, repeated additions of fresh, high-quality green manures to the soil are strongly favourable to soil quality and crop performance. Although land tenure in the state is predominantly in the hands of the small-scale family farmer, technological change and economic liberalization are changing the face of the agricultural sector. As in many other places, farmers here are trying to negotiate an uneasy passage from traditional production practices to newer and, hopefully, more sustainable ones.

Most small-scale farmers use draft animals or mini-tractors for traction. Local manufacturers produce a variety of conservation tillage machinery for this market, and its availability has permitted the spread of reduced tillage in the region. The farmer's management expertise is crucial to the success of no-till systems, particularly in complex, multi-year rotations that include a variety of different crops. In some parts of the state, highly successful conservation tillage systems have become well established for certain key crops such as maize (*Zea mays* L.), beans (*Phaseolus vulgaris* L.), tobacco (*Tabacum nicotianum* Berchtold & Opiz.) and onions (*Allium cepa* L.). Usually, numerous local variations of a given cropping system include different elements such as minimum or zero tillage, animal or mechanical traction, or variations in the species composition and sequencing of the rotation.

M. Eilittä et al. (eds.), Green Manure/Cover Crop Systems of Smallholder Farmers, 1-36.

Overall, four trends are clear in the agricultural sector. First, the hilly landscapes and granular pattern of land ownership contributed to a shift in production of many field crops (e.g. soybean - *Glycine max* [L.] Merr., wheat – *Triticum sativum* Lam. and potatoes - *Solanum tuberosum* L.) from Santa Catarina to other regions where greater returns to scale are possible. Second, output has grown in specialized commodities such as processed meats, flue-cured tobacco and horticultural products. Third, the emergence of a modern, dynamic sector exposes the vulnerability of traditional socio-economic structures in the state. In order to compete in the new markets, successful farmers must invest in capital improvements to improve productivity and efficiency. From 1985 to 1995, struggling farmers sold or abandoned their properties in a surge of rural exodus that continues today, and average farm size is slowly increasing as the more financially consolidated properties snap up cheap land.

The fourth trend is that within the specialized agricultural commodities mentioned above, no-till management of green manure/cover crops is increasingly adopted on the small farm. Majority opinion in the state, backed by good data, holds that there are three primary reasons for the adoption of this technology by farmers. First, the technology reduces labour requirements and soil cultural operations. Second, the technology controls erosion. And third, as soil quality improves, so do crop performance and financial returns.

INTRODUCTION

In the State of Santa Catarina, Brazil, soils are intensively cultivated: although it represents just over 1% of Brazilian national territory, the state accounts for about 13% of commercial agricultural output nationally. The social basis of agricultural production, the small-scale family farm, stands out relative to the rest of Brazil, where larger or estate farms typically predominate. A variety of agro-ecological zones spans the state and agricultural production is diverse. Processed pork and poultry, tobacco (*Tabacum nicotianum* Berchtold & Opiz.), onions (*Allium cepa* L.), apples (*Malus pumila* Mill.), garlic (*Allium sativum* L.), bananas (*Musa* spp. L.), rice (*Oryza sativa* L.), potatoes (*Solanum tuberosum* L.), tomatoes (*Lycopersicon esculentum* Mill.), wheat (*Triticum sativum* Lam.) and soybeans (*Glycine max* [L.] Merr.) are important export crops; while beans (*Phaseolus vulgaris* L.) and cassava (*Manihot esculenta* Crantz) are staple food crops in the state. Despite the volume and diversity of commercial products, about half of the area devoted to annual cropping is planted to maize (*Zea mays* L.), which is pivotal to the domestic economy of rural households for home consumption, as an animal feed and as a cash crop.

Most arable soils in the state are rocky and shallow and occur on hilly topography. Production increases since the 1950s have been based on clearing new areas and on expanding the use of industrial inputs and mechanical tillage. Yet this came with a heavy cost: soil erosion and degradation are the primary threat to sustainable production and are the chief factors behind declining yield trends observed in many fields. In consequence, farmers, extension agents, researchers, municipal governments and industry representatives joined forces to seek effective technological alternatives that are adapted to the soil, climate and socio-economic conditions of the state. After 20 years of using and experimenting with green manure/cover crops (GMCCs) and conservation tillage, farmers in the region increasingly leave GMCCs on the soil surface to minimize erosion and weed growth, to build fertility and to reduce fluctuations in soil humidity and temperature. Farmers and researchers report beneficial and multi-faceted results for soil physical, chemical, biological and economic properties.

This flowering of applied research and extension involved actors from diverse segments of the commodity chain and the government. Together they produced a wealth of technological options for recovering and sustaining the productive capacity of arable soils. Technical alternatives exist for GMCC germplasm for summer and winter crop cycles, for different rotations and cropping systems, for fertilization and for conservation tillage. However, many problems remain unsolved or poorly understood and despite robust adoption of conservation systems in Santa Catarina as a whole, many regions, crops and communities have seen little progress in this regard. The adoption of conservation methods is uneven and fragmentary across the different regions and the financial security of many small farms is still precarious at best. Hence, the need for further experimentation and adaptation is widely felt, propelling continued collaboration between diverse groups working on rural development.

BIOPHYSICAL FEATURES

Santa Catarina is the smallest of the three states in Brazil's southernmost region (Figure 1). It has a humid subtropical climate with generous rainfall (1200-2370 mm) evenly distributed throughout the year (Box 1). Given its varied topography and soil types, the state hosts a diversity of landscapes and vegetation sequences. The principal groups are tropical Atlantic rainforest along the coast, subtropical rainforest in the west and savannah-type grasslands mixed with forests in the high middle plains. Exuberant native rainforests were decimated, leaving agricultural land and extensive areas of secondary regrowth. Old forest fragments are tiny and rare.

Depending on the region, some 40 to 80% of agricultural soils are moderately to steeply sloping. Together, the hilly topography and high rate of rainfall cause intense erosion when soil cover is inadequate. In uncultivated soils, organic matter varies between 3 and 4%, pH <5.0, cation exchange capacity (CEC), K levels are low and P is negligible. Because of the low fertility and the hilly topography, about 60% of the state's soils are not well suited to annual cropping, requiring lime, fertilizers and conservation measures for satisfactory production. Despite this, many of these soils are cultivated year round (Santa Catarina 1973).

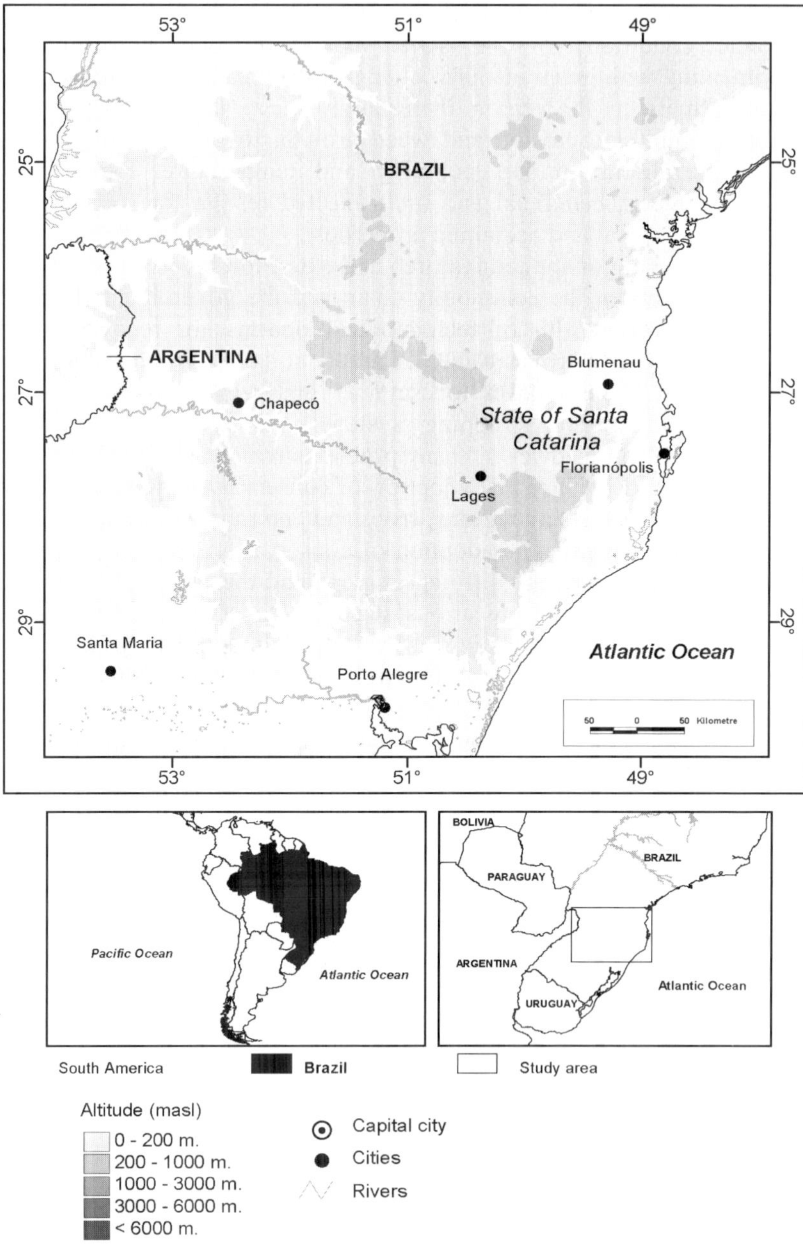

Figure 1. Location of case study, Santa Catarina, Brazil.

***Box 1.* Geography and climate of Santa Catarina state**

Location: Southern Brazil, between parallels 25°57'S and 29°23'S and meridians 48°19'W and 53°50'W.

Surface area: 95 925 km^2, or 1.13% of national territory.

Relief: Three major zones: (1) coastal plains, (2) coastal mountain range and (3) western plateaus.

Drainage basins: Two main watershed areas, one for the western plains and a group of smaller, independent watersheds along the coast.

Water resources: 80% of the state's water sources are threatened by the presence of heavy metals, agricultural pollutants or urban or industrial effluents.

Climate: Koppen's classification – Cf, humid subtropical mesothermic. In the west and along the coast, two warm season cropping cycles, whereas in the high middle plains, only one warm season cropping cycle.

- Average annual temperature: Between 18 and 22 °C.
- Average annual rainfall: Between 1220 mm in the southern part of the state and 2370 mm in the west.
- Rainfall distribution: In general, well distributed throughout the year, with no well-defined rainy or dry seasons.
- Average relative humidity: Between 73% in the west and 85% along the coast.
- Freezes: Only sporadic occurrences in the western plateaus.

The primary soil groups in the state are Cambisols (40% of total area), Ultisols (17%), Mollisols (16%) and Podzols (14%) (Pundek 1994). Table 1 presents the characteristics of a typical Cambisol.

Table 1. Upper profile characteristics of a typical Cambisol in Santa Catarina, Brazil (*Source*: Santa Catarina 1973).

Horiz	Depth (cm)	Sand	Silt (%)	Clay	pH	Org C (%)	Ca+Mg	CEC (meq 100 g^{-1})	Al sat. (%)	K (ppm)
A1	0-15	51	30	19	4.0	3.0	0.6	19.6	83	54
A3	15-48	50	26	24	4.4	1.8	0.6	18.4	86	15
B1	48-80	49	26	25	4.6	1.8	0.6	19.0	84	11
B2	80-125	50	21	29	5.0	0.7	0.5	13.8	81	7

AGRICULTURE IN SANTA CATARINA

Land Tenure and Use

Europeans colonized Santa Catarina in two stages. The Portuguese crown initially sent Azoreans to settle along the coast in the 17th and 18th centuries. Later, from the mid-19th to early 20th century, national authorities sponsored the colonization of the interior regions of southern Brazil by central Europeans, mainly Germans and northern Italians, who brought with them the tradition of green manuring. The settlement companies typically distributed around 100 ha to each family and today a property of up to 50 ha is considered a small-scale family farm. Properties up to this size represent about 90% of the registered holdings in the state and about 43% of the total area (Table 2). Hence, the state is somewhat unusual in Brazil in that land tenure is predominantly in the hands of small-scale family farmers.

Table 2. Distribution of farm properties, surface area and agricultural employment by farm size in Santa Catarina in 1985 and 1995 (*Source*: Zoldan 1998).

	Farm size (ha)					
	< 10	10 – 50	50 – 100	100 – 1000	> 1000	Total
Share of no. of properties (%):						
1985	39	51	6	4	0.2	234,851
1995	36	54	6	4	0.2	203,347
% change[a]	-21	-9	-9	-7	-11	-
State area devoted to agriculture (%):						
1985	6	35	12	30	17	7,419,535
1995	9	34	12	30	15	6,882,845
% change[a]	-19	-9	-9	-8	-17	-
No. of people employed:						
1985	277,247	479,515	65,717	55,542	9,093	887,114
1995	215,164	395,307	51,492	44,454	12,045	718,462
% change[a]	-22	-18	-22	-20	+32	-19

[a] Percentage change of absolute values (not shown) from 1985 to 1995.

A government study (Santa Catarina 1973) indicates that only 30% of state soils are apt for annual cropping, while 40% are apt for permanent crops, with a moderate aptitude for annual cropping in limited areas. Altogether, 24% of the state area is apt only for permanent crops, pastures or forestry. In terms of actual land usage, Moser et al. (1994) concluded that 50% of the state area was being used in accordance with its aptitude; 30% of state area was being used in accordance with its aptitude, so long as adequate conservation measures were taken; and 18% of state area was being used in a manner inconsistent with its aptitude, presenting a clear threat to soil and water quality.

Industrialization of Agriculture

Like many other places, Santa Catarina is undergoing an uneasy shift from traditional agriculture, with few manufactured inputs, to an industrially based agriculture. Between these two poles are farmers at every point along the continuum. The penetration of industry into agriculture varies greatly according to geographic region, crop type, farm size, farmer strategy and even municipal politics. The following 1995 indicators reflect the access of state farmers to modern production factors (Zoldan 1998):

- While 86% of properties had electricity, only 25% had at least one mini-tractor or tractor.
- While 85% of farmers used mineral fertilizers, only 7% used irrigation.
- Of all farmers, 65% received some type of technical assistance, of which 57% was from government agencies, the remainder coming mostly from the tobacco multinationals.
- Only 19% purchased inputs on credit from suppliers and only 6% received credit from financial institutions.

Trends in Agricultural Production

Agricultural production is diverse in the state, both between and within the major regions. In contrast, within enclaves where a specific crop is particularly adapted, the tendency is towards greater specialization and homogenization of production practices. Small farm output accounts for 75-95% of most commodities, with the exception of soybean, wheat and apples, which larger scale farmers dominate.

The emergence of a modern, dynamic sector in agriculture exposes the vulnerability of traditional socio-economic structures in the state. The period from 1985 to 1995 witnessed a surge of rural exodus (Table 2) that continues today. From 1985 to 1995, the total number of farm workers fell from 887 000 to 718 000, or an average reduction of about 20%. The exception to this trend was in the largest properties.

Among the forces propelling the transformation of rural society in Santa Catarina, two major structural elements seem to cut in different directions. On the one hand, the hilly topography and atomized land structure limit the introduction of large-scale, mechanized agriculture. On the other hand, Brazil's integration with the regional trading organization, Mercado Común del Sur (MERCOSUR), as well as the globalization of many commodities, force greater modernization and homogenization of production practices, increased integration with the agro-industrial complex and the consolidation of larger farms.

The area planted to field crops such as soybean, wheat, beans and potatoes has dwindled recently. Production of these crops has shifted to other areas where topography and land tenure are more conducive to mechanization and returns to scale. In contrast, for those commodities where a small production scale and the availability of family labour are a relative advantage, such as tobacco or intensive horticultural crops, acreages have increased. Examples include onion, garlic, tomato, cauliflower (*Brassica oleracea* L. var. *botrytis*), peppers (*Capsicum annuum* L.), and beets (*Beta vulgaris* L. subsp. *vulgaris).*

In the western part of the state, pork and poultry commodity chains supply national and regional markets. In these chains, small farms are tightly integrated with the agro-industrial complex. The production practices permit reasonable economies of scale to be realized in properties of up to about 100 ha. Hence, there is a trend for the average farm size to increase. This sector has seen significant increases in productivity and output over the past 20 years, although growth in farm income has not kept pace.

Tobacco production, also controlled by agro-industrial concerns, is somewhat the same and yet different. Flue-cured tobacco is a characteristic crop of the smallest and poorest family farms. Its harvest and curing requires intensive manual labour involving the entire farm family. Pesticide use and the risk of exposure are high in production and processing activities. Two Anglo-American firms dominate the commercialization of this product. Returns to the farmer are often low but it is a popular crop because the companies will provide industrial inputs (improved seeds, fertilizers and pesticides) on credit to contracted farmers. Because of the intense demand on family labour, farm size has not tended to increase and tobacco remains an economic anchor for the poorest and most vulnerable farm families.

The well-distributed rainfall and hilly topography endow the state with a natural aptitude for perennial agriculture. Most permanent crops have seen their acreage and production expand over the past 10 years, with apples, *erva-mate* (*Ilex paraguariensis* A. St.-Hil., a tree whose leaves are used to produce a green tea) orange (*Citrus sinensis* [L.] Osbeck), peach (*Prunus persica* [L.] Batsch var. *persica*) and forest products leading the way (Zoldan 1998). Exotic tree species such as eucalyptus and pine do very well and have been promoted by state agencies, while native species have been less well studied and are generally promoted by non-governmental organizations (NGOs). Mixed agroforestry systems have received scant attention from state officials and most research funds are directed towards annual cropping systems.

SOIL CONSERVATION AND THE INSTITUTIONAL CONTEXT

Background

Interest in green manures in Brazil followed a pattern familiar in other parts of the world: curiosity grew in the early part of the 20th century but the green revolution provoked a rush to industrialize farming and this research was all but abandoned. The recent efforts to promote soil conservation techniques in Santa Catarina began in the 1960s, with an emphasis on terracing. This technique, at best, reduces the effects of erosion, without addressing the cause: raindrop impact on bare soil. It is not very well adapted to the topography or socio-economic conditions in the region, so results were minimal. The 1970s saw a peak in the expansion of both the agricultural frontier and the use of mechanized tillage on hillsides. With these developments came the alarming rates of erosion that mobilized agronomists and farmers to seek solutions.

The Instituto Agronômico do Parana (IAPAR), the state agricultural research and extension agency of the neighbouring state of Paraná, began looking at green manures again in 1975. In 1978, IAPAR's counterpart in Santa Catarina, the Empresa de Pesquisa Agropecuária e difusão de tecnologia (EPAGRI), adopted a new strategy that focused on proven appropriate technologies: (1) contour strip buffers with vegetation and rocks and (2) GMCCs with reduced tillage. The agency launched its integrated programme for soil and water management and conservation, which mapped out the agronomic potential of cover crops in the state and encouraged their adoption by farmers in target municipalities. The first demonstration units for these technologies were installed on farmers' properties. A key limitation, seed supply, was overcome with the seed exchange programme, whereby EPAGRI distributed 150 kg of common vetch (*Vicia sativa* L.) in small quantities to farmers, who had to return double the quantity they received for several years. EPAGRI also began screening summer and winter species for cover cropping, green manuring and soil recovery at one of its regional experiment stations. Over the next 12 years, this initiative was expanded to all of EPAGRI's experiment stations.

In 1983, the Brazilian Enterprise for Agricultural Research (EMBRAPA, the Portuguese acronym) organized the first national meeting on green manures, hoping to stimulate exchange between researchers and extension agents across Brazil. The quantity and quality of research on this topic has continued to grow nationwide since this catalytic event. In the mid-1980s, throughout the southern region of Brazil, the tobacco companies began to promote GMCCs on demonstration plots and to contract for the production of GMCC seeds. In 1987, the federal government launched its national programme for micro-watershed management, which became the model for subsequent state conservation programmes.

Throughout the 1980s, the planning and execution of soil conservation programmes was *de facto* the purview of federal and state agencies. The 1990s saw a new situation emerge, whereby diverse institutions from both the public and private sphere contribute to different aspects of agricultural development. These institutions include agricultural cooperatives, producer unions, private companies and agro-industry boards, agricultural vocational schools and universities, municipal governments and NGOs. This broad-based institutional context was pivotal for advancing environmental awareness and the use of GMCCs and other conservation measures in the landscapes of Santa Catarina.

The Santa Catarina Micro-watersheds Project

From 1991 to 1999, these various efforts were consolidated and given greater focus by Santa Catarina's micro-watershed basin management project. Among other roles, the micro-watershed project served as the backbone for conservation initiatives undertaken with local entities and various governmental units. This programme was co-financed by the World Bank and administered by EPAGRI.

This project had a major impact on the conservation and management of natural resources in the state, involving numerous state agencies, agricultural cooperatives and commercial entities. Its primary objective was to increase the amount of soil cover for erosion control. The principal lines of action were to:

- Increase soil cover and control surface run-off;
- Encourage soil use in accordance with aptitude;
- Protect water sources, water courses and other natural resources;
- Increase agricultural production; and
- Train farmers in technical skills and conduct environmental education.

With a budget of about US$71.6 million, split between the World Bank and the state government, the programme included a component, Prossolo, offering financial incentives to farmers for adopting conservation methods.[1]

Promotion of Soil-conserving Technologies

During this period, diverse institutions sponsored different strategies and events to promote soil-conserving technologies.

Farmer Training Courses

At 11 training centres across the state, EPAGRI offers 64 courses to farmers, extension workers and agricultural technicians to continually improve the technical capacity in the agricultural sector. Themes range from commodity-based courses, food processing and packaging, machinery operation and maintenance, to more cross-cutting topics, such as soil and water conservation, conservation tillage, farm enterprise administration and administration of rural cooperatives.

Agricultural Cooperatives

Starting in the 1990s, producer unions began to take a more active role in the technical training of their members, and demonstration units and field trips became common. Today, diverse regionally and sectorally based cooperatives come together to sponsor trade fairs focusing on sustainable agriculture throughout the state.

Efforts of the Tobacco Companies

As soils became denuded and production costs increased, the tobacco companies came to focus on the problem of soil conservation, first through their research departments and then through extension efforts with their contracted farmers. They began projects for cover crop introduction, screening, seed production, demonstration units and field days for tobacco producers. They also promoted the development of an animal-drawn direct seeder for such row crops as maize, beans and soybean, facilitated its purchase by groups of contracted farmers, and began research on reduced tillage for tobacco cropping systems.

Municipally Sponsored Technical Field Days

During the second half of the 1990s, the municipal governments of larger towns organized regional seminars and trade fairs in conservation tillage and sustainable agriculture. Farmers, extension agents, cooperatives, agribusiness firms, merchants, agricultural labour unions and professional organizations have all participated in these events.

The Rural Extension Component of the Micro-watersheds Project

Extension agents from EPAGRI, municipal governments, private enterprises and agricultural cooperatives participated in the programme. Before farmer training and implementation activities, the extension agents received specialized training in one or several of the following:

- Concepts and application of micro-watershed planning,
- Soil identification,
- Aptitude classification and management,
- Soil fertility and conservation practices,
- Reforestation,
- Farm planning and administration,
- Education,
- Training and
- Communication in rural extension and environmental education.

The extension component was the principle vehicle for reaching the goals stipulated in the programme. The following activities were emphasized: (1) land use capability classification and planning at farm-level, (2) use of GMCCs, (3) mechanical erosion control practices, (4) reforestation, (5) sewage and effluent treatment and (6) conservation tillage. Activities in the programme were grouped in three phases. Phase one was motivational: meetings and field trips with leaders and key community stakeholders, and technical excursions with farmer groups. Phase two was for micro-watershed planning and individual farm planning. Phase three was for execution: formation of the micro-watershed commissions, with farmer training in participatory planning and monitoring of planned activities, technical training for farmers based on planning results and farmer cooperative organization for collective acquisition and use of machinery and equipment.

Agro-industrial Initiatives

Consortia led by multinational and domestic agro-industrial enterprises launched various related initiatives. The crop residue project (*Pró-palha*) sought to consolidate no-till planting in maize, bean and soybean cropping systems. Research projects evaluate and make recommendations on fertilizer sources and methods of application, residue management, crop rotation, liming and other topics, while the extension component focuses on extension agents from EPAGRI, municipal governments, cooperatives and NGOs. The *Milho-Feijão* (Maize-Beans) Project focused on improving the productivity and quality of bean production through improved varieties, agrochemical inputs and GMCCs with direct seeding. Increasing the productivity of subsistence crops, such as maize and beans, enhances the food security of farmers and relieves the cash wage requirements that the agro-industrial firms must pay to contracted farmers in the tobacco, pork and poultry commodity chains.

MANAGEMENT OF GMCC SYSTEMS

Monegat (1991) and da Costa et al. (1993) provide a wealth of information on the management of green manures, their incorporation into regional cropping systems and the advantages and limitations of different species for different purposes. Derpsch et al. (1991) and CEPA-SC (1999) provide additional information on machinery and cropping practices. This section employs information from these texts and other observations to consider briefly the basic principles and issues involved. The sections below address different strategies for incorporating green manures into a production system, management of the GMCC biomass, the variety of equipment used by farmers for this purpose, and the seed production of GMCCs in these systems.

Different Ways of Using Green Manures

Because agricultural production is so diverse in the state, a variety of strategies are used for integrating green manures into existing farms (da Costa et al. 1993). The most important are:

- Spring/Summer Sole-crop Green Manuring
  Although this is reputedly the oldest form of green manuring in the region, warm season species generally compete with economic crops, so this option is mostly used when the economic crop is a fall/winter crop. The advantages include abundant biomass and excellent N fixation, as well as soil cover during the rainiest part of the year. Most commonly, *Mucuna* spp.Adans are used.
- Fall/Winter Sole-crop Green Manuring
  Cool season species take the place of annual fallowing. Multiple uses for the green manure, such as seed or forage production, are common. This is the most usual use of GMCCs in the state. Most often, black oats (*Avena strigosa* Schreb), oilseed radish (*Raphanus sativus* L., var. *oleiferus*) and vetches (*Vicia* spp. L.) are used.
- Green Manures Relay-cropped with Annual Crops
  Some farmers practice relays to ensure that the soil is adequately covered at all times. For example, systems where GMCCs are relayed into maize are common in some areas. A disadvantage is that sowing the GMCCs and/or harvesting the economic crop can be more difficult, although this is less of a problem if the cover crop is seeded relatively late, for example, after maize tasselling. Both winter and summer species are used. This option is recommended with GMCC species that are long seasoned or initially slow growing. The most common system involves maize intercropped with *Mucuna* spp. Adans.
- Contour Strip Farming
  The GMCC and economic crop are planted along the contour in strips, usually two rows wide. The following year, the two are switched. This system can be used where soil quality is adequate and the farmer does not want to completely forego the economic crop. Most commonly, maize is grown with *Mucuna* spp. or jackbean (*Canavalia ensiformis* [L.] DC.).
- Orchard Floor Green Manuring
  Non-aggressive species, some of which re-seed themselves, are managed to limit competition with a banana, citrus or apple crop. Species include vetches, black oats, rye grass (*Lolium multiflorum* Lam.), *Pisum sativum* L., some *Mucuna* spp., *Crotalaria* spp. L., pigeon pea (*Cajanus cajan* [L.] Millsp.) and *Arachis* spp. L. Most commonly, hairy vetch (*V. villosa* Roth), black oats or rye grass are followed by *Mucuna* spp. or *Crotalaria* spp. L.

- Improved Fallows
  Normally used for degraded soils, summer and winter green manures are cropped in succession until the soil quality has improved. Both annual and perennial species are used. For annuals, species that re-seed themselves are preferred.

These categories are not necessarily mutually exclusive. For example, some farmers combine strip contours with relay-cropping and reduced tillage: while the GMCC is still green, the soil is ploughed in narrow strips so that an economic crop can be planted.

For each of these uses, only certain cover crops are appropriate, depending on the season, the agro-climatic region and the specific objectives of the farmer. Currently, EPAGRI recommends the species in Table 3 for use in the different regions of Santa Catarina, although farmers are also using others. Yet other species are under study and some of these have been disseminated to farmers to evaluate on-farm performance.

Table 3. Recommended winter and summer green manure and cover crop species for use in Santa Catarina (*Source*: Wildner et al. 1995).

Region[a]	Species used
Coast (20 °C, 1250 mm, < 500 m)	Winter: *Avena strigosa* Schreb, *Vicia sativa* L., *Pisum sativum* var. *arvense*, *Lathyrus sativus* L., *Lupinus luteus* L., *L. angustifolius* L., *L. albus* L., *Raphanus sativus* L. var. *oleiferus*, *Spergula arvensis* L. Summer: *Mucuna pruriens* (L.) DC., *M. aterrima* (Piper & Tracy) Merr., *M. deeringiana* (Bort.) Merr., *Crotalaria spectabilis* Roth, *C. pallida* Aiton, *Cajanus cajan* (L.) Millsp., *Canavalia ensiformis* (L.) DC., *Can. brasiliensis* Mart ex Benth.
Itajaí Valley (18 °C, 1500 mm, 500-700 m)	Winter: *A. strigosa*, *V. sativa*, *P. sativum*, *Ornithopus sativus* Brot., *Lath. sativus*, *R. sativus*, *S. arvensis* Summer: *M. nivea* (Roxb.) D.C., *Can. ensiformis*, *C. pallida*, *V. unguiculata* (L.) Walp., *Caj. cajan*, *Lablab purpureus* (L.) Sweet
High Plains (16 °C, 1500 mm, > 800 m)	Winter: *A. strigosa*, *S. arvensis*, *R. sativus*, *Lath. sativus*, *V. sativa*, *V. villosa* Roth Summer: *C. retusa* L., *C. virgulata* subsp. *grantiana* Klotzsch (Harv.) Polh., *C. spectabilis*, *Can. ensiformis*, *Caj. cajan*, *M. deeringiana*, *M. nivea*
West (18 °C, 1500 mm, <500 m)	Winter: *A. strigosa*, *Secale cereale* L., *S. arvensis*, *R. sativus*, *V. sativa*, *V. villosa*, *P. sativum*, *Lath. sativus* Summer: *M. deeringiana*, *M. nivea*, *Can. ensiformis*, *C. retusa*, *C. spectabilis*, *C. virgulata*

[a] Information on average temperature, rainfall and altitude is given in parentheses.

GMCCs and Reduced Tillage in Beans and Maize

Both these crops are cultivated throughout the state, although the area planted to maize (some 750 000 ha) is about three times the area planted to beans. Maize predominates because of its importance as feed, food and a cash crop. Beans are the first cash crop of the year for many farmers and are important to the family diet but their principal downside is the difficult manual harvest. Typically, a farmer will plant at least some land to both these crops each year. Numerous variations for these two crops are possible with different GMCCs. The species most often used with beans are black oats and oilseed radish, whereas with maize, vetch (*Vicia* sp. L.) and *Mucuna* are also common. The maize or beans are generally planted using direct seeding or minimum tillage, while some farmers use a disc harrow to incorporate the green manure. State-wide, the average yield for maize is about 4 t ha^{-1}.

GMCCs and Minimum Till in Onions and other Horticultural Crops

During the 1970s and 1980s, the area devoted to onion production rose greatly in the Upper Itajaí River Valley, which features steep slopes and fine-textured soils. The conventional system of soil management, which used the disc harrow and rotary hoe in succession, led to acute soil losses in the onion crop, whose foliage provides scant cover for the soil. Consequently, GMCCs were introduced and machinery was adapted for a minimum till system.

In some parts of the Valley, minimum tillage is used on 60-70% of the properties. Onions are cultivated on small-scale family farms on plots generally no larger than 2 ha. Typically, a maize-GMCC-onion rotation is repeated annually, with black oats or oilseed radish as the GMCC. Vetch is also used, either alone or, more commonly, intercropped with black oats or radish. Some farmers use *Mucuna* or *Mucuna* plus maize. *Mucuna*, *Canavalia* and *Crotalaria* species yield good results as a summer GMCC but most farmers are unwilling to forego the maize or bean crop. After a period of rest in which the green manure crop dries and begins to decay, a special narrow rotary hoe is passed over the soil for each crop row, so that only about one third of the surface area is actually tilled, and the rest remains covered by the GMCC residues. An outlet for applying mineral fertilizer precedes each rotor, so that three or four rows are prepared in a single pass. As the C/N ratio of black oats is wide, heavy doses of mineral fertilizer are frequently applied, from 500 to 1500 kg ha^{-1} or more of 5-20-10. The onion seedlings are then transplanted manually and frequently a pre-emergent herbicide is applied.

Use of family labour during transplanting and harvest is intense. Maize follows the onion crop, benefiting from the residual fertilizer applied to the cash crop. The method has proven effective for minimum till in many different horticultural crops, including tobacco. Adaptations exist for animal traction, micro tractor, and full-size tractor. Recently, a mechanical seedling transplanter appeared on the market that can be adapted to either draft animals or a mini-tractor. Onion yields in this region average about 10 t ha^{-1} fresh weight of marketable bulbs. Although it may take several years for a yield advantage to materialize, research results indicate that this system may have favourable impacts.

GMCCs and Minimum Till in Tobacco

Flue-cured tobacco, also a labour-intensive crop, can represent a significant portion of the family's annual income, particularly among the smaller scale farmers. The main limitation on production is the availability of family labour for harvest and curing. As with onions, use of family labour is intense during transplanting and harvest of the tobacco crop. The multinationals that dominate the market have intensely promoted the use of GMCCs. The most common winter species used by farmers are *Lupinus angustifolius* L., *A. strigosa, Vicia* spp. and *R. sativus*, although some use of *Mucuna* spp. has been reported. Average tobacco yield in the state is about 2 t ha^{-1}.

Cultural Operations

The GMCCs can be sown before the economic crop is harvested (relay planting), after it is harvested or after the crop residue is flattened, depending on the particular situation and the desired effect. Seeds of other GMCC species require a light incorporation at planting. Medium- to large-seeded species are generally direct seeded. In some regions, it is possible to rely on natural re-seeding of certain cover crops, such as *V. sativa, Spergula arvensis* L., *Lol. multiflorum, Melilotus albus* (Mediki), *Ornithopus sativus* Brot., *Trifolium subterraneum* L., certain varieties of *T. incarnatum* L. and some *Crotalaria* spp.

The timing of cultural operations can be crucial in determining the success of the subsequent crop and requires, most of all, good husbandry skills on the part of the farmer. Numerous factors come into play, including:

- The cover crop species,
- Its growth stage at the desired planting date of the economic crop,
- Cover crop growth habit and density,
- Season of the year,
- Recent weather,
- Whether seeds or seedlings are to be subsequently planted,
- Type of machinery to be used for levelling/killing the cover crop and planting the subsequent crop, and
- Whether or not desiccants are used to kill the cover crop.

Managing the cover crop generally requires flattening it with a roll-chopper (see the section on machinery, below) so as to kill it before planting the economic crop. This is usually done between full flowering and maturation of the cover crop, because this provides the greatest biomass and most nutrients to the soil. Most farmers apply a herbicide to the cover crop before flattening, which makes subsequent mechanical management easier and more effective. With black oats, for example, the cover crop is often flattened when the grain is in the milky stage, before maturation. It is important not to wait too long, because the stem of most plants begins to lignify after pollination. This widens the C/N ratio of the plant biomass, retards the decay and release of nutrients, may hamper the effectiveness of the roll-chopper and the operation of the planting equipment, and may turn mature GMCC seeds into weeds in the next cycle. On the positive side, lignified GMCCs lead to longer-lasting soil cover.

Most farmers observe a waiting period after the GMCC is flattened in the field and before the subsequent crop is planted. This seems to have several purposes, including: (1) insect herbivores flee the decaying crop field; (2) planting operations are easier after the cover crop has begun to decay; and (3) the decaying foliage may at first release volatiles or leachates unfavourable to the growth of the subsequent crop. (This alleopathic effect varies greatly both by GMCC and by main crop [Almeida 1988]). The required length of waiting period varies from species to species, depending on the C/N ratio of the mulch and on factors listed above. In general, when legumes or oilseed radish are flattened, the waiting period is 1 to 2 weeks, while winter cereals may require 3 to 4 weeks.

The green manure biomass is managed in one of three ways for subsequent planting of the economic crop as described below.

Full Incorporation of Plant Biomass with Conventional Tillage

This option has become less common over the past 20 years. The cover crop is flattened and then incorporated with a plough or a disc harrow.

Partial Incorporation of Plant Biomass with Minimum Tillage

In this case, the soil is prepared by opening a furrow for planting seeds or transplants, thereby incorporating some 30% of the cover crop. There are four variations, based on the cover crop species and its growth stage at the time of planting the economic crop.

The first is to minimum till once, before flowering of the cover crop: When the cover crop growth habit is prostrate and of slow initial growth or long seasoned (e.g. *V. villosa*, *Lathyrus sativus* L.), the furrow line is ploughed when the cover crop reaches 100% soil cover. This allows early sowing of the main crop, while the cover crop plants between the furrow lines are allowed to mature their seed. This variation can be used for *T. incarnatum*, *Lens culinaris* subsp. *culinaris* and *S. arvensis*. In some cases, after the cover crop has set seed, a secondary crop can be planted between the rows of the main crop in relay fashion.

The second possibility is to minimum till twice, first before flowering and then during full bloom of the cover crop: in this variation, maximum cover crop biomass is the objective. When plant growth is robust and the farmer uses animal traction, the cover crop may impede the ploughing of the furrow, or subsequent germination of the economic crop. Hence the furrow is ploughed twice: once when the cover crop reaches 100% soil coverage, and again when the cover crop is in full bloom, for the actual seeding of the subsequent crop if *S. arvensis* or species of *Vicia* or *Lathryus* are used. After the cover crop matures its seed, a secondary crop can be direct seeded in relay fashion.

The third variation is to minimum till after winter grain harvest: When the cover crop is a winter grain and the grain is to be harvested, such as wheat, black oats, triticale, or rye, the furrow is ploughed and the succeeding crop is sown immediately after harvest of the grain.

The fourth variation is to minimum till after flattening the green manure: This is typical of summer cover crops, such as *Mucuna*, but can also be done for certain winter crop systems, such as vetches. The cover crop is first flattened, with or without prior use of a desiccant. After it has begun to decay, the furrow is ploughed.

With animal traction, the minimum till system is designed for crops with relatively large spacing between the rows, around 1 m, such as maize, cassava and tobacco. These systems can present difficulties with ploughing the furrow (due to, for example, plant mass and resistant stems), increased incidence of rats and certain soil pests and greater susceptibility to particular weeds. In addition, animal traction is difficult to use with crops that use less than 1 m between rows.

No Incorporation of Plant Biomass with Zero-tillage

Zero tillage management of cover crops requires specialized, although low-tech, machinery for direct seeding and flattening of cover crops. This option is becoming increasingly popular as better and cheaper equipment comes on the market and as farmers in the region gain experience in managing GMCCs. Because the soil is not moved, this system requires the least effort of the three, although it arguably requires the greatest management expertise. If soil cover is adequately managed, erosion is minimal in this variation. The effects of zero-tillage on soil quality and economic returns are discussed later under Impacts of GMCC Systems.

Machinery Adapted to Work with GMCCs

The private companies that settled German immigrants in Santa Catarina in the late 19th century took care to recruit skilled craftsmen and professionals. The metalworking and mechanical talents of the colonists for adapting farm implements is still observable today. When erosion revealed the need for new methods of soil management, farmers, researchers and craftsmen worked together to devise farm tools appropriate for the task. Numerous machines are available from private companies that specialize in equipment for zero tillage. Most small-scale farmers in Santa Catarina own draft animals or mini-tractors, and small local firms build a variety of machines for this market. The development and availability of these implements have been crucial factors in the adoption of GMCC systems in Santa Catarina. For full-sized tractors, machinery from national and multinational firms is available.

Machines have been developed for the different cultural operations in the GMCC systems, both for the flattening of the GMCC and for the seeding of the main crop. For flattening the cover crop, two different types of machines are available:

(1) An implement called *rolo-faca* (i.e. rolling stalk chopper, roll-chopper, roller-crimper or knife roller) is most common. It consists of a cylinder or drum with blunt blades set across it lengthwise; the drum rolls around an axle, which is pulled along behind the power source. The blades on the drum crimp and crush the plant shoots as it rolls over them.

(2) For plants with a trailing growth habit, the *rolo-disco* (i.e. disc roller), which consists of a row of discs, is used to cut through the vines and stems. This machine is mainly used in the state of Santa Catarina.

Although both of these implements work quite well even on steep slopes, neither works well in rocky soils and they may require the field to be levelled with a disc harrow in the first year to work properly. Some farmers use a disc harrow at a superficial depth, leading to partial incorporation of the cover crop, which hastens decay and nutrient release and yet leaves some soil cover.

The seeds of the economic crop can be planted in a variety of ways, including manual, mechanized and combined planting systems. If the crop is planted from seed, the manual option is generally to use the *matraca* (i.e. hand jab planter), which is a hand-held, direct seeder for large-seeded crops, such as maize and beans. Alternately, the *sulcador*, a special animal-drawn ripper fitted with a coulter in front, cuts a 40-cm wide furrow through the cover crop, leaving the GMCC residue on either side undisturbed. The farmer then uses the jab planter to seed along the prepared furrow line. When this implement is used on the contour with a

1-m spacing between crop rows, the result is contour strips with and without mulch cover.

The mechanized options include a *semeadora* (seeding machine), which is normally used with animal traction or mini-tractors. Although there are numerous variations, the seeder generally consists of a chassis with a coulter to cut through cover crop stems, followed by a tine, or forward-facing, converging double discs, to create a slot for the seed, and a guide for placing the seed and fertilizer. Rearward-facing, converging double discs, or a weighted wheel, then close the furrow and compress the soil. A simpler arrangement is to have an implement for opening the furrow, which consists of just the coulter and inward-facing, double discs or tines and sometimes a guide for applying fertilizer. Seeding is then done manually or with a *matraca*.

For transplanting seedlings, two options exist: The manual option is to fashion a seedling hole in the cover crop mulch with a hoe. Most commonly, however, a combined system is used for seedlings: a minimum-till cultivator or adapted moldboard plough prepares the furrow line into which the seedlings are manually transplanted. This is a common scenario for tobacco, onions and most vegetables. The minimum-till machines for tractors or mini-tractors use a set of narrow rotary tillers to cultivate tillage lines for planting the seed or seedling. Local firms have recently produced mechanized seedling transplanters but they are an expensive novelty.

The seeding machine and hand-jab planter are usually built to deliver fertilizer at the same time as seeding is performed. Different versions of some machines exist for the different power supplies: tractors, mini-tractors and animals. Poorer farmers simply use an ox to haul a log or heavy chain over the cover crop after applying a desiccant and then use a *matraca* to plant the seed, with or without the modified moldboard plough.

Seed Production of Green Manure/Cover Crops

The availability of seed is one of the principal factors limiting the adoption of GMCCs in Santa Catarina. Availability differs dramatically for different species, in different regions and localities, and from year to year. When the weather conditions are favourable to seed production, many farmers simply reserve the area with the most abundant growth of the cover crop for seed production. In good years, it may be reasonably easy to find relatively high-quality seed of a certain species, whereas in bad years, the most highly sought-after species, generally legumes, can be simply unavailable. A few farmers specialize in GMCC seed production, and some integrated commodity chains (i.e. pork, poultry, tobacco) contract for the production of desired GMCC seeds, as do some producer cooperatives.

Numerous other problems stem from the lack of established 'best practice' guidelines for seed production of the different species and cultivars. The lack of certification results in the marketing of low-quality, sometimes pathogen-bearing, seeds. The lack of standards, quality control and market grading leads to considerable confusion about the precise seed species and quality. All of these irregularities can lead to yield losses in the economic crop. Although the state and the private sector have made some advances in sorting through these issues for the most important species, much progress remains to be made. Currently, EPAGRI has several projects in GMCC quality control, certification and plant breeding. Monegat (1991) and da Costa et al. (1993) discuss different aspects of seed production in detail and present copious data.

Because of problems with seed availability, many farmers have opted for black oats or oilseed radish, two species with rapid soil cover and robust biomass, which also reliably produce abundant seed. In Santa Catarina, these two are legendary for their benefits to economic yield, soil cover and weed suppression, and they are equally important as forage crops. Because the C/N ratio of black oats is wide, its use is invariably tied to use of mineral N, which is heavily applied to subsequent cash crops, such as vegetables and tobacco. Black oats can also be intercropped with *Vicia* spp. Another option in parts of the state is to rely on *Brachiaria plantaginea* (Link.) Hitchc., a naturalized grassy weed, which re-seeds itself yearly.

Other Considerations

The use of GMCCs has made it possible to establish cropping systems that maintain soil cover year round. Because GMCCs are more difficult to manage with just mechanical/cultural operations, most farmers opt for the practicality of herbicides, generally glyphosate. The species of the GMCC also factors in; for example, oilseed radish lends itself well to strictly mechanical management, whereas chemical control is more efficient for black oats and the vetches. Certain fields may need to be prepared for reduced tillage by removing rocks and stumps or harrowing the soil so that the various implements, such as the knife roller or direct seeding equipment, work well.

Numerous authors stress the importance of combining crop rotations and reduced tillage to reap the full economic and soil conservation benefits of GMCCs. Equally important for soil protection is the soil conservation infrastructure present on each farm. These measures, such as land-use capability planning, contour tillage and vegetated strip buffers, reforested or grassed waterways, and proper site selection and construction of roads, should be adopted in parallel to the use of GMCCs.

IMPACTS OF GMCC SYSTEMS

Over the last 20 years, GMCC cropping systems have been the object of exhaustive study by researchers in southern Brazil. Derpsch et al. (1991), Monegat (1991) and da Costa et al. (1993) are three valuable reviews of this large body of research. Most studies document favourable impacts of GMCCs on soil quality, crop productivity, labour requirements and financial returns. CEPA-SC (1999) compared adopting farmers' opinions regarding conventional versus reduced tillage systems, the latter being typically associated with GMCC systems, and their responses were largely in line with scientists' findings. A synopsis of research results is presented below, including some negative impacts of GMCC systems.

Labour-Saving Attributes of Reduced Tillage

Soil preparation under conventional tillage requires ploughing and then harrowing the soil. In animal-drawn systems, these are intensely fatiguing physical tasks. Monegat (1991) states that farmers will not adopt any technology that increases manual labour, whether in man-hours or the arduousness of the task. Under conventional tillage, the use of green manures increases overall labour requirements: GMCC seeds must be produced, harvested and processed and the crop must be turned under. Reduced tillage, however, reduces the number of soil operations and eliminates the most arduous tasks—ploughing and harrowing the soil. Hence, it decreases the labour days and animal-days or tractor-hours required to cultivate 1 ha (Table 4). Reduced labour for ploughing and harrowing is one of the single most important factors in farmer adoption of reduced tillage GMCC systems: CEPA-SC (1999) found that 87%, and Maciel et al. (1998) found that 100%, of farmers interviewed cited lower labour requirements as a reason for adopting zero-tillage. In mechanized systems, reduced tillage brings fewer tractor hours, lower fuel costs and less wear on equipment.

Table 4. Comparison of labour and animal or tractor usage for 1 ha in conventional-tillage (CT) and zero-tillage (ZT) systems (*Source*: CEPA-SC 1999).

Crop	Traction source	Labour (d)		Use of traction (animal-d or tractor-h)		Savings from using ZT (%)	
		CT	ZT	CT	ZT	Labour	Use of traction
Maize	Animal	7.22	3.23	6.82	2.32	55	66
Bean	Animal	12.93	5.37	5.48	2.97	59	46
Maize	Mech.	3.81	1.85	7.28	5.92	51	19
Soybean	Mech.	0.86	0.77	7.59	5.56	10	27
Onion	Mech.	34.74	29.29	30.98	21.99	16	29

Reduced tillage with cover crops affects farm labour and the overall ease of operations in another important aspect: the faster infiltration of water and its greater availability afterward bring a larger window of time during which soil conditions are favourable to planting. This reduces labour demand peaks and makes the farm family much more self-sufficient in terms of labour (Derpsch et al. 1991; CEPA-SC 1999).

Impacts on Soil Quality

The use of GMCC systems with reduced tillage improves a broad spectrum of soil properties. The single most important factor is probably the protection of the soil from raindrop impact, the catalyst for erosion. Research in Brazil clearly shows that well-managed GMCC systems can reduce erosion to less than 5% of that observed under conventional soil tillage (Castro 1956; IAPAR 1981; Fundação Cargill 1984). Beyond the effects of protective cover, the recurrent addition of sufficient quantities of fresh high-quality organic matter to the soil surface promotes a notable series of improvements in soil physical, chemical and biological properties. Table 5 presents research reports available on the subject.

Table 5. Research results on soil properties that improve under reduced tillage and GMCC systems.

Property	Source
Physical:	
Bulk density	Derpsch et al. 1986
Porosity	Derpsch et al. 1986
Infiltration	Miyasaka 1984; Derpsch et al. 1991
Soil water retention	Derpsch et al. 1991
Available water	Roth 1985
Soil moisture	Derpsch et al. 1991
Hydraulic conductivity	Roth et al. 1986
Soil temperature	Miyasaka et al. 1966; Derpsch et al. 1985; Sidiras and Pavan 1986b
Aggregate stability	Sidiras et al. 1982
Soil cover from residue	IAPAR 1984; Roth et al. 1986; Monegat 1991
Chemical:	
C, N, P, K, base saturation, pH	Sidiras and Pavan 1986a
NO_3^-	Heinzmann 1984; McGuire 1999
Biological:	
Earthworms	Voss 1986; Kemper and Derpsch 1981
Arthropods	Kemper and Derpsch 1981
Cellulose degradation	Kronen 1984
Respiration	Kronen 1984
Rhizobial nodulation	Voss and Sidiras 1985
Nematode control	Sharma et al. 1982
Weed control	Derpsch et al. 1982; Almeida and Rodrigues 1985; Almeida 1988
Soil microbial biomass	Monegat 1998; McGuire 1999; Pereira 1999

Black oats, for example, are particularly renowned among farmers for their ability to enhance soil structure (Antonio Uberti, personal communication. 1999). In general, the improvements to soil quality greatly enhance the ability of crops to surmount adverse climactic conditions, particularly dry weather. Conversely, it appears that in very humid periods, excessive moisture may hamper crop development in reduced tillage GMCC systems, and cooler soil temperatures may delay the germination of certain warm-season crops. Also, the effect of freezes may be intensified by the presence of soil cover.

Impacts on the Yield and Quality of the Economic Crop

When compared to a fallow, the residual effect of GMCCs on the productivity of the subsequent economic crop can be positive or negative. The impact varies according to a host of factors, especially the specific GMCC-economic crop combination in question, soil and climate characteristics and the management given to the crops. When these variables are favourable, the literature documents yield increases of up to 75% over a fallow control, with figures around 30% commonly observed. With maize, the following results have been obtained.

The state-wide average for maize yield is about 4000 kg ha^{-1}. CEPA-SC (1999), based on yields reported by farmers, found that maize yields were higher with direct seeding and GMCCs than with conventional soil management by 30% for mechanized systems and 20% for animal-drawn systems. Working with summer GMCC species, Calegari (1990) found that, compared to a control, maize yields after *Crotalaria spectabilis* Roth were 22% higher, after *Pennisetum americanum* (L.) Leeke, 55% higher, and after *C. pallida* Aiton, 61% higher. Working with winter GMCCs, Derpsch et al. (1991) found that, compared to a winter fallow, maize yields were 25% higher after two legumes, *L. albus* and *V. villosa*, and 13% higher after *R. sativus*, but the yield impact was negative after winter grains such as black oats, wheat, or rye. Calegari (1990) found that maize yields after *V. villosa* were higher than after a winter fallow + 120 kg ha^{-1} N. In a 3-year study, Mendes (1938) found that *Mucuna* relay cropped into maize improved yields by an average of 14% compared to a maize monocrop and that maize yields after *Pisum* spp. were 23% higher than after a winter fallow. In a 10-year study at three locations, Viégas et al. (1960) found that *Mucuna* relay cropped into maize improved yields by an average of 11 to 36%, depending on site, compared to a maize monocrop. They also found that maize yields with *Mucuna* increased over time, such that yields in the latter years at two of the locations were significantly higher than the 10-year average. Kanthack et al. (1991) found that maize yields after lupine did not respond to additional nitrogen.

For beans, the average yield is about 500 kg ha^{-1}. Research results include the following. CEPA-SC (1999) found that bean yields were 37%

higher with GMCCs and direct seeding than with conventional soil management, while Bassi (1999) reported that bean yields improved by only 3%. Derpsch and Calegari (1985) found that bean yields after *A. strigosa* were 85% higher, after *R. sativus*, 67% higher, and after *L. albus*, 30% higher, compared to after a winter fallow. Mascarenhas et al. (1967) found that bean yields in eight localities averaged 31% higher after *C. juncea* L. than without.

With diverse crops, results include the following. Comparing production of onion in GMCC/minimum till systems versus systems with conventional soil management, CEPA-SC (1999) found that yields were 26% higher with minimum tillage and Tassinari (1989) found yields to be between 15% and 30% higher with minimum tillage. IAPAR (1984) found that soybean yields were 37% higher after black oats than a winter fallow. Passos et al. (1981) found that yields of maize and cotton (*Gossypium hirsutum* L.), without added nitrogen, were greater following soybean than fallow plus added nitrogen.

In tobacco and maize cultivation, the case of one farmer, Mr. Roland Ristow of Ibirama, is particularly interesting. For half a century, his family cultivated tobacco on a small property with acid soils on a 20-45% slope. In 1964, Mr. Ristow began using an annual tobacco/*Mucuna*/maize rotation that continues today. In 1974, he adopted minimum tillage and in 1984, he moved on to zero tillage. During the tobacco harvest, *Mucuna* is planted between the rows of tobacco plants. The seeds of *B. plantaginea*, a rapid-growing naturalized grass, germinate along with the *Mucuna*, and the result is a very well-adapted cereal-legume GMCC association: the *Mucun*a provides nitrogen to the system and the grass provides durable soil cover. Soil quality on his property is so high that his tobacco yield in 1997, 2.7 t ha^{-1} (35% higher than the regional average), was obtained using 50% of the recommended fertilizer rate and 50% less man-hours than with conventional tillage (Pereira 1999). His maize yields have also risen with time. His yield in 1997, 5100 kg ha^{-1}, was 82% higher than the average yield in his township, 2800 kg ha^{-1}, even though Mr. Ristow does not apply mineral fertilizers to his maize.

According to Monegat (1991), the residual effect has a cumulative component, such that while a jump in yield may or may not be observed the first year, the full impact may not occur for 10 or more years. Many farmers growing crops with stringent grading standards, such as tobacco and other horticultural crops, note that the crop quality is improved in reduced tillage GMCC systems. Since plants suffer less water stress when soil cover is abundant, the yield advantage of reduced tillage GMCC systems is greater in low rain years. Accordingly, yield stability from year to year is greater under reduced tillage. Over the long term, the greatly decreased erosion under zero tillage and the effects on soil quality clearly have significant impacts on yield (Pereira 1999).

Yield-reducing residual effects may be due in part to allelopathic properties of the cover crop for the subsequent planting. Beyond trial and error, there seem to be few concepts or tools to predict which specific

GMCC-economic crop combinations will be successful in increasing yield, but some trends are clear:

- Abundant production of plant biomass by the GMCC is a necessary condition.
- Rapid soil cover is needed to suppress weeds, although slower growing species can be successful if they are relay cropped into the previous crop cycle, giving them an early start.
- Even when considering just leguminous GMCCs, the above two factors seem to outweigh the importance of relative nitrogen-fixing ability.
- A gramineous cover crop is not advisable before a non-leguminous economic crop.

GMCC residues in the soil interact favourably with applied mineral fertilizers. The use of GMCCs can lead to remarkably increased fertilizer use efficiency by the economic crop (Kumwenda et al. 1996; Rosemeyer 1996; McGuire 1999). The most skilful farmers combine green manures and mineral fertilizers in such an artful way that they need apply only one third to one half of the recommended fertilizer rate to achieve the same yields obtained under conventional tillage without a cover crop (Pereira 1999). Findings by McGuire (1999) suggest that the impact of fresh organic matter on soil microbial dynamics plays an important role in mediating nutrient availability to the economic crop.

Weeds and Herbicides in Reduced Tillage-GMCC Systems

Do reduced tillage systems with cover crops reduce or increase herbicide use? This question elicits contrasting opinions, apparently conflicting data and a fair amount of confusion. Recent data from Paraguay suggest that when using appropriate GMCC, weed pressure can be significantly reduced in the no-till system (Vallejos et al. 2001). Instituto CEPA-SC (1999) concludes that weed incidence and herbicide use are greater in reduced tillage, while Monegat (1991) and Pereira (1999) affirm that careful management of soil cover with appropriate rotations leads to reduced weed incidence.

It is a complex issue with, perhaps, no simple answer. Many factors are involved, such as soil characteristics, weather conditions, the species involved, the specific crop sequences used, the machinery employed and even the farmer's previous experience with herbicides (Roland Bunch, personal communication, 2000). Some factors are time dependent, such as the degree of recovery of soil quality, the degree of adaptation of machinery to local conditions, or the amount of knowledge in a community on managing a system that is new, versus one that is well established. Field observations suggest that the crop husbandry skills of the farmer are crucial: the 'best' farmers are able to reduce herbicide use, while the 'average' farmer may use more.

The use of desiccants to kill the cover crop is common among farmers, in part to ease the mechanical operations and in part to eliminate cover crop resurgence. Once the cover crop has been flattened, the blockage of the sun's rays inhibits the germination of many weed species. Others may be affected by the allelopathic properties of certain cover crops, such as black oats or rye. On the other hand, the mulch interferes with pre-emergent herbicides, so selective contact herbicides are often employed in the economic crop. In Brazil, multinational agrochemical companies have invested in research and extension on the management of GMCCs. In their view, conservation tillage is inextricably linked to the use of herbicides, although they claim that reduced tillage systems require less herbicide than conventional tillage.

Economic Impacts

Reduced tillage systems with green manures affect the cost structure of crop production in several ways. The presence of green manures increases fertilizer use efficiency and crop productivity, as well as lowers N requirements when a legume is part of the rotation. Requirements of P and Ca are lower when mulch is abundant and the soil is not disturbed (Derpsch et al. 1991). While some studies report a reduction of overall operating costs by around 10% in reduced tillage systems, other studies find that lower soil preparation costs are offset by increased expenditures for herbicides. A consistent finding is that improved crop productivity with GMCCs brings greater returns than conventional tillage.

Evaluating a system that combined crop rotations, GMCCs and zero tillage, Sorrenson and Montoya (1984; 1989) found returns can be as much as 100% higher compared to repeated soybean-wheat or maize-fallow cycles with conventional soil management. CEPA-SC (1999), looking at maize and soybean, found returns to zero tillage systems to be about 65% higher than returns to conventional systems based on mechanized tillage and up to 90% higher than returns to conventional systems based on animal traction. Frasson (1997) found that, compared to conventional tillage, direct-seeded maize with animal traction reduced costs by 9% and increased returns by 16%, and direct seeded maize with mechanized traction reduced costs by 12% and increased returns by 16%.

However, reduced tillage can bring unfavourable economic results in the first few years, as farmers gain experience in managing the cover crops, soil conditions improve and the machinery is adapted to local conditions. Capital requirements for specialized machinery can be a barrier to adoption among poorer farmers. This problem has been partly addressed in some areas of Santa Catarina with the collective purchase of equipment by groups of farmers.

Hydrographic Impacts

Bassi (1999) monitored water quality and river discharge in the Chapecó micro-watershed in western Santa Catarina. From 1986 to 1996, under the auspices of the micro-watershed basin management project, the area cropped with cover crops in this micro-watershed increased from 365 to 1963 ha, while the area under reduced tillage increased from 0 to 1465 ha. During this time, both average and minimum discharge rates increased, hence water became more available. Maximum discharge rates remained relatively stable, suggesting less run-off and an increase in subsurface flow. Over the 10-year period, turbidity fell by 61% and the sediment load by 70%, so water quality improved measurably.

ADOPTION OF GMCC SYSTEMS

Given the collaborative efforts, institutional support and the contributions from formal research related above, the use of GMCCs and conservation tillage has increased enormously in the state over the past 20 years. In what follows, we first describe farmer adoption of these systems and then discuss the factors that both foster and limit adoption.

Trends in the Use of GMCC Systems

Results abstracted from different studies by EPAGRI are brought together in Table 6. Because the methodologies employed in these studies were not uniform, the precision of these data is open to question but the overall trend is nevertheless clear.

Table 6. Trends in area cropped to green manure/cover crops (GMCCs) in Santa Catarina (*Source*: ACARESC 1987; Freitas et al. 1998).

Year	Area with GMCCs (ha)	% of arable cropland
1987	112 000	5
1994	332 000	18
1997	793 000	44

Several studies by EPAGRI sought to uncover the extent of reduced tillage in the state. The area under conservation tillage was estimated at 123 000 ha in 1994 and at 685 000 ha in 1997 (Freitas et al. 1998). Of those 685 000 ha, 36% were managed with minimum tillage and 64% with zero tillage. Freitas (1997) conducted a study of one particular microcatchment basin, Ribeirão das Pedras, where extension efforts were intense over a 10-year period. In 1984, GMCCs were used in 18% of arable cropland, whereas in 1994, the figure was 83%. During this time, the use of reduced tillage rose from 0 to 80% of arable soils.

Recent data from one tobacco company that promotes GMCCs and reduced tillage indicate that the percentage of contracted farmers using GMCCs continues to increase (Table 7). Among these farmers, zero tillage is rising, minimum tillage is declining and conventional tillage has disappeared.

Table 7. Trends in the use of green manure/cover crops (GMCCs) and tillage regime among 7150 farmers in contract with a tobacco company in western Santa Catarina (*Source*: Alceu Cericatto, personal communication, 1999).

Indicator	1998	1999
Area cropped to tobacco (ha)	7913	7794
Farmers using GMCCs (%)	71	92
Black oat seeds financed (t)	330	540
Area cropped with GMCCs (%)	61	82
Of which:		
zero till (%)	31	65
minimum till (%)	61	35
conventional till (%)	8	0

Developments in GMCC technology in Santa Catarina are part of wider changes in soil management sweeping through the Southern Cone countries, where zero-tillage is growing rapidly. According to Derpsch (2001), in the MERCOSUR countries (Brazil, Argentina, Paraguay and Uruguay), the technology has expanded 20-fold between 1987 and 1997, compared to a fivefold increase of the area in the USA in the same period. In just 1 year, 1997, the land under no-till increased by 28%, compared to 4% in the U.S.

Factors Enabling Adoption and Diffusion

Several intertwining factors fostered the widespread adoption of conservation tillage GMCC systems in Santa Catarina. While farmer-driven technology may be at the core of this phenomenon, institutional factors clearly had multiple impacts.

Endogenous Development of Appropriate Technology

Different local actors developed a variety of equipment adapted to diverse social strata. Farmers, local metalworkers, technicians and businessmen collaborated to make this technology available for the poorest farmers with no source of traction as well as for animal-drawn machines, mini-tractors and full-size tractors. For all but the full-size tractors, the machinery is well adapted to operate along the contour of even steep hillsides.

Labour-saving and Cost-cutting Impact of Conservation Tillage Machinery

Farmers adopt these conservation technologies less from a heartfelt need to preserve the environment than because they make farm labour less exhausting and reduce the man-hours involved. This humanizes the work and frees up time for other activities. In some systems, and depending on the husbandry skills and infrastructure of the particular farmer, cash outlays can be reduced significantly for a given volume of product or level of net income.

Private Sector Involvement

Local entrepreneurs who believe in the sustainability of the technology produce the machines in workshops and small factories, making them available to farmers. State-wide and multinational agro-industrial concerns promoted their diffusion and use for soil conservation.

Dedicated Agricultural Research and Extension Agencies

Each of the three states in southern Brazil has an extensive and far-reaching state agency for agricultural research and extension—IAPAR in Paraná, EPAGRI in Santa Catarina and Empresa de Assistência Técnica e Extansao Rural (EMATER) in Rio Grande do Sul. They all seek technical solutions for soil erosion, actively promote conservation tillage GMCC systems and collaborate extensively in their search for solutions. These efforts are complemented by agencies in other states and by EMBRAPA, the national agricultural research agency.

Shifts in National Agricultural Policy

In the 1970s the Brazilian government initiated a programme to modernize agriculture with industrial inputs through credit subsidies and extension efforts. This led most farmers and researchers to abandon the use of green manures. At the same time, rotary hoes for soil preparation became common. In the 1980s, the oil crisis led the government to eliminate the credit subsidy for agrochemicals. In parallel, soil erosion became a motivating factor, demonstrating the cumulative effects of years of soil management based on mechanical tillage and chemical fertilizers. Farmers and agriculturists then returned to GMCCs for cost-effective recovery of soil health.

Favourable State and Municipal Agricultural Policy

The strong tradition of family farming in Santa Catarina, rooted in a smallholder land tenure regime, makes state and municipal governments sympathetic to the small-scale farmer. Hence, successive administrations fund diverse and extensive state agricultural agencies and numerous research and extension projects.

External Development Organizations

The micro-watershed basin management project, co-financed by the World Bank, gave greater focus and impetus to the diverse actors and initiatives promoting GMCCs and enabled sustained efforts toward clearly established goals in education, training, research and extension.

Awareness among Farmers and Technicians

To explain why some farmers adopt conservation technologies, extension agents often invoke this factor—the individual is simply more ‘conscientious’ or concerned about conservation issues. At the same time, agricultural professionals became more aware of conservation issues as well, illustrated by trends toward a farming systems or holistic approach, environmental conservation, appropriate technology and farmer-driven research.

Factors Limiting Adoption and Diffusion

Rabelo (1991) studied why farmers fail to adopt GMCC systems in Santa Catarina. Among the reasons he cites are:

- Availability of GMCC seeds is poor or irregular.
- Little information is available on how to manage the cover crop.
- Appropriate equipment is lacking for handling the cover crop.
- The technology conflicts with the existing production structure or work calendar of a particular farm.
- The technology does not reflect the objectives or perceived needs of the farmer.
- The technology does not provide an immediate economic benefit to farmers.
- Farmers, technicians or administrators are not convinced of the need for them.

RESEARCH WITH GREEN MANURE/COVER CROPS

Since the early 1980s, studies with GMCC species have became widespread in southern Brazil. In Santa Catarina, these efforts received generous financial support and greater focus from the micro-watersheds project, which included a component for study of GMCC species and their integration into local cropping systems. The research of EPRAGI on these crops can be divided into four areas, although not every issue is exhaustively studied, nor is the work necessarily done in sequence.

(1) Introduction and initial screening: phenology, soil coverage, biomass production, seed production and pest and disease incidence.
(2) Crop ecology: seed and plant eco-physiology, planting density and planting date, seed storage, nitrogen fixation and weed suppression.
(3) Cropping systems research: crop rotations, intercropping, relay cropping, crop residue management, integration with existing cropping systems, effects on soil properties and residual effects on economic crops.
(4) Machinery: conservation tillage, cultural operations, refining existing equipment and adapting them for use in new crops.

Experiment station research sometimes follows in the path of creative farmers when trying to assess the widespread applicability and downfalls of their innovations. There can be a commodity-oriented emphasis to a particular research project, for example, trying to fit a new species into existing cropping systems (e.g. maize, beans, cassava, tobacco and vegetables). Efforts are also directed at integrating GMCCs into new cropping systems such as apples and passion fruit (*Passiflora edulis* Sims).

CONCLUSION

People in Santa Catarina have a cultural patrimony of small farm husbandry and local craftsmanship. There are several important manufacturing clusters in the state and incomes are slowly rising in the cities. The farmers are reasonably well educated and technical training is available in many parts of the state. This human capital explains in part the success of some rural regions in carving out an economic niche based on specialized agricultural commodity production. Competitive economic pressures continue to take their toll and the rural population continues to dwindle, while new specialized crops are appearing in some regions. The need for new economic options and sustainable production techniques is widely felt, propelling continued collaboration among farmer cooperatives, state officials, industry groups and small businesses in the evolution of the agricultural sector in this state. In light of the state's natural aptitude for perennial agriculture, state officials have been slow to explore the economic potential of forestry with native species and mixed agroforestry systems with native and exotic germplasm.

NOTE

1. While conservation practices have certainly grown in the state, one study by Fontana (1998) found scant evidence for a cause and effect relationship between the financial incentive and farmer behaviour. The author concluded that the policy was not as effective as hoped and other factors were also at work that had a greater impact on adoption.

REFERENCES

ACARESC - Associação de Crédito e Assitência Técnica e Extensão Rural de Santa Catarina (1987) Relatório interno de conservação do solo e da água. ACARESC, Florianópolis, Brazil. Unpublished document, available from the Empresa de Pesquisa Agropecuária e difusão de tecnologia de Santa Catarina (EPAGRI), Brazil.

Almeida F S (1988) A alelopatia e as plantas. Instituto Agronômico do Paraná (IAPAR), Circular 53, Londrina, Brazil. 60 p.

Almeida F S, Rodrigues B N (1985) Guia de herbicidas: Contribuição para o uso adequado em plantio direto e convencional. Instituto Agronômico do Paraná (IAPAR) Londrina, Brazil. 468 p.

Bassi L (1999) Impactos sociais, econômicos e ambientais na microbacia hidrográfica do Lageado São José, Chapecó, SC – Estudo de caso. Empresa de Pesquisa Agropecuária e difusão de tecnologia de Santa Catarina (EPAGRI), Chapecó, Brazil. 48 p.

Calegari A (1990) Plantas para adubação verde no sudoeste do Paraná. Instituto Agronômico do Paraná (IAPAR), Boletin Técnico, no prelo. Londrina, Brazil.

Castro F S (1956). Conservasión de suelos. Salvat, Barcelona, Spain. 298 p.

CEPA-SC – Instituto de Planejamento e Economia Agrícola de Santa Catarina (1999) Avaliação do Projeto Microbacias: Análise comparativa do plantio direto frente ao sistema convencional de manejo do solo em sistemas de produção de lavouras em Santa Catarina Relatório de Avaliação Final. CEPA-SC, Florianópolis, Brazil. 37 p.

Costa M B da, Calegari A, Mondardo A, Wildner L do P, Alcântara P B, Miyasaka S, Amado T J (1993) Adubação verde no sul do Brasil. Assessoria e Serviços (AS)-Projetos de Tecnologias Alternatives (PTA) Rio de Janeiro, Brazil. 346 p.

Derpsch R (2001) Frontiers in conservation tillage and advances in conservation practice. *In:* Stott D E, Mohtar R H, Steinhardt G C (Eds.) Sustaining the global farm. Selected papers from the 10th International Soil Conservation Organization Meeting held 24-29 May 1999. National Soil Erosion Research Laboratory, Purdue University and the United States Department of Agriculture Agricultural Research Service (USDA-ARS), USA. p. 248-254.

Derpsch R, Calegari A (1985) Guia de plantas para adubação verde de inverno. Instituto Agronômico do Paraná (IAPAR), Documentos 9, Londrina, Brazil. 96 p.

Derpsch R, Hoogmoed W, Sidiras N, Almeida F S (1982) A escarificação como alternativa de preparo e conservação do solo. Presented at IV Encontro Nacional de Pesquisa Sobre Conservação de Solos. Campinas, Brazil.

Derpsch R, Sidiras N, Heinzmann F K (1985) Manejo do solo com coberturas verdes de inverno. Pesqui. Agropecu. Bras. 20(7):761-773.

Derpsch R, Sidiras N, Roth C H (1986) Results of studies made from 1977 to 1984 to control erosion by cover crops and no-tillage techniques in Paraná, Brazil. Soil & Tillage Res. 8:253-263.

Derpsch R, Roth C H, Sidiras N, Kopke U (1991) Controle da erosão no Paraná, Brasil: Sistemas de cobertura do solo, plantio direto e preparo conservacionista do solo. Deustsche Gesellschaft fur Technische Zusammenarbeit (GTZ) - Fundação Instituto Agronômico do Paraná (IAPAR), Paraná, Brazil. 268 p.

Fontana R B (1998) O comportamento dos produtores rurais frente aos incentivos financeiros para a adoção de práticas conservacionistas (um estudo do Projeto Microbacias na região Sul de SC). Masters Thesis, Universidad Federal de Santa Catarina (UFSC), Florianópolis, Brazil. 170 p.

Frasson Z (1997) Aspectos econômicos do plantio direto na pequena propriedade. Presented at 5o Encontro Nacional do Plantio Direto na Palha, Goiânia, Brasil.

Freitas V H de (1997) Impactos econômicos, sociais e ambientais em duas microbacias hidrográficas de Santa Catarina: Ribeirão da Pedras e Rio Macaco. Empresa de Pesquisa Agropecuária e difusão de tecnologia de Santa Catarina (EPAGRI) Documentos 188, Florianópolis, Brazil. 84 p.

Freitas V H de, Bet M, Gubert R (1998) Situação da adubação verde/plantas de cobertura em Santa Catarina. *In:* Reunião Centro-Sul de Adubação Verde e Rotação de Culturas, 5, 1995. Resumos. Empresa de Pesquisa Agropecuária e difusão de tecnologia de Santa Catarina (EPAGRI), Chapecó, Florinaópolis, Brazil. p. 111-122.

Fundação Cargill (1984) Adubação verde no Brasil. Fundação Cargill, Campinas, Brasil. 363 p.

Heinzmann F K (1984) Auswertung von untersuchungen zur N-dynamik in abhängigkeit von fruchtfolge und bodenbearbeitung sowie zur nährstoffaufnahme von soja auf einem Latossolo Roxo Distrófico. Deustsche Gesellschaft fur Technische Zusammenarbeit (GTZ), Eschborn. Unpublished mss. 33 p.

IAPAR - Instituto Agronômico do Paraná (1981) Relatório técnico anual, 1980. IAPAR, Brazil. 232 p.

IAPAR - Instituto Agronômico do Paraná (1984) Relatório técnico anual, 1982. IAPAR, Brazil. 326 p.

Kanthack R A D, Mascarenhas H A A, Castro O M, Tanaka R T (1991) Nitrogênio aplicado em cobertura no milho após tremoço. *In:* Pesqui. Agropecu. Bras. 26(1):99-104.

Kemper B, Derpsch R (1981) Results of studies made in 1978 and 1979 to control erosion by cover crops and no-tillage techniques in Paraná, Brazil. Soil & Tillage Res. 1:254-267.

Kronen M (1984) Der Einfluβ von Bearbeitungsmethoden und Fruchtfolgen auf die Aggregatstabilität eines Oxisols. Kulturtech. Flurbereinig. 25:172-180.

Kumwenda J D T, Waddington S D, Snapp S, Jones R B, Blackie M J (1996) Soil fertility management research for the maize cropping systems of smallholders in southern Africa: A review. Natural Resources Group (NRG) Paper 96-02. Centro Internacional de Mejoramiento de Maíz y Trigo (CIMMYT), Mexico. 36 p.

Maciel A A, Melo I B, Spada V R (1998) Viabilização do sistema plantio direto tração animal na pequena propriedade. *In*: Encontro Latino Americano Sobre Plantio Direto na Pequena Propriedade, III. Programmea e Resumos. Instituto Agronômico do Paraná (IAPAR), Pato Branco, Brasil. 31 p.

Mascarenhas H A A (1967) Efeito da adubação verde do feijoeiro 'da seca' Com *Crotalaria juncea* L. empregando-se a toda vegetação ou retirando-se do campo as hastes despojadas de suas folhas. S.I, Bragantia, Brasil: 26(17):217-234.

McGuire M E (1999) Efeitos e manejo do solo sobre fisiologia vegetal e incidência de pragas e doenças na cebola. Masters dissertation, Universidade Federal de Santa Catarina (UFSC), Florianópolis, Brasil. 62 p.

Mendes C T (1938) Adubos Verdes. Secretaria da Agricultura, Indústria e Comércio do Estado de São Paulo. Directoria de Publicidade Agrícola (Boletin). São Paulo, Brasil. 80 p.

Miyasaka S (1984) Histórico de estudos de adubação verde, leguminosas viáveis e suas características. *In:* Adubação Verde no Brasil. Fundação Cargill, Campinas, Brazil. p. 64-123.

Miyasaka S, Freire E S, Mascarenhas H A A, Nery C, Campana M P, Sordi G (1966) Efeito da adubação verde com uma gramínea e quatro leguminosas sobre a produção do feijoeiro da seca em terra roxa misturada. *Bragantia*, s.l., 25(25):277-289.

Monegat C (1991) Plantas de cobertura do solo: Cararacterísticas e manejo em pequenas propriedades. Centro Internacional de Información sobre Cultivos de Cobertura (CIDICCO), Tegucigalpa, Honduras. 337 p.

Monegat C (1998) Avaliação multidimensional do desempenho do manejo do solo no sistema do pequeno agricultor. Masters dissertation, Universidad Federal de Santa Catarinal (UFSC), Florianópolis, Brazil. 144 p.

Moser J M, Gama A M R C, Justus A R M (1994). Aptidão agrícola, padrões de uso e cobertura do solo e aptidão agrícola × uso atual de Santa Catarina. Instituto Brasileiro de Geografía e Estatísticas (IBGE)-Directoria de Geociências (DGC) e Secretaria de Agricultura e Abastecimento de Santa Catarina (SAA-SC), Florianópolis, Brazil. 42 p.

Passos S M G, Trani P E, Ferreira S (1981) A matéria orgânica e o algodoeiro. Boletim Técnico 56. Coordenadoria de Assistência Técnica Integral (CATI), Campinas, Brazil. 13 p.

Pereira J C (1999) Rolando Ristow: Uma contribuição ao estudo da agricultura sustentável. Masters dissertation, Universidade Federal de Santa Catarina (UFSC), Florianópolis, Brasil.

Pundek M (1994) Principais solos de Santa Catarina. Unpublished document, available from the Empresa de Pesquisa Agropecuária e difusão de tecnologia de Santa Catarina (EPAGRI), Brazil.

Rabelo R R (1991) Levantamento da situação atual de rotação de culturas e adubação verde na região Centro-Sul. *In:* Reunião Centro-Sul de Adubação Verde e Rotação de Culturas, 3.Organizacão das Cooperativas do Estado do Paraná (OCEPAR), Cascavel, Brazil. p. 53-90.

Rosemeyer M E (1996) Efficiency of phosphorus application to the slash/mulch and conventional systems of bean cultivation over three years. *In*: Araya R, Beck D (Eds.) Proceedings of the workshop on tolerance to low phosphorus in beans, November 1995, San José, Costa Rica. University of Costa Rica Press, San José. p. 53-90. (In Spanish with English abstract)

Roth C H (1985) Infiltrabilität von Latossolo-Roxo-Böden in Nordparaná, Brasilien, in Feldversuchen zur Erosionskontrolle mit verschedenen Bodenbearbeitungssystemen und Rotationen. Göttinger Bodenkundliche Ber. 8:101-104.

Roth C H, Meyer B, Frede H G, Derpsch R (1986) The effect of different soybean tillage systems on infiltrability and erosion susceptibility of an Oxisol in Paraná, Brazil. Z. Acker Pflanzenb. 157:217-226.

Santa Catarina (1973) Levantamento de reconhecimento dos solos do Estado de Santa Catarina. Universidade Federal de Santa Maria (UFSM), Santa Maria, Brazil. 2 vols.

Sharma R D, Pereira J C, Resck D V S (1982) Eficiência de adubos verdes no controle de nematóides associados à soja nos cerrados. Empresa Brasileira de Pesquisas Agropecuárias (EMBRAPA) - Centro de Pesquisa Agropecuária do Cerrados (CPAC), Boletim de Pesquisa, 13. EMBRAPA, Planaltina, Brazil. 30 p.

Sidiras N, Pavan M A (1986a). Infuência do sistema de manejo do solo no nível de fertilidade. Rev. Bras. Ci. Solo. 9:249-254.

Sidiras N, Pavan M A. (1986b). Infuência do sistema de manejo na temperatura do solo. Rev. Bras. Ci. Solo. 10:181-184.

Sidiras N, Derpsch R, Henklain J C (1982) Vergleich von drei Bodenbearbeitungsverfahren in bezug auf einige physikalische Eigenschaften, Boden- und Wasserkonservierung und Erträge von Soja und Weizen auf einem Oxisol. Z Acker Pflanzenb. 151:137-148.

Sorrenson W J, Montoya L J (1984) Implicações econômicas da erosão do solo e de práticas conservacionistas no Paraná, Brasil. Instituto Agronômico do Parana, (IAPAR) -Deustsche Gesellschaft fur Technische Zusammenarbeit (GTZ) Eschborn, Londrina, Brazil. 321 p.

Sorresonson W J, Montoya L J (1989) Implicações econômicas da erosão do solo e do uso de algumas práticas conservacionistas no Paraná. Instituto Agronômico do Parana (IAPAR), Boletim Técnico No. 21, Londrina, Brazil. 110 p.

Tassinari M H (1989) Plantio direto na cultura da cebola. Rev. Agropecuária Catarinense 2(1):11-13.

Vallejos F, Kliewer I, Florentin M I, Casaccia J, Clegari A, Derpsch R (2001) Abonos verdes y rotación de cultivos en siembra directa. Sistemas de producción tractorizados. Proyecto 'Conservación de Suelos', Ministerio de Agricultura y Ganadería (MAG) - Deutsche Gesellschaft für Technische Zusammenarbeit (GTZ), San Lorenzo, Paraguay. 92 p.

Viégas G P, Freire E S, Fraga Jr., C G (1960) Adubação do milho. XIV: Ensaios com mucuna intercalada e adubos minerais. Bragantia s.l. 19(57) 909-941

Voss M (1986) População de minhocas em diferentes sistemas de plantio. Plantio Direto 17:6-7.

Voss M, Sidiras N (1985) Nodulação da soja em plantio direto em comparação com plantio convencional. Pesquisa Agropecuária. Brasileira 20:775-782.

Wildner L do P, Alexandre A D, Almeida F S, Mondardo E, Lavina M L, Pereira J C (1995) Adubos verdes. *In:* Recomendação de cultivares para o Estado de Santa Catarina 1995-1966. Empresa de Pesquisa Agropecuária e difusão de tecnologia de Santa Catarina (EPAGRI) Boletim Técnico, 92, Florianópolis, Brazil: p. 7-11.

Zoldan P (1998) Evolução da agropecuária catarinense – análise intercensitária. *In*: Síntese Anual da Agricultura de Santa Catarina – 1997. Instituto de Planejamento e Economia Agrícola de Santa Catarina (CEPA-SC), Florianópolis, Brazil. p. 11-20.

Chapter 2

Slash-and-Mulch System: *Fríjol Tapado* in Costa Rica

G. Meléndez

Centro de Investigaciones Agronómicas, Costa Rica

SUMMARY

Traditionally a pre-Columbian system, the *fríjol tapado*, or slash-and-mulch system, is a system of migratory agriculture, which allows 3 months of production in 1 year. The ensuing fallow lasts for the rest of the year, or for several years, depending on the climatic conditions and availability of land in the area. The system is found especially on hillsides, in areas with low population pressure and high precipitation (800-4000 mm). It is almost exclusively practiced by smallholder farmers and is principally utilized for bean (*Phaseolus vulgaris* L.) production.

Fríjol tapado management consists of first selecting appropriate land and then slashing paths through the vegetation to create access for subsequent planting, broadcasting at high rates (25 to 40 kg of seed ha^{-1}) and slashing of fallow vegetation over the bean seeds. The fields are then left untouched until harvest. Typically, thick mulch results in low bean germination and survival, and therefore low yields (at 0.49 t ha^{-1} they are less than the national average of 0.68 t ha^{-1}).

Agronomic research efforts on *fríjol tapado* have concentrated on increasing bean yields and on better understanding the system's ecological function. The system's economics have also been studied in comparison to a dibble-stick system promoted by government extension. The socio-economic importance of the *fríjol tapado* system is rooted in the domestic production of beans, in the minimum utilization of external inputs and labour, and in the provision of additional income to the household, as about 40% of the production is sold. Moreover, the system poses little risk and requires almost no initial investment.

The *fríjol tapado* system presently accounts for 24% of national bean production, compared to 80% in 1980. Reasons for the decline include globalization/market liberalization, rural-to-urban migration, increasing

M. Eilittä et al. (eds.), Green Manure/Cover Crop Systems of Smallholder Farmers, 37-64.

land use pressure, competition with other crops and enterprises, increased risk due to changing climate and poor availability of capital to make improvements. Some factors tend to favour the continued practice of *fríjol tapado*, in particular, strong cultural traditions placing a high premium on food security and positive impacts of recent research and extension programs. Future technological improvement of the system must therefore confront the socio-economic factors that form its niche and include farmers as participants.

INTRODUCTION

Throughout the tropics, a large percentage of food for domestic consumption is produced on hillsides where productivity is generally poor and soil erosion is high. In Central America, hillsides are increasingly used for agricultural production because of the changing structure of agriculture and mounting population pressure. Commercial agricultural production, both of livestock and export crops, such as coffee (*Coffea* L.) and horticultural and ornamental crops, is on the rise and is pushing subsistence-oriented production to more marginal lands. Because of population pressure, lack of rural and urban employment and the scarcity and unequal distribution of suitable land for agriculture, small-scale farmers have migrated, especially to forested hillsides and mountains. During the 1980s, an estimated 64% of the area deforested by cutting and burning of trees in Central America was utilized to plant two staple crops, maize (*Zea mays* L.) and beans (*Phaseolus vulgaris* L.) (Lindarte and Benito 1991).

Traditionally, a pre-Columbian system, the *fríjol tapado*, or slash-and-mulch system, is used to produce beans in a sustainable manner on hillsides in Costa Rica. This system is essentially the same low-input one that has been practiced in various forms all over Mesoamerica, from Guatemala to northern Colombia (Thurston 1994). A system of migratory agriculture, it allows only 3 months of production in 1 year. The ensuing fallow may last for the rest of the year or for several years, depending on the climatic conditions and availability of land in the area. The system is found especially on hillsides, in areas with low population pressure and high precipitation. It is almost exclusively practiced by smallholder farmers.

The *fríjol tapado* system is principally utilized with beans but also, to a minor extent, with maize and rice (*Oryza sativa* L.); formally, there was some utilization with other legumes besides beans.

In Costa Rica, beans are a principal source of protein among low-income families (23% of the dietary protein in rural areas) and are consumed by 96% of the population (Castro et al. 1991; Castro 1996). The *fríjol tapado* system accounts for 24% of national bean production (CNP 1998). However, the proportion of total beans produced in the *tapado* system has decreased since 1980, when it was practiced by 80% of all

Costa Rican bean farmers and represented 47% of the national production and 63% of the total area planted (Alfaro 1994). This decrease is due to changes in the macro-economy and policies, including trade liberalization, decreased bean prices, increased cultivation of export crops and expansion of mechanized production. However, given the importance of beans for family food security as well as for national consumption, and the fragility of the lands on which they are grown, economically and ecologically viable methods are needed for confronting these changes while assuring production. Farmers in Mesoamerica have produced beans with the *fríjol tapado* system for generations, attesting to the system's resilience and sustainability.

This case study describes the current state of knowledge on the *fríjol tapado* system, including recent research efforts. A relatively large amount of research has been conducted on the system and much of it verifies its long-term sustainability, although it has some disadvantages. Lessons learned about the functioning of a time-proven, mulch-based system, such as *fríjol tapado*, will hopefully guide future development efforts of similar systems elsewhere.

CONTEXT OF *FRÍJOL TAPADO*

Biophysical Context

In Costa Rica, the areas of Acosta, Puriscal, Pérez Zeledón, La Cruz, Santa Cruz, Upala, Coto Brus, Turrialba and indigenous reserves produce considerable quantities of *fríjol tapado* beans for consumption and sale. These areas have steep slopes and rainfall ranging from 800 to 4000 mm, with soils typically of low to medium fertility. A large part of the research on *fríjol tapado* described below has taken place in two areas, Coto Brus and Acosta (Figure 1).

At Coto Brus, the total annual rainfall is about 4000 mm. Soils are relatively deep (100 cm or more to the C horizon) and fertile (effective cation exchange capacity [ECEC] of 3-15 cmol L^{-1}), with pH values in the arable layer ranging from 5.0 to 6.0. Soil organic matter levels are intermediate (4 to 21%) and aluminium saturation is 1 to 45%. Typically, phosphorus (4 to 8 ppm) and at times potassium are limiting. Most of the soils are classified as Acroxic Hapludands and Typic Haplohumults.

In comparison, Acosta, located in the Central Valley of Costa Rica, receives about 1800 mm of rainfall annually. Soils tend to be slightly more fertile, with greater ECEC and lower percentage of Al saturation but they are less deep (<60 cm). Soils in Acosta are classified as Dystropepts or Ustorthents (Mata et al. 1999).

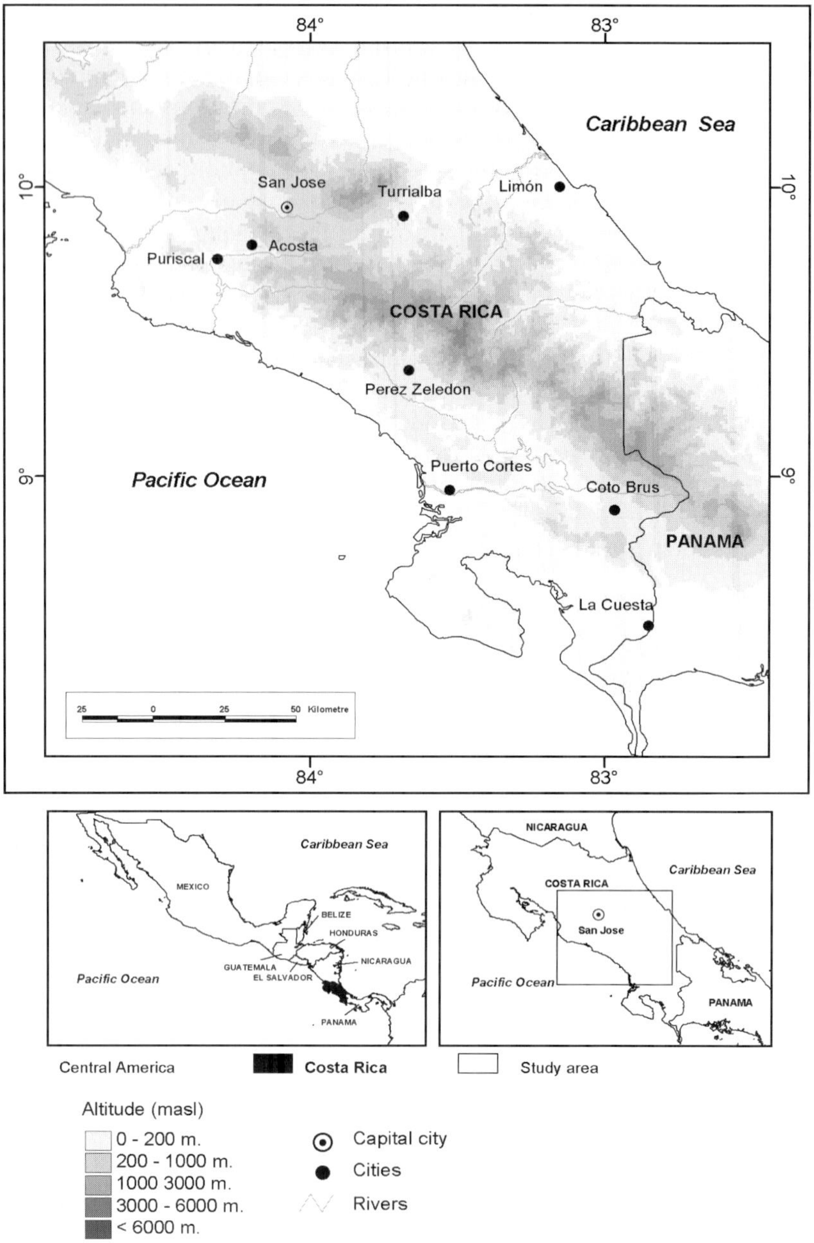

Figure 1. Location of study area, Costa Rica.

Socio-economic and Policy Context

In Costa Rica, during the past decade, a number of factors have led to a general decrease in bean production, increased importance of mechanized production systems and a reduction in the number of *fríjol tapado* farmers. At the national level, the trade of basic grains has been liberalized and the number of institutions assisting smallholder farmers, the main producers of basic grains, has been reduced. In addition, minimum producer prices have been eliminated, maximum selling prices for consumers established and restrictions on external trade eliminated. Finally, the Consejo Nacional de Producción (CNP), an institution that aided producers in commercializing grain production, has been discontinued. In the agricultural sector, the macroeconomic changes (liberalization and opening of markets) have decreased the production of primary food crops, such as maize and beans, for the domestic market. These measures have also brought about a disincentive to the *fríjol tapado* system that is mainly practiced by small-scale farmers and for food security.

At the local level, the expansion of export-oriented crops has resulted in competition for better land, and small-scale farmers have often been forced to occupy the less favourable and more fragile areas of hillsides. Their decreased access to land leads to reductions in the fallow period, which threatens to destabilize any system requiring fallow, including *fríjol tapado*.

FRÍJOL TAPADO SYSTEM

Overview

The *fríjol tapado* system is based on the use of short-term fallow and is typically found in humid areas. Its characteristic components include a thick mulch of secondary vegetation, which prevents soil erosion and loss (Bellows 1992), and a fallow-cropping cycle, which serves to maintain soil productivity during extended periods of time. The system, however, differs from other slash-and-mulch systems in that:

- The fallow period is short (ranging from 1 to 4 years typically but only 1 to 2 years in Costa Rica), hence total biomass accumulation is low and the biomass accumulated is less woody than in forest-based fallows.
- The cut fallow vegetation is not burned, resulting in a slow nutrient release, a more rapid regrowth of cut vegetation and continual soil coverage.
- The above-mentioned characteristics result in low bean germination and survival because of physical barriers to germination and increased competition from the newly growing vegetation. As a result, the average bean yields of *fríjol tapado* (0.49 t ha^{-1}) are less than the national average of 0.68 t ha^{-1} (Gonzalez and Araya 1994; CNP 1996).

Management

The *fríjol tapado* practice consists of, first, the selection of appropriate land. Criteria used to select sites include considerations such as: land use history and past performance, duration of the fallow, diversity of the vegetation present, apparent fertility of the soil, slope, timing of planting and the bean cultivars used. The use of traditional knowledge related to biodiversity is a particularly important aspect of the system. Throughout *fríjol tapado* areas, farmers identify fallow species as 'good', 'bad' or 'neutral' (Jiménez 1985; Araya and González 1986) (see below). In addition to the presence of certain plants that indicate the field's suitability for *tapado*, fields are often selected based on plant density. East-facing slopes are often preferred because of their more favourable microclimatic conditions, which reduce the incidence of pests, diseases and slugs (*Vaginulus plebeius* F.) (Thurston 1984; Arias and Amador 1990; Bellows 1992).

Paths are first slashed through the vegetation to create access for subsequent planting. Seed is then broadcast into the vegetation at a rate of 25 to 40 kg seed ha^{-1} (von Platen and Rodríguez 1982; Campos-Alvarado 1983; CEDECO 1992), although up to 75 kg seed ha^{-1} are sometimes used. These seeding rates are extremely high and equivalent to about 180 000 to 400 000 seeds ha^{-1} (Araya and González 1986). Bean seeding is timed so that harvest occurs during relatively dry periods. In many areas (especially those where bean producers own the land they farm), diverse local bean varieties can be found in the *tapado* fields.

Immediately after bean seeding, the fallow vegetation is slashed to provide a more or less compacted mulch layer through which the bean seedlings emerge. Cutting height can vary from soil surface to 20 cm or more above the surface, depending on the bean cultivar used and local practices. Cutting height affects both the rate of regrowth of secondary vegetation and bean germination and survival.

The fields are then left untouched until harvest. Beans are harvested when they reach 70% to 85% of physiological maturity. The plants are pulled up, left to dry in the field for 2 to 8 days and then threshed (Araya and González 1986). In some cases, livestock are allowed to graze the bean residues left in the field.

Following 2 to 3 years of production, the system requires a fallow of 1 to 4 years. When sufficient land is available, a farmer often divides the farm into three or four lots and rotates *tapado* production among them. Recently, however, fallow periods have been reduced, and in some cases eliminated, because of increasing land use pressure.

Socio-economic Issues

To meet the labour demands of the system, on most farms, a combination of family and reciprocal labour exchange is used. Although relatively little labour is needed, most work involved is heavy because of the steep slopes on which *tapado* is practiced. The role of women in the system is primarily limited to seed preparation and selection, storage, marketing and to participation in decision making regarding management. Their participation in fieldwork is often limited because the fields are distant from the homestead and hard physical labour is needed to work on steeply sloping land. However, if women are heads of families, or if *tapado* fields are close to the homestead, they may also participate in the field activities.

Land tenure can affect *fríjol tapado* management. For example, in Coto Brus, where landholdings are large and lands have been recently settled, farmers have sufficient land to maintain a fallow-based system. In contrast, in Acosta, where more than 80% of the land used in *fríjol tapado* is rented from a small number of owners, fallow periods are often shorter and farmers have few incentives to invest in measures or inputs that would increase land use sustainability.

Beans produced in the system are generally used for family consumption, although a good crop may generate some cash income as well. *Fríjol tapado* is a favoured practice among many small-scale farmers because of its low production cost, small risk of capital investment (generally no fertilizer or pesticide inputs), low soil losses to erosion, and its provision of food security while using one of the few resources readily available to the family – its labour (Bellows 1992).

CURRENT KNOWLEDGE ON THE SYSTEM

During the last 10 years, a variety of institutions have been involved with research or extension efforts on *fríjol tapado*. These include:

- Centro Agrónomico Tropical de Investigación y Enseňanza (CATIE), a regional, internationally funded research centre;
- National universities (the University of Costa Rica and the National Autonomous University);
- Foreign universities (Cornell University and the University of Georgia); and
- Local non-governmental organizations (NGOs), such as the Corporación Educativa para el Desarrollo Costarricense (CEDECO) and the Instituto para el Desarrollo y Acción Social (IDEAS).

Most of the research has been directed at alleviating biophysical constraints to *fríjol tapado* production, although the NGOs also have been involved in applied research and extension of post-harvest processing and commercialization techniques. A more recent effort has been a collaborative project among the University of Costa Rica, University of Guelph and various NGOs and Costa Rican producer organizers, 'Improvement of Cover Systems: *Fríjol Tapado*.' The project has conducted integrated and interdisciplinary research on the system, focusing not only on the biophysical aspects but also on the socio-economic characteristics and opportunities for improvement. As mentioned earlier, most of the research has taken place in two locations, Coto Brus and Acosta.

It should not be forgotten, however, that the *fríjol tapado* system is in fact a result of centuries of farmer-led research, farmer dissemination, the adaptation of external contributions through long processes of collective experimentation and the gradual validation of the system's requirements. This has permitted the development of intricate systems of traditional knowledge on the use and conservation of local bean varieties, on their combination with medicinal plants, on the selection of areas with more diverse vegetation, on the use of certain plants as indicators of soil fertility and on the use of family labour, among others. Examples of such processes are the simultaneous conservation of traditional bean varieties with the experimentation of improved varieties and the conservation of desired plant species in the fallow.

In the following, results of researcher-led investigations from the past two decades are discussed.

Agronomic Aspects

Research on *fríjol tapado* has generally concentrated on increasing bean yields and in better understanding the ecological function of the system. Bean production in Latin America is generally nutritionally limited by the availability of phosphorus and nitrogen. Many of the research projects on *fríjol tapado* have focused on overcoming these nutritional constraints. Common research themes, discussed below, include: methods to increase the availability of P and other nutrients via P fertilizers, lime or mycorrhizal inoculation; bean inoculation with improved strains of *Rhizobium*; better seed handling; impact of plant biodiversity in fallow and the substitution or enrichment of the natural fallow with other species.

Fallow Dynamics and Bean Yields in *Fríjol Tapado*

Soil fertility is one of the least sustainable resources of agro-ecosystems (Fresco and Krooneberg 1992) because of the small quantities of nutrients added annually from natural sources and the potentially large and rapid nutrient losses that can occur once a natural system is brought into agriculture. From this point of view, it is important to examine nutrient dynamics in *fríjol tapado*.

In both Coto Brus and Acosta, the physical properties of the superficial horizon are good despite various years of cultivation under *fríjol tapado*. Bulk density is low (0.5 to 0.9 Mg m^{-3}), moisture retention is high (about 70%) and hydraulic conductivity is greater than 20 cm h^{-1} (Mata et al. 1999). The maintenance of good soil physical properties is probably due to the volcanic mineralogy of these soils, the large additions of organic residues, continual root growth and continual cover that inhibits soil compaction caused by rainfall impact, animals or human beings.

In both Acosta and Coto Brus, the quantity of biomass accumulated is relatively high, especially in long fallows (Table 1). The slow decomposition suggested by the litter data may be due to the frequent additions of low quality fallow residues that are dominated by a fern, *Pteridium aquilinum* (L.) Kuhn and, in Coto Brus, perhumid conditions.

Table 1. Live and dead biomass (t ha^{-1}) of fallows, and bean yields (t ha^{-1}) in consecutive years following fallow slashing in *fríjol tapado* system at Acosta and Coto Brus, Costa Rica (*Source*: Meléndez and Szott 1999).

Site	Fallow age (years)[a]	Biomass		Bean yield		
		Living	Dead	Year 1	Year 2	Year 3
Acosta	3	9.75	1.45	0.44	0.50	0.49
	2	7.28	1.78	0.60	--	--
	1	5.27	1.80	0.62	--	--
Coto	4	17.46	2.75	0.26	0.45	0.43
Brus	1	2.47	11.87	0.34	--	--

[a] For fallows > 1 year old, n = 1 at Acosta and Coto Brus. For 1-year-old fallows, n = 8 at Acosta and n = 11 at Coto Brus.

Nutrient Balances

The construction of nutrient balances can often help identify key components of the system's sustainability. A nutrient balance for *fríjol tapado* in Acosta is shown in Table 2. Clearly, except for P, (1) most of the nutrients are stored in the soil, not in the vegetation and (2) the system's sustainability is largely due to the small quantities of nutrients removed or lost (a result of low yields and continuous protection of the soil surface) relative to the large quantities present. Data from plots cultivated under the system for up to 80 years suggest that it has maintained soil organic matter and plant production (Meléndez and Szott 1999).

Table 2. Nutrient stocks in various components of the *fríjol tapado* system, Acosta, Costa Rica (*Source*: Meléndez and Szott 1999).

Component[a]	Mass ($t\ ha^{-1}$)	Nutrients ($kg\ ha^{-1}$)[b]				
		N	P	K	Ca	Mg
Fallow biomass (DM)	5.0	47.4	7.7	102.5	16.0	11.3
Litter (DM)	2.0	6.0	1.4	3.8	5.7	3.2
Subtotal	7.0	53.4	9.1	106.3	21.7	14.5
Soil	-	4500	6.6	315	186	408
Total	-	4553.4	15.7	421.3	207.7	422.5
Grain	0.5	19	2.5	8	2.5	1.5
(% exported)	-	(0.4)	(15.9)	(1.9)	(1.2)	(0.4)
Stover	0.24	1.7	1.2	n.d.	0.5	n.d.

[a] DM = dry matter.
[b] n.d. = not determined.

The data also suggest that P is the weak link in this system since the quantity exported is large relative to the stocks present, although there are reserves of P in the soil that can buffer the quantities removed. The differences in soil P and plant production between Acosta and Coto Brus, as well as the positive response to rock P applications, also lend support to the limiting role of P. Because of the key role of P in the system, changes in system management that result in increased losses or export of P will easily have a negative impact on the future sustainability of *fríjol tapado*.

Bean Yields in Relation to Fallow

Bean yields under traditional management in Acosta varied between 0.40 and 0.90 t ha^{-1}, with most of the values between 0.40 and 0.50 t ha^{-1}. In Coto Brus, yields varied between 0.25 and 0.70 t ha^{-1}, with most of the values between 0.30 and 0.45 t ha^{-1} (Table 1 shows partial results). The lower yields in Coto Brus compared to Acosta are because of lower soil fertility, less fallow growth and a rainier climate. In both regions, crops are sometimes lost because of unfavourable climatic conditions or pests.

In 1-year-old fallows in Acosta, bean yields usually increase with fallow mass (Figure 2). In Coto Brus, there was little relation between total fallow biomass (i.e. live and dead) and bean yields but bean yields and live biomass of the fallows were related. Interestingly, the relationship between fallow biomass and bean yields does not hold for fallows 3 to 4 years old. In these systems, yields immediately following fallow slashing tend to be less than in following years, because of the large quantity of non-decomposed material that affects bean germination and survival (Table 1).

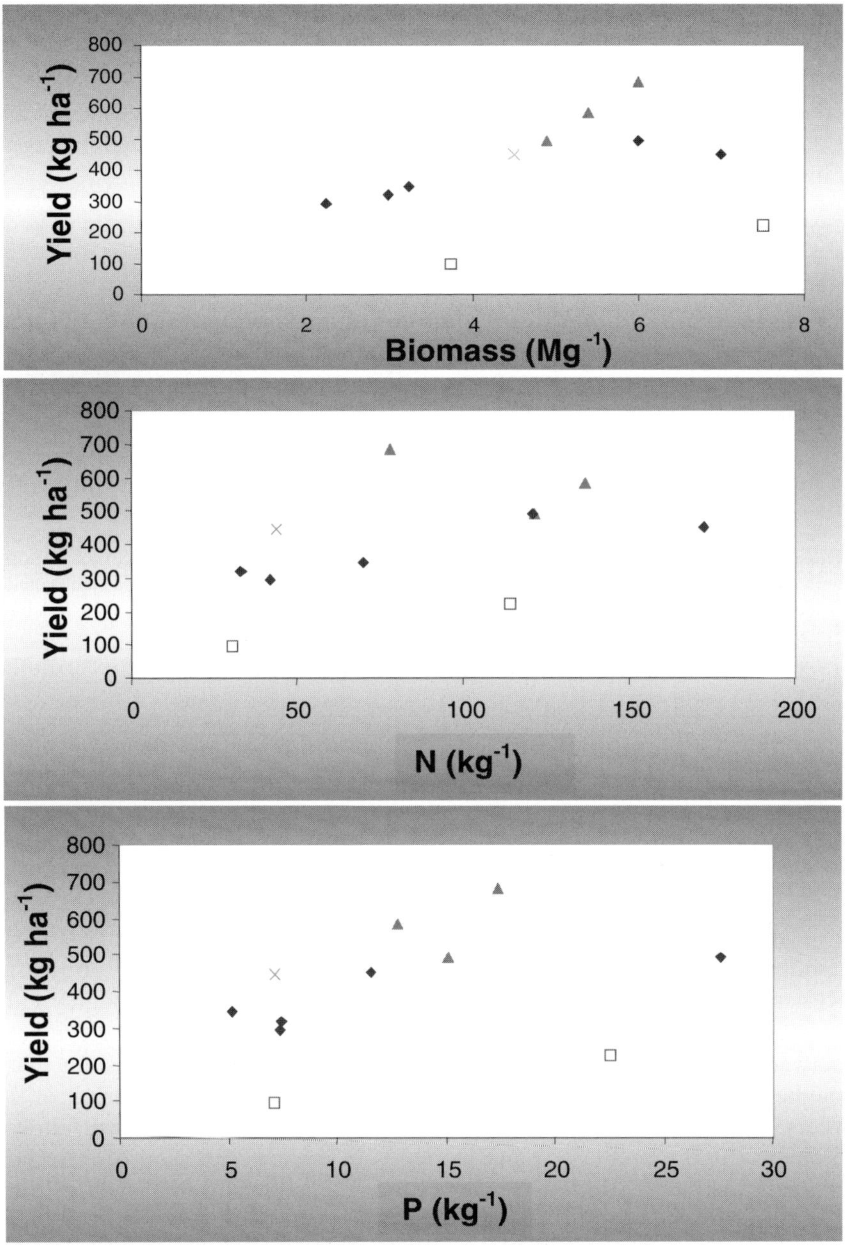

Non-favourable ▲ Mixtures × Natural fallow control ◆ Favourable □

Figure 2. Impact of biomass quantity and nitrogen and phosphorous application on yield in natural fallow, mixtures and in vegetation with favourable and non-favourable species (*Source*: Adapted from Meléndez and Szott 1999).

Addition of Phosphorus and Lime

The addition of P in the form of rock phosphate (RP) has a consistently large and positive effect on bean yields both in Acosta and in Coto Brus (Tables 3 and 4). Increases in yields due to RP application range from 20 to 140% compared to the non-fertilized control.

Table 3. Immediate and residual effects of the application of rock phosphate (216 kg ha^{-1}) and inoculation with *Rhizobium* on the bean yields in *frijol tapado*, Acosta, Costa Rica (*Source*: Gadea and Briceño 1999).

Treatment	Bean yield (t ha^{-1})			
	Immediate	Year 1	Year 2	Year 3
Control	0.87	0.86	0.75	0.53
Rock phosphate (RP)	1.03	1.23	0.87	0.59
Rhizobium	0.82	0.77	0.82	0.55
RP+*Rhizobium*	0.77	0.68	0.76	0.52

Table 4. Immediate and residual effects of the application of rock phosphate, inoculation with *Rhizobium*, and lime on bean yield in Coto Brus, Costa Rica (*Source*: Gadea and Briceño 1999).

Treatment	Bean yield (t ha^{-1})			
	1993	1994	1995	1996
Control	0.85	0.80	0.67	0.71
Rock phosphate (RP)	2.05	1.45	1.38	1.10
Rhizobium	0.69	1.05	0.82	0.73
RP + *Rhizobium*	1.75	1.20	1.06	0.98
Lime	1.55	0.75	0.72	0.73
Lime + *Rhizobium*	1.49	1.33	1.05	0.81

Often, the effect of RP is greater in the year following fertilizer application than in the year in which it is applied. The residual effect of RP lasts until the third or fourth year following application, although yield benefits decrease with time (Tables 3 and 4). Few differences are evident between a one-time application and annual applications, suggesting that the benefits from RP can be achieved with one application every 3 or 4 years. The beneficial effect of RP appears to stem from increased fallow biomass (by 0.5 to 2 t ha^{-1}) and increases in the concentration of P in the biomass and soil. Additional research has demonstrated that most of the first year's yield benefit comes from P and not from calcium. In Coto Brus, yield with lime only was higher than the control yield but not as high as with RP, presumably because of deficiency in the soil.

Inoculation with *Rhizobium* and Mycorrhizae

Inoculating beans with improved strains of *Rhizobium* had either no effect on bean yields or its effect was negative (Table 3). This could be caused by a number of factors. Local strains of *Rhizobium* may be more efficient than improved strains, which may even represent a physiological drain on the beans. Alternatively, Rosemeyer (1994b) found that low levels of N did not increase bean yields in an Andisol in Coto Brus. This suggests that other nutrients limit N_2 fixation and/or that levels of inorganic N in soil are relatively high. Research results exist to support both hypotheses. Positive response in nodule biomass and bean yields to P additions support the former hypothesis. The high concentrations of organic matter in these soils and evidence from Rosemeyer (1994b), showing that N fertilization reduces nodulation and N_2 fixation in soils under the *fríjol tapado* system, support the latter hypothesis.

The results related to the importance of mycorrhizae are partial and inconclusive. A survey of mycorrhizal infection of fallow vegetation indicated that most of the plants (some *Poaceae* are the exceptions) had high levels (>80%) of infection. However, it is not clear whether the mycorrhizae present are those needed by beans. A greenhouse study showed that there was little response in bean growth to sterile soil inoculated with *Glomus* mycorrhizae or to non-sterile soil, unless P was added. However, a different study showed that the percentage of bean roots infected by mycorrhizae decreased with increases in the quantity of P applied to a *fríjol tapado* system (Rosemeyer 1994a).

Seed Quality

Low bean yields and the need to plant large quantities of seed are partially related to poor seed quality. Under sterile laboratory conditions, the germination of three local bean varieties varied between 56 and 81%, and 9 to 44% of the seed was infected, the majority with *Aspergillus*, *Fusarium*, or *Rhizoctonia*. Field sampling at 15 days after planting (DAP) indicated that diseases were present in 38 to 73% of the bean plants. At 30 DAP, 27 to 57% of the beans were infected with *Fusarium*, while 12 to 24% were infected with *Rhizoctonia*. Although it has been reported that the mulch formed by the slashed fallow can reduce the level of diseases due to less propagation through raindrop splash and changes in microclimate (Galindo 1994), more work is needed on diseases in beans produced under *fríjol tapado* conditions.

With regard to the effect of seed quality or type on bean yields, most previous attempts to improve the yields of *fríjol tapado* using improved seed were unsuccessful since the yields of local varieties were equal to or greater than those of improved seed (Alfaro 1994). The most probable explanation for this result is that the improved varieties have different characteristics (such as rapid and indeterminate growth, and tall or climbing growth habits) than those needed for the *fríjol tapado* system (Smith 1994).

Conditions in the fallow can also reduce bean germination, growth and yield. Data from Bellows (1994) and Shenk (1994), respectively, suggest that good contact between the seed and soil can increase yields by 0.15 and 0.40 t ha^{-1}, and that the temporary elimination of competitive secondary vegetation by herbicides can result in another increase of about 0.20 t ha^{-1}. In total, this represents a relative increase of 80 to 600% compared with traditional management. Methods of slashing the vegetation (i.e. at the soil level, as in Acosta, or 20 cm above the soil surface as in the Bellows research) has a big effect on bean germination and on competition from secondary vegetation and therefore on bean yields.

Plant Biodiversity in Fallow

Various studies also have described and analyzed the diversity of plants present in the *fríjol tapado* system. A vegetation survey of plant diversity in these systems in four agro-ecological zones indicated high diversity (Ocampo 1985; De la Cruz 1994). For example, 59 to 89 species were found in 8 to 10 plots per zone. Plants in the *Asteraceae*, *Poaceae* and *Euphorbiaceae* families were found in all the zones, and representatives of the *Solanaceae*, *Fabaceae* and *Labiatae* families were present in three of the four zones. With 10 to 20 species per zone, *Asteraceae* and *Poaceae* families had the greatest number of species. Studies also indicate that some fallow species are particularly rich in nutrients. *Melampodium divaricatum* (Rich.) DC. (*botoncillo*) is rich in N, K and Mg; *Bidens pilosa* L. (*moriseco*) is rich in Mg; *Melinis minutiflora* P. Beauv. (*calinguero*) is rich in N; and *Elvira biflora* (L.) DC. (*prima hermana*) is rich in N and Mg.

Bellows (1992) noted that a decrease in the fallow period led to a reduction in the diversity of what farmers regarded as beneficial plant species. In areas that have been in continuous production for less than 2 years, grasses tend to dominate the secondary vegetation, whereas mixtures of broad-leaved herbs, shrubs and small trees dominate areas with 2 to 3 years of fallow (Bellows 1992). In a study conducted in five locations in Costa Rica, Araya and González (1986) observed that the appearance of 'weedy' species was associated with certain land use practices and soil texture, with certain levels of soil fertility and with the amount of post-harvest cover of fallow residues. Fields under intense use suffered more from the presence of plants characterized as 'bad' secondary vegetation species.

Farmers have been surveyed also for their views on species present in the fallow. According to the farmers, 'good' species indicate fertile soils, are easy to cut, decompose rapidly or interact favourably with beans. In contrast, 'bad' species resprout rapidly, produce heavy shade, attract pests such as slugs, indicate infertile or acid soils, or decompose slowly.

Species recognized as 'good' or 'bad' differed among zones. Both the *Asteraceae* and the *Poaceae* have many species classified as favourable or non-favourable by the farmers. Generally, species that farmers use to indicate fertile sites are: *Tithonia diversifolia* (Hemsl.) A. Gray (*mirasol, girasol*), *Melanthera aspera* (Jacq.) Small (*paira*), *Calea urticifolia* (Mill.) DC. (*jaral*), *Spermacoce verticillata* L. (*botón blanco*) and *E. biflora*. It needs to be determined if these patterns are causes (i.e. the plants selectively accumulate certain nutrients) or effects (i.e. tissue nutrient concentrations reflect the nutritional status of the soil). It has been suggested that what made the weedy herb *paira* or *podrax* (*M. aspera*), a 'good' weed was its ability to accumulate relatively high levels of the limiting nutrient, P, and the subsequent return of P upon decomposition.

Fallow Enrichment

Some attempts have been made to measure the effect on bean yields of partially enriching the natural fallow vegetation or substituting it completely. In the former case, bean yields after fallows enriched with *T. diversifolia*, *Mucuna* spp. Adans or *Canavalia ensiformis* (L.) DC. were 10 to 53% greater than after the natural fallow (Figure 2). However, bean yields after seven sole-cropped fallow species were similar to, or significantly less than, those after natural fallow (Table 5). Low yields were found also for sole-cropped species that farmers considered favourable (Figure 2, Table 4). The low yields were seemingly caused by a number of factors. For example, *Ageratum conyzoides* L. produced only a small quantity of biomass, *Rottboellia cochinchinensis* (Lour.) Clayton added only low amount of nutrients, *P. aquilinium* residues mineralized slowly, *T. diversifolia* was infested by pests and *Mucuna* spp. acted as a refuge for bean-devastating slugs.

Table 5. The effect of enrichment or substitution of the natural fallow on bean yields in the *frijol tapado* system, Acosta, Costa Rica (*Source*: Meléndez and Szott 1999).

Fallow	Fallow biomass (t DM ha^{-1})[a]	Bean yield (t ha^{-1})
Sida rhombifolia L.	3.00	0.32
Elvira biflora (L.) DC.	3.25	0.34
Ageratum conyzoides L.	2.25	0.29
Tithonia diversifolia (Hemsl.) A. Gray	6.00	0.49
Pteridium aquilinium (L.) Kuhn	7.50	0.22
Rottboellia cochinchinensis (Lour.) Clayton	3.75	0.10
Canavalia ensiformis (L.) DC. (120 days)	7.00	0.45
Mucuna spp. Adans (90 days)	5.50	---
Natural fallow (NF)	4.50	0.45
NF + 20% *T. diversifolia*	6.00	0.68
NF + 50% *Mucuna* spp.	4.90	0.49
NF + 60% *C. ensiformis*	5.40	0.58

[a] DM, dry matter.

On the other hand, data suggest that although the mixtures had lower nutrient concentrations than the pure patches of the same species, their concentrations were greater than those of the natural fallow. These results indicate that the greater yields observed in the enriched fallows may be because of a combination of greater quantities of nutrients in the vegetation (compared to the natural fallow) and a greater protection against pests (compared to the mono-cropped fallows). Figure 2 presents some results regarding fallow enrichment.

Socio-economic Aspects[1]

The socio-economic importance of the *fríjol tapado* system is rooted in the domestic production of beans, in the minimum utilization of external inputs and labour, and in the provision of additional income to the household, as about 40% of the production is sold (CEDECO 1992). Moreover, the system poses little risk and requires almost no initial investment.

The economics of the system have been studied in comparison to a dibble-stick system promoted by the government extension. The structure of the costs is very different between the two systems (Table 6). For example, in the dibble-stick system, 27% of the production costs come from labour compared to 43% in *fríjol tapado*. Nevertheless, the total amount of labour devoted to *fríjol tapado* production is considerably less (at 41 hours) than in the dibble-stick system (61 hours). Moreover, labour use in *fríjol tapado* is concentrated at seeding and harvesting (Figure 3). This permits the family to use its labour in the intervening time for other on- or off-farm activities. In the free time between seeding and harvest, farmers usually tend coffee or pastures.

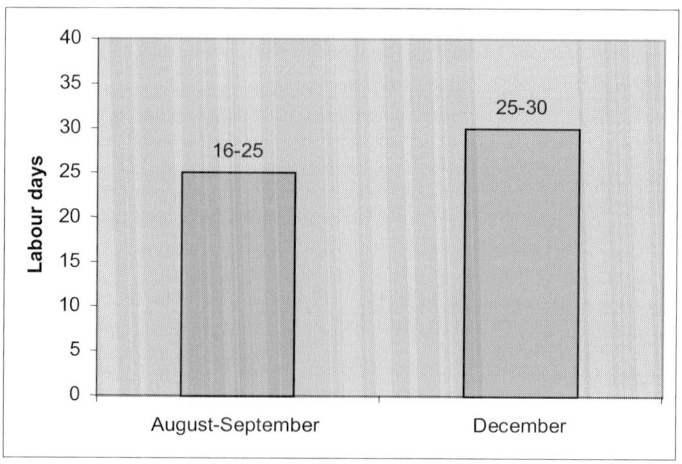

Figure 3. Distribution of labour in the *fríjol tapado* system (*Source*: Amador and Briceño 1999).

Table 6. Comparison of production costs of the dibble-stick and *fríjol tapado* systems (*Source*: Arias et al. 1999).

Item	Dibble stick[a]		*Tapado*[a]	
	Amount (colones)	Share of total cost (%)	Amount (colones)	Share of total cost (%)
Manual labour:				
Slashing	0	2.0	27000	43.9
Pulling up	24000	262.0	22500	36.6
Removing pods	4500	4.9	3000	4.9
Threshing, drying	7500	8.2	6000	9.8
Transport to the house[b]	-	-	1500	2.4
Transport to the field[b]	-	-	1500	2.4
Weeding	18000	19.7	0	0.0
Planting and first fertilization	27000	29.5	0	0.0
Herbicide application	4500	4.9	0	0.0
Pesticide application (slugs)	1500	1.6	0	0.0
Fungicide application	1500	1.6	0	0.0
Foliar fertilization	3000	3.3	0	0.0
Subtotal	91500	67.5	61500	82.0
Materials:				
Seed	9331	21.2	12586	93.0
Sacks	1169	2.6	923	7.0
Fertilizer (10-30-10)	14022	31.8	0	0.0
Pesticide (for slugs)	2365	5.4	0	0.0
Benomyl-Benlate	7872	17.8	0	0.0
Floazifob-butil	3198	7.3	0	0.0
Paraquat (Gramoxone)	4428	10.0	0	0.0
Adhesive	1722	3.9	0	0.0
Subtotal	41742	32.5	13509	18.0
Production costs	135607	88.4	75009	91.5
Others[c]:				
Incidentals	6780	4.4	3750	4.6
Depreciation	11066	7.2	3242	4.0
Subtotal	17846	11.6	6992	9.0
Total costs	153453		82001	

[a] Costs, US$1 = 250 colones. Share of total cost: for items within subtotals, % of the subtotal. For subtotals of manual labour and materials, % of the production costs. For subtotals of production costs and other costs, % of the subtotal from the total costs.
[b] These items are reported only for the *tapado* system. Unlike in the dibble-stick system, the fields in *tapado* are usually located far from the homestead.
[c] Note that following were not included: opportunity cost of land, and administrative and financial costs.

In contrast, in the dibble-stick system, a larger share of the costs comes from buying of materials (73%) than in the *fríjol tapado* system (57%). The most important expenses in the dibble-stick system are fertilizers, labour for land preparation, weeding and planting. Reducing these costs would result in better profitability. However, a reduction of 10% in the weeding costs increases the profitability of the system by only 2.6% and reduces the probability of financial losses by 5%. Only if the reduction of costs implies also an increase in productivity can one expect a better response in profitability and risk. For the *fríjol tapado* system, seed cost is the highest of the production costs and impacts the system's profitability and risk. Decreasing the amount of seed by 10% increases the system's profitability by 2.4%, while the risk decreases much more. For this reason, a study on the impact of planting density on the costs and productivity of the system would be particularly useful.

The profitability of the dibble-stick system is poorer and its risk greater compared to the *fríjol tapado* system. In addition to being more profitable (¢1535 per ha in contrast to ¢55776 per ha for the dibble-stick system[2]), it is less risky (possibility of losing money is 53% compared to 93% for the dibble-stick system). Although the coefficient of variation for earnings is higher for the *fríjol tapado* system than for the dibble-stick system, indicating a larger variability in earnings, the range in earnings is also more favourable in the *fríjol tapado* system. For example, with a 67% probability, the earnings from *fríjol tapado* are between ¢79376 and ¢32447, while for the dibble-stick system, earnings range from ¢91395 to ¢20158, which includes the possibility of no profits at all. Because of the low capital costs in the *fríjol tapado* system, return to assets is 2.05% in comparison to -8.38% for the dibble-stick system. The continuation of the latter can be explained in no small part by the subsidies given.

For bean production in Nicoya, Guanacaste, Borbón (1989) calculated a marginal rate of return for *fríjol tapado* of 531% compared to 96% for the somewhat more intensified dibble-stick system (one of the systems supported by the Ministry of Agriculture).

Clearly, the results of the economic analysis are very favourable for the *fríjol tapado* system. Not only is it more profitable but also its risk is almost zero and its sensitivity to fluctuations in, for example, yield or prices, is small. The *fríjol tapado* system gives socio-economic benefits for the families that practice it, especially by minimizing risk of financial loss. The system does this in two ways. First, it allows the farmer to undertake other activities to generate income during the growing period. Second, it produces a subsistence crop that reduces the farmer's exposure to markets where prices and supply are constantly fluctuating. Moreover, in addition to the direct ecological benefits to the farmer, such as low pest and disease incidence, the system gives indirect benefits for the society, such as plant and animal diversity; such benefits may be immediate or may be achieved in the medium-or long-term.

It should be noted that the economic results depend greatly on the supposed yields, and therefore soil and climatic conditions, as well as the farming practices of each producer. The many differences among the farmers and production zones are caused by differences in soil, climate, technological level and cultural practices; such differences are not well documented and could help us better understand the benefits of each system. The preference of each producer in using a particular system depends greatly on his/her conditions as well as attitude to financial risk.

Explaining the Logic of *Fríjol Tapado*

It is important to understand the *fríjol tapado* system as an expression of the smallholder farmer's logic and rationality, which focus on achieving the most efficient use of resources within the production unit available to him or her. This rationality explains many aspects of *fríjol tapado*, such as conservation of the region's and family's bean varieties, primary importance of subsistence production, integration of medicinal plants in the system, selection of areas with highest diversity of plant species, use of certain plants as soil fertility indicators and reliance on labour inputs.

This same rationality causes smallholder farmers to seek a margin of security against the threats caused by macroeconomic fluctuations. Whenever possible, small-scale farmers attempt to diversify their enterprises as much as they can to avoid buying more than the minimum from the market. For that reason, smallholder farmers in most regions where *tapado* is practiced sell only the surplus and not the totality of their production. Similarly, this rationality causes farmers to utilize the factor of production they have most of—labour. It may be that for the smallholder farmer, the technicians' repeated phrase of low yields in *fríjol tapado* system is of little importance. The farmer's main interest is to obtain sufficient yield for family consumption and for the following year's seed, as well as some surplus for selling to enable the purchase of some necessities. In that respect, the farmers are seeking to eliminate any risk related to the family's minimum food requirements.

ADVANTAGES AND DISADVANTAGES OF *FRÍJOL TAPADO*

As evident in the highlights presented above, research has helped us understand the *fríjol tapado* system at an increasingly detailed level, including its advantages and disadvantages, which are presented in a summary form below.

Advantages

In the *tapado* system, the continual presence of a mulch or litter layer and growing vegetation favour nutrient recycling. Since many of the soils under *fríjol tapado* in Costa Rica are deficient in P (Rosemeyer and Gliessman 1992), mineralization of P from the mulch/litter layer can provide important amounts of P to the beans (Schlather 1998). The presence of a mulch/litter layer, the growing vegetation and the improvements in soil structure help minimize erosion losses (von Platen and Rodríguez 1982; Tapia and Camacho 1988). These characteristics, in association with tight nutrient cycling, have permitted the system to be used for more than 80 years on the fragile hillsides of Costa Rica.

Biomass additions result in permanent soil cover and improve soil structure and decrease rainfall impact, thus reducing compaction of the soil surface. Studies of water infiltration show greater rates of infiltration under *fríjol tapado* than under the clean-tended bean systems (Roth et al. 1991). The soil under *fríjol tapado* is also less resistant to penetration and has larger and more stable aggregates than soil under clean-tended bean systems (Bellows 1992).

Despite no use of synthetic chemicals, the *fríjol tapado* system tends to be less affected by pests, with the exception of slugs, than other bean production systems (Borbón 1989). The function of the fallow residues as a mechanical barrier has been suggested as the explanation for the lower incidence of web blight (*Thanatephorus cucumeris* [A.B. Frank] Donk; Galindo et al. 1983) and *Xanthomonas campestris* p.v. *phaseoli* (Smith) Dye (Tapia and Camacho 1988), thus diminishing or eliminating the use of fungicides. The biodiversity in *fríjol tapado* fields is much higher than in the clean-tended fields, because of the presence of the fallow and the absence of pesticide and herbicide use. For example, there were 30 times more detritivorous insects in *fríjol tapado* fields than in clean-tended bean fields (Cook et al. 1993).

Weed pressure is reduced because of the reduced germination of weed seeds caused by the physical barrier of the mulch (Borbón 1989). Araya and González (1986) reported that the fallow slash served to inhibit regrowth of other vegetation and that non-bean species remained largely in the seedling stages until the beans reached the pod-forming stage. Weed growth has been shown to vary inversely with the thickness of the mulch layer (Eckert et al. 1991).

Under certain conditions, nodule weight is greater in *fríjol tapado* than in clean-tended beans (Rosemeyer 1994b). Moreover, effective strains of *Rhizobium* appear to be present locally, because there is usually little or no positive effect associated with *Rhizobium* inoculation (Huntington et al. 1986; Rosemeyer 1990; Meléndez et al. 1994).

The *tapado* system has low capital costs and poses little risk. It requires labour inputs only during planting and harvesting, and machinery, agrochemicals and other active forms of management such as weeding are seldom used. In terms of production per unit of labour or capital, rather than unit area, *fríjol tapado* is relatively efficient (von Platen et al. 1982). Furthermore, if one considers the real costs of land and labour, the benefit: cost ratio is better for the *fríjol tapado* system than for either semi-mechanized or highly mechanized bean production systems (CIAT 1986). Labour use in *fríjol tapado*, which is almost entirely by the family, is concentrated at seeding and harvesting, permitting the family to use its labour in the intervening time for other on- or off-farm activities (Pachico and Borbón 1987; Ribier et al. 1988).

Finally, the system helps maintain traditional knowledge since it has been used over a long period of time to satisfy a large part of national bean consumption (Alfaro and Waaijenberg 1992; Rosemeyer 1994a).

Disadvantages

Perhaps the chief disadvantage of the *tapado* system is that bean yields and income are lower than in higher input bean systems. Yields under *fríjol tapado* range usually from 200 to 500 kg ha^{-1} (von Platen et al. 1982; Araya and González 1986; CIAT 1986; Tapia and Camacho 1988) but in some cases can reach 1400 kg ha^{-1}. In more humid areas, the mulch from the slashed fallow may serve as a refuge for slugs, which results in further yield decreases (Hallmann and Andrews 1989). In comparison to these figures, yields of clean-tended beans are usually around 500 to 1055 kg ha^{-1} (PRIAG 1993). Statistically valid comparisons of *fríjol tapado* with higher input bean systems, however, are limited in number. Recent decreases in bean prices have also reduced income.

As a result of the need for a fallow period (9 months to 4 years, with an average of 2 years), *fríjol tapado* fields need to be often out of production. This is a disadvantage when land is limited. Under such conditions, producers are often forced to farm continuously, resulting in soil degradation.

Land tenure patterns often discourage investment in sustainable forms of land use. When rented land is used for *fríjol tapado*, landowners often permit livestock to graze the bean residues following the crop harvest, which results in greater nutrient exports and soil physical problems, such as compaction. Moreover, high land rents and lack of year-to-year security act as disincentives to investments in inputs or adequate fallow period.

Farmers practicing *fríjol tapado* receive limited technical or financial assistance (Araya and Gonzáles 1986; Borbón 1989) and often remark on the lack of respect with which officials of the Ministry of Agriculture and other institutions treat them. This lack of support has been justified by the need for structural adjustment and the system's low production (harvest per unit area) and smaller average seed weight in comparison to higher-input bean production systems.

Finally, heavy physical labour is required for planting and harvesting. Competition for labour by other simultaneous and more profitable activities, for example, coffee picking, may reduce the quantity of labour available for *fríjol tapado*.

ADOPTION OF THE *FRÍJOL TAPADO* SYSTEM

Biophysical and socio-economic factors determine the niche of *fríjol tapado*. As seen above, the system is environmentally friendly, reduces risk and is highly efficient in terms of labour, capital and natural resources, at the cost of reduced crop yields. It works well where labour is limited, sufficient land is present to allow some fallow, relatively great importance is placed on bean self-sufficiency (e.g. in situations of limited capital, difficult access to markets or great cultural importance assigned to bean production) and other productive activities can be undertaken when labour requirements to the system are low.

Constraining Factors

As discussed above, in the last several years, the system has lost some of its former importance and presently only accounts for 24% of national bean production (CNP 1998), in comparison to 80% in 1980 (Alfaro 1994). There are a number of reasons for the decreased importance of the *fríjol tapado* system.

First, during recent years, structural adjustment programs and government institutions that tend to favour monocultures of export-oriented crops have acted as disincentives for the practice of *fríjol tapado*. Policies have tended to liberalize the commercialization of beans at both the international and national levels and virtually no technical assistance or credit is available for *fríjol tapado* production. Within Costa Rica, price supports as well as the direct purchase of beans from farmers by the government have been eliminated. Declining prices, decreased bean production in other Central American countries and increased competition by Argentina, Canada and other countries are increasingly resulting in the importation of beans and the loss of regional food security.

Second, migration from rural to urban areas, especially of younger family members, caused by changes in regional employment and cultural mores, is affecting the amount of family labour available for *fríjol tapado*. The large fluctuations and generally lower bean prices, hard physical labour associated with the system and employment opportunities with multi-national companies act as incentives for the sons of bean producers to look for other, easier sources of off-farm employment. Don Rodrigo Arias, a farmer from the Bajo Los Arias de Acosta community, mentions that: 'The youth of today no longer want to *tapar* beans. They prefer to look for work outside, even if it means being a policeman in order to earn enough money to buy beans.'

Third, the increasing pressure on land is forcing the use of inadequate fallow periods, resulting in declining soil fertility and increased land degradation. Lack of land availability and the lack of capital to increase farm area have obligated many *fríjol tapado* farmers to shorten the length of fallow, resulting in a net loss of nutrients, soil erosion, a general degradation of the soil and accompanying loss of production (Bellows 1992). Land tenure systems and changes in land use have reduced access to lands suitable for *fríjol tapado* and act as disincentives to the sustainable management of these systems.

Fourth, farm resources have shifted to more profitable crops and enterprises and away from *fríjol tapado*. Favourable coffee prices have had a large effect on labour availability for *fríjol tapado* in Coto Brus, in particular. In recent years, the number of fields dedicated to the system has tended also to decrease as owners of these lands convert them to pastures.

Fifth, climate change is perceived as having increased the risk of crop losses in the system. Typically, bean plants are left to dry on the fields for about 1 week during a time when tropical storms can greatly harm the beans. Losses caused by climate are perceived as having increased in recent years.

Finally, research has shown that P fertilization can increase the biological and economic productivity of the *fríjol tapado* system but problems of access to capital restrict the use of inputs.

Factors Favouring the System

Some factors tend to favour the continued practice of *fríjol tapado*. In particular, strong cultural traditions placing a high premium on food security are common among many farmers. The important role that *fríjol tapado* plays within this value system is one of the reasons for its continued persistence. As Alvaro Loaiza, a farmer from Toledo de Acosta, mentions in regard to the possibility of no longer practicing *fríjol tapado*: 'While I'm alive, I'm going to *tapar* beans, even if only just for the food (that they provide).'

There is little actual promotion of research results for the *fríjol tapado* system. The NGOs currently working with *fríjol tapado* farmers mainly concentrate their activities on post-harvest processing, bean collection and distribution, and commercialization of organic beans. Viewed against this panorama, the chief positive impacts of the research and extension programs on farmer adoption have been:

- Improved low-cost methods for selection and multiplication of high quality local seed,
- Seed sanitation procedures,
- The use of improved fallow species in *fríjol tapado* as well as other systems,
- A greater awareness of the value of plant biodiversity and
- Post-harvest bean processing techniques that permit farmers to tap into markets for organic products.

ALTERNATIVES AND IMPROVEMENTS

The principal biophysical limitations to *fríjol tapado* include climatic interactions, pests and disease losses, low P levels in many soils and problems with the germination and initial growth of beans. These factors suggest that the following areas merit future attention: seed quality and sanitation, selection of suitable bean varieties, use of RP in small quantities and broadcast over mulch (Schlather 1998), inoculation of beans with adapted *Rhizobium* strains, improvements in soil-seed contact (e.g. by substituting broadcast planting with dibble stick planting) and enrichment of fallow vegetation.

Table 7 shows some of the estimated costs and benefits associated with these techniques. Because of the characteristics of the *fríjol tapado* system, any alternative must have low capital costs.

Table 7. Estimated costs and benefits associated with various techniques to improve bean production in the *fríjol tapado* system (US$1 = 250 colones) (*Source*: Meléndez and Szott 1999).

Technique	Marginal cost (price and/or labour spent)	Expected marginal yield[a] (kg ha^{-1})
Artesian methods of disinfecting seed	c/250 + 1 h	100-200
Use of bean varieties selected for *tapado* conditions	0	?
Rock phosphate:		
Purchase	c/2200 + 1 d	-
Transport	c/500	-
Application	0.5 d	200-400
Rhizobium		
Purchase	c/3600 + 1 d	?
Planting with a dibble stick	9 d	300-600
Fallow enrichment:		
Seed collection	2 d	100-300
Broadcasting seed	1 d	

[a] In the *fríjol tapado* system, no varietal selection and inoculation are used; question mark indicates the uncertainty of their possible impact on yield.

Another alternative that has been tested in order to solve the problem of reduction in the fallow period includes the enrichment of the fallow with alley-cropped leguminous trees. The trees are pruned during the mulching of the fallow vegetation and provide fruit, fuel wood and N. In Las Tumbas, Perez Zeledon, the natural fallow is being substituted by kudzu (*Pueraria phaseoloides* [Roxb.] Benth.). In other areas around Ciudad Neily, various legume species are being evaluated as a means of fallow enrichment. Farmers in the Acosta-Puriscal region are currently experimenting with leguminous trees planted as fallow enrichment species. Finally, the intermediate use of inputs (dibble planting and the selective application of chemical fertilizers and pesticides) is also being researched.

Any improved technological alternative, however, must confront the socio-economic factors that form the *fríjol tapado* niche: low capital and labour requirements, low risk and cultural traditions. Such alternatives also need to take into account the factors threatening its existence, that is, low income, competition for labour and changing cultural mores. Therefore, any future program of technology development must include farmers as co-participants in order to assure that the technology is in harmony with their needs and resources.

NOTES

1. This section is mainly based on the work of Amador and Briceño (1999).
2. US$1 = 250 colones.

REFERENCES

Alfaro, R M (1984) Logros de la investigación sobre fríjol común (*Phaseolus vulgaris* L.) en Costa Rica. *In*: Congreso Agronómico Nacional IV, San José, Costa Rica. Programa Ministerio de Agricultura y Ganadería (MAG)-Universidad de Costa Rica (UCR)- Consejo Nacional de Producción(CNP)-Centro Internacional de Agricultura Tropical (CIAT), Costa Rica. p. 155-175.

Alfaro, R M (1994) Improving the slash/mulch *fríjol tapado* system. *In:* Thurston H D, Smith M, Abawi G, Kearl S (Eds.) Tapado/slash mulch: How farmers use it and what researchers know about it. Cornell International Institute for Food, Agriculture, and Development (CIIFAD), Ithaca, New York, USA. p. 209-214.

Alfaro R M, Waaijenberg H. (1992) El cultivo de fríjol tapado en Costa Rica: Un resumen de investigaciones 1978-1991. Technical Series, Technical Report No. 190, Report No. 28. Centro Agronómico de Investigación y Enseñaza (CATIE), Turrialba, Costa Rica. 12 p.

Amador M, Briceño J (1999) Aspectos socioeconómicos en el sistema de fríjol tapado. *In*: Meléndez G, Vernooy R, Briceño J (Eds.) El fríjol tapado en Costa Rica: Fortalezas, opciones y desafíos. Asociación Costarricense de la Ciencia del Suelo (ACCS), San José, Costa Rica. p. 103-114.

Araya R, Gonzalez W (1986) El fríjol bajo el sistema tapado en Costa Rica. Centro Internacional de Agricultura Tropical (CIAT), Cali, Colombia. 272 p.

Arias F, Amador M (1990) Fríjol tapado, un sistema ventajoso para el pequeño productor. Corporación Educativa para el Desarrollo Costarricense (CEDECO), San José, Costa Rica. Avances en Investigación 4:11-15.

Arias J, Chacón A, Meléndez, G (1999) Valoración financiera y de riesgo de los sistemas de siembra de fríjol, tapado, y espequeado. *In*: Meléndez G, Vernooy R, Briceño J (Eds.) El fríjol tapado en Costa Rica: fortalezas, opciones y desafíos. Asociación Costarricense de la Ciencia del Suelo (ACCS), San José, Costa Rica. p. 129-155.

Bellows B C (1992) Sustainability of steep land bean (*Phaseolus vulgaris* L.) farming in Costa Rica: An agronomic and socio-economic assessment. Ph.D. dissertation, University of Florida, Gainesville, USA. 282 p.

Bellows B C (1994) Fríjol tapado, fríjol espeque and labranzo cero: A socio-economic and agroecological comparison of bean production in Costa Rica. *In*: Thurston H D, Smith M, Abawi G, Kearl S. (Eds.) Tapado/slash mulch: How farmers use it and what researchers know about it. Cornell International Institute for Food, Agriculture, and Development (CIIFAD), Ithaca, New York, USA.

Borbón E (1989) Evaluación socioeconómica de la producción de fríjol (*Phaseolus vulgaris* L.*)* en el Cantón de Nicoya, cosecha veranera, 1987. Tesis de Licenciatura. Universidad de Costa Rica, San José, Costa Rica. 145 p.

Campos-Alvarado W (1983) Manual de recomendaciones de cultivos agrícolas de Costa Rica. Ministerio de Agricultura y Ganadería de Costa Rica, Boletín Técnico No. 62.

Castro L (1996) Efecto del uso agrícola y el barbecho sobre los contenidos de biomasa microbiana en ultisoles y andisoles de Costa Rica. *In*: II Congreso Nacional de Suelos, Memoria Vol. III, Costa Rica. 95 p.

Castro G, Meza N, Alfaro R M, Rodríguez N, Rojas K (1991) Encuesta nacional sobre consumo aparente de alimentos 1989. Ministerio de Salud, Dirección Nacional de Estadísticas y Censos, Comité de Acción de Apoyo al Desarrollo Económico y Social de Centroamérica-Comunidad Económico Europea (CADESCA-CEE), San José, Costa Rica. p. 150-158.

CEDECO - Corporación Educativa para el Desarrollo Costarricense (1992) Memoria del 1 encuentro nacional campesino de fríjol tapado, San Ignacio de Acosta, 12-14 de Julio de 1991. San José, Costa Rica. 126 p.

CIAT – Centro Internacional de Agricultura Tropical (1986) Bean research payoff in Costa Rica. CIAT International 5:3-5.

CNP – Consejo Nacional de Producción (1996) Diagnóstico de la actividad fríjolera y propuesta de políticas. Dirección de Planificación, CNP, San José, Costa Rica. 26 p.

CNP – Consejo Nacional de Producción (1998) Dirección de mercadeo y agroindustria. Sistemas de información de mercados. Estadística de producción y comercialización de granos básicos. CNP, San José, Costa Rica. 35 p.

Cook S, Dyhrmann S, Ingram K (1993) The abundance of detrivores in fríjol tapado vs. espequeado plots. *In*: Bansak T S, Seandel M (Eds.) Dartmouth Studies in Tropical Ecology 1993. The Department of Biological Sciences, Dartmouth College. p. 96-98.

De la Cruz R (1994) La utilidad de la diversidad de malezas en fríjol tapado: Dificultades con el uso de herbicidas. *In:* Thurston H D, Smith M, Abawi G, Kearl S (Eds.) Tapado/slash mulch: How farmers use it and what researchers know about it. Cornell International Institute for Food, Agriculture, and Development (CIIFAD), Ithaca, New York, USA. p. 261-266.

Eckert G, Schellhorn N, Fernald S, Geiger C, Ramstad K, Treviño L, Reeves M (1991) Microenvironment of the fríjol tapado system. *In:* Tropical managed ecosystems. 1991 Course Book, Organization for Tropical Studies (OTS), Costa Rica. p. 89-92.

Fresco L O, Krooneberg (1992) Time and spatial scales in ecological sustainability. Land Use Policy 7:155-167.

Gadeo A, Briceño J (1999) Efecto de insumos externos en el rendimiento de fríjol tapado. *In*: Meléndez G, Vernooy R, Briceño J (Eds.) El fríjol tapado en Costa Rica: Fortalezas, opciones y desafíos. Asociación Costarricense de la Ciencia del Suelo (ACCS), San José, Costa Rica. p. 185-193.

Galindo J J (1994) Incidence of web blight in beans grown under the *fríjol tapado* system in Costa Rica. *In:* Thurston H D, Smith M, Abawi G, Kearl S (Eds.) Tapado/slash mulch: How farmers use it and what researchers know about it. Cornell International Institute for Food, Agriculture, and Development (CIIFAD), Ithaca, New York, USA. p. 109-116.

Galindo J J, Abawi H D, Galvez, H G (1983) The effect of mulching on beans in Costa Rica. Phytopathology 73:610-615.

González W, Araya R (1994) Agro-economic study of beans (*Phaseolus vulgaris* L.) grown under the 'tapado' slash/mulch system in Costa Rica. *In:* Thurston H D, Smith M, Abawi G, Kearl S (Eds.) Tapado/slash mulch: How farmers use it and what researchers know about it. Cornell International Institute for Food, Agriculture, and Development (CIIFAD), Ithaca, New York, USA. p. 263-272.

Hallmann G, Andrews K L (1989) 'Fríjol.' *In:* Andrews K L, Rutilio Q J (Eds.) Manejo integrado de plagas insectiles en la agricultura: Estado actual y futuro. Escuela Agrícola Panamericana (El Zamorano), Honduras. p. 523-545.

Huntington T G, Smith M, Thomas G W, Blevins R L, Pérez A (1986) Response of *Phaseolus vulgaris* to inoculation with *Rhizobium phaseoli* under two tillage systems in the Dominican Republic. Plant Soil 95:77-85.

Jiménez G (1985) Levantamiento de malezas en el cultivo de fríjol tapado en la zona de Upala. *In:* Soto A, Koch W, Jùrgens G, García J (Eds.) Resúmenes del Seminario Integrado de Malezas. Plits 3(2):213-224.

Lindarte E, Benito C (1991) Agricultura sostenible de ladera en América Central: instituciones, tecnología, políticas. *In*:Taller sobre agricultura sostenible en las laderas centroamericanas: Oportunidades de colaboración interinstitucional, 13-16 agosto 1991. Instituto Interamericano de Cooperación para la Agricultura (IICA), San José, Costa Rica. p. 77-188.

Mata R, Uribe L, Reeves M, Salas E (1999) Suelos característicos de tapaderos. *In*: Meléndez G, Vernooy R, Briceño J (Eds.) El fríjol tapado en Costa Rica: Fortalezas, opciones y desafíos. Asociación Costarricense de la Ciencia del Suelo (ACCS), San José, Costa Rica. p. 41-78.

Meléndez G, Szott L (1999) El barbecho y el funcionamiento biofísico del fríjol tapado. *In*: Meléndez G, Vernooy R, Briceño J (Eds.) El fríjol tapado en Costa Rica: Fortalezas, opciones y desafíos. Asociación Costarricense de la Ciencia del Suelo (ACCS), San José, Costa Rica. p. 159-184.

Meléndez G, Briceño J, Gadea A (1994) Mejoramiento de sistemas con cobertura: fríjol tapado. Informe anual 1995. Centro Internacional de Investigaciones para el Desarrollo (CIID)-Universidad de Costa Rica (UCR)-University of Guelph, Costa Rica. 120 p.

Ocampo R (1985) Metodología para la identificación de plantas indeseables en el cultivo de fríjol tapado en Costa Rica. *In:* Soto A, Koch W, Jurgens G, García J (Eds.) Resúmenes del Seminario Integrado de Malezas. Plits 3(2): 105-114.

Pachico D, Borbón E (1987) The adoption of improved bean varieties: A case study in Costa Rica. Centro Internacional de Agricultura Tropical (CIAT), Cali, Colombia. 18 p.

PRIAG - Programa Regional de Reforzamiento a la Investigación Agronómica sobre los Granos en Centro América (1993) Síntesis de los estudios nacionales. El impacto del ajuste estructural sobre los sistemas de producción de granos básicos en Centroamérica. PRIAG, San José, Costa Rica.

Ribier V, Barahon M, Damais G, Mora H, Munguía S, Sáenz C (1988) Estudio sistemático de la realidad agraria de una microregión de Costa Rica: Nicoya-Hojancha IV. Sistema de granos básicos. Universidad Nacional, Heredia, Costa Rica. p. 40-51.

Rosemeyer M E (1990) The effects of different management strategies on the tripartite symbiosis of bean (*Phaseolus vulgaris* L.) with *Rhizobium* and vesicular-arbuscular mycorrhizal fungi in two agroecosystems in Costa Rica. Ph.D. dissertation, University of California-Santa Cruz, Costa Rica. 160 p.

Rosemeyer M E (1994a) Avances en investigación de fríjol tapado con énfasis en fuentes alternativos de fósforo. *En*: Memoria de 1 Encuentro Nacional Campesino de Coberturas, 4-7 octubre 1993, Corporación Educativa para el Desarrollo Costarricense (CEDECO), San José, Costa Rica. p. 45-57.

Rosemeyer M E (1994b) Yield, nodulation and mycorrhizal establishment in slash/mulch vs. row-cropped beans. *In:* Thurston H D, Smith M, Abawi G, Kearl S (Eds.) Tapado/slash mulch: How farmers use it and what researchers know about it. Cornell International Institute for Food, Agriculture, and Development (CIIFAD), Ithaca, New York, USA. p. 169-178.

Rosemeyer M E, Gliessman S R (1992) Modifying traditional and high-input agro-ecosystems for optimisation of microbial symbioses: A case study of dry beans in Costa Rica. Agric. Ecosyst. Environ. 40:61-70.

Roth D, Olson D, Brown N, Kettler J, González R, Webb T, Rosemeyer M E (1991) Physical characteristics in soil in different management systems. *In:* Tropical managed ecosystems, 1991. Course Book, Organization for Tropical Studies (OTS), Costa Rica. p. 85-88.

Schlather K J (1998) The dynamics and cycling of phosphorus in mulched and unmulched bean production systems indigenous to the humid tropics of Central America. Ph.D. dissertation, University of Cornell, New York. 134 p.

Shenk M (1994) Possible modifications of planting and mulch management schemes for fríjol tapado. *In:* Thurston H D, Smith M, Abawi G, Kearl S (Eds.) Tapado/slash mulch: How farmers use it and what researchers know about it. Cornell International Institute for Food, Agriculture, and Development (CIIFAD), Ithaca, New York, USA. p. 129-132.

Smith M (1994) Crop variety improvement for slash/mulch production systems. *In:* Thurston H D, Smith M, Abawi G, Kearl S (Eds.) Tapado/slash mulch: How farmers use it and what researchers know about it. Cornell International Institute for Food, Agriculture, and Development (CIIFAD), Ithaca, New York, USA. p. 239-248.

Tapia M, Camacho A (1988) Manejo integrado de la producción de fríjol basado en labraza cero. Deutsche Gesellschaft für Technische Zusammenarbeit (GTZ), Managua, Nicaragua. 131 p.

Thurston H D (1984) Tropical plant diseases. The American Phytopathological Society Press, St Paul, Minnesota, USA.

Thurston H D (1994) Historia de los sistemas de siembra con cobertura muerta o sistemas de tumba y pudre en América Latina. *In:* Thurston H D, Smith M, Abawi G, Kearl S (Eds.) Tapado/slash mulch: How farmers use it and what researchers know about it. Cornell International Institute for Food, Agriculture, and Development (CIIFAD), Ithaca, New York, USA. p. 1-4.

von Platen H, Rodríguez G (1982) La producción de fríjol tapado en la región de Acosta-Puriscal, Costa Rica. Documento presentado en la XXVII Reunión Anual del Programa Cooperativo Centroamericano para el Mejoramiento de Cultivos Alimenticios (PCCMCA) en San José, Costa Rica.

von Platen H, Rodríguez G, Lagemann J (1982) Sistemas de finca en Acasta Puriscal, Costa Rica. Centro Agronómico Tropical de Investigación y Enseñanza, Costa Rica (CATIE), Deutsche Gesellschaft für Technische Zusammenarbeit (GTZ)-Ministerio de Agricultura y Ganadería (MAG). CATIE Technical Report No. 30, Turrialba, Costa Rica. 146 p.

Chapter 3

Mucuna Use by Hillside Farmers of Northern Honduras

B. Triomphe[1] and G. Sain[2]

[1] *Centre de Coopération Internationale en Recherche Agronomique pour le Développement, France - Centro Internacional de Mejoramiento de Maíz y Trigo, Mexico (CIRAD-CIMMYT)*
[2]*CIMMYT*

SUMMARY

In the humid lowlands of northern Honduras, thousands of small-scale hillside farmers decided in the early 1980s to replace their traditional maize (*Zea mays* L.) cropping systems with a rotational one that included maize and *Mucuna* (*Mucuna pruriens* [L.] DC.). This new system consists of rotating a short-term *Mucuna* fallow during the main rainy season with a dry-season maize crop that is planted in the thick *Mucuna* mulch at the end of the rainy season. *Mucuna* re-establishes itself via natural reseeding towards the end of the maize cropping cycle and is always slashed prior to maize planting.

This case study describes the system, recounts the story of its adoption and disadoption (by the mid-1990s, many adopters had abandoned the *Mucuna* system) and discusses its strengths and limitations from a variety of angles. From the agro-ecological viewpoint, this system exemplifies the multiple benefits that cover crops, no tillage and soil mulching may bring to hillside maize farming. Such benefits include erosion control, improvement of soil fertility (increases in organic matter and better infiltration), better moisture retention, weed control, nutrient capture (via N fixation), synchrony between nutrient supply and demand, and a higher productivity without the need to apply chemical fertilizer. Agronomic problems, on the other hand, stem from the poor re-establishment of the *Mucuna* stand via natural reseeding and the consequent aggressive establishment of weeds such as *Rottboellia cochinchinensis* (Lour.) Clayton (itchgrass). From the socio-economic viewpoint, when compared to the traditional maize cultivation, the maize- *Mucuna* rotation brought

M. Eilittä et al. (eds.), Green Manure/Cover Crop Systems of Smallholder Farmers, 65-97.

significant benefits in terms of labour savings and economic returns per hectare. However, it did not compare well to other alternatives, such as common bean (*Phaseolus vulgaris* L.) production, Tabasco chilli peppers (*Capsicum frutescens* L.) or livestock production. From the adoption viewpoint, land tenure did not constitute an obstacle to adoption of the *Mucuna* system. After its introduction in the region by migrant workers, adoption was basically spontaneous, through farmer-to-farmer diffusion. The main reasons for adoption included labour-saving, productivity boost, lower risk of crop failure and a higher market price of the dry-season maize. Abandonment, on the other hand, stemmed from the loss of productivity related to itchgrass infestation and a general tendency for maize to be displaced by more lucrative enterprises, such as meat and milk production. Although there are different ways of improving the performance of the *Mucuna*-maize system, the future niche for the system is unknown.

INTRODUCTION

Why did thousands of small-scale farmers in the Honduran hillsides decide in the early 1980s to substitute their traditional maize (*Zea mays* L.) cropping systems based on shifting cultivation for one in which maize was grown in rotation with *Mucuna pruriens* (L.) DC. Is the Spanish name given to this system (*abonera*, a name derived from *fríjol abono*, or fertilizer bean, the farmers' name for *Mucuna*) really justified? And why have adoption rates of this system declined in the 1990s although they had risen just a decade earlier?

Certainly, many important lessons can be learned from this case of spontaneous adoption and disadoption of an archetypical green manure/cover crop (GMCC) system. The *Mucuna* system in Honduras is here discussed from several complementary viewpoints: agronomic, economic, social, institutional and political. No original research has been conducted to write this account. Instead, it is a compendium of information taken mainly from Buckles et al. (1992), Humphries (1994), Sain et al. (1994), Triomphe (1996), Buckles et al. (1998), Buckles and Triomphe (1999) and Neill and Lee (2001). Figure 1 shows the location of the area under study.

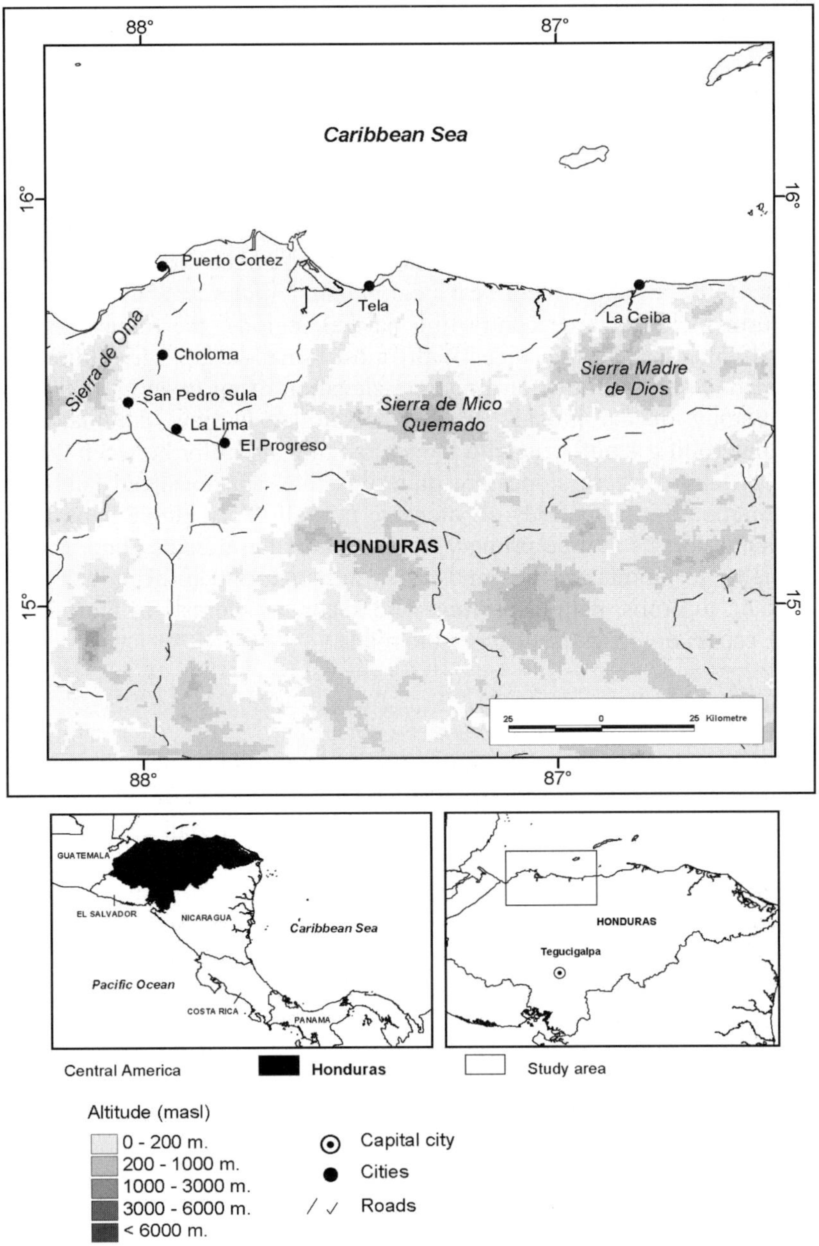

Figure 1. Location of study area in Honduras.

REGIONAL CONTEXT

Biophysical Environment

Topography and Vegetation

Sedimentary materials from the ocean floor were pushed up during the Tertiary era to form the Nombre de Dios mountain range, which runs parallel to the coastline with peaks culminating at almost 2500 masl. This landform includes three contrasting natural regions: the mountains, the coastal plain and (the area with most small-scale agriculture) an intermediate hillside area of moderate elevation of up to about 600 masl. Topography of the hillside area is mixed but largely dominated by irregular rolling landforms with slopes ranging typically between 20 and 100%. Many of these slopes are quite unstable, and localized landslides are common during periods of intensive rainfall. Very humid subtropical deciduous forests are the primary vegetation of the hillside zone. Forests still tend to dominate the landscape, although much of it has been deflected to crops, natural pastures and secondary forests in the past 30 years, especially in areas closer to the roads.

Climate and Soils

The climate of northern Honduras is classified as humid tropical. Under the influence of the Nombre de Dios Mountains, the hillsides receive between 2000 and 3000 mm of annual rainfall in a bimodal distribution, with some rain during virtually every week of the year (Figure 2). Relatively wetter periods occurred during the late 1970s and early 1980s, while relatively drier periods occurred during the mid-1970s, mid-1980s and, particularly severely, in the early 1990s. Although rainfall has become possibly less reliable than previously, farmers still estimated in the early 1990s that there were fewer than 2 'bad years' out of every 10 with respect to the climate; and many even disputed the very idea of climatically bad years. Indeed, were it not for the occasional storms and hurricanes (e.g. Hurricane Mitch, which devastated the region in late 1998), northern Honduras could almost be considered a haven for rain-fed agriculture from a water balance perspective.

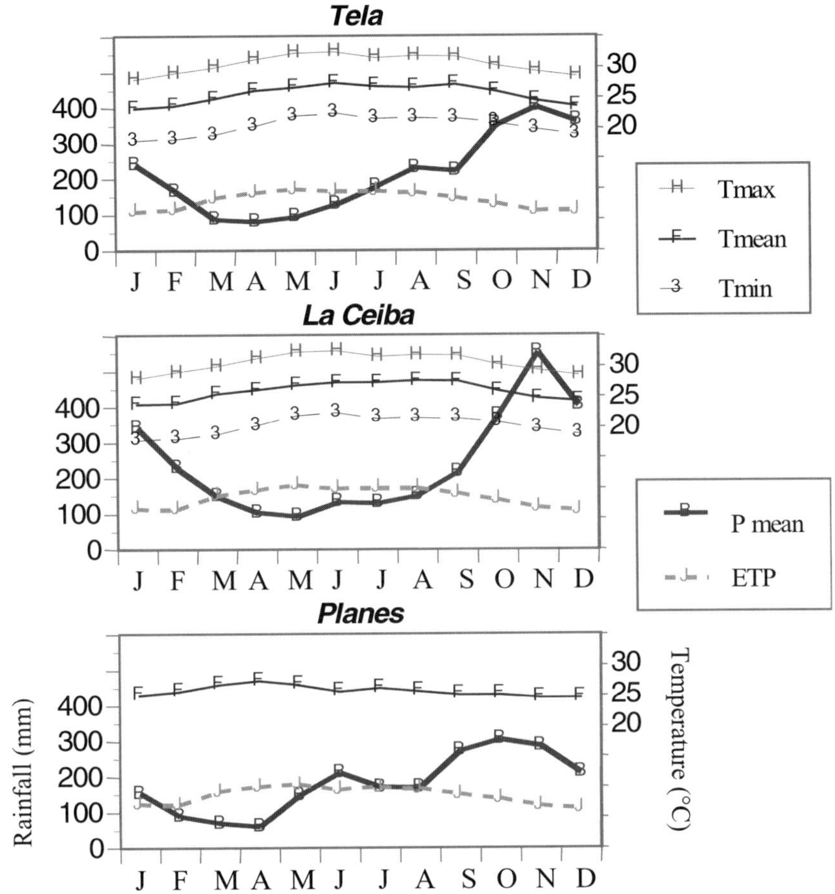

ETP evapotranspiration; P precipitation; T temperature.

Figure 2. Average annual rainfall in three communities, northern Honduras (Triomphe 1996). Monthly rainfall averages were calculated over 38 years and monthly minimum and maximum temperatures over 12 years (Hargreaves 1980).

The first rains usually begin in June, establishing the *primera* or first season. The heaviest and most consistent rainfall on the Atlantic Coast coincides with the last portion of the year (September-December), initiating a second major cropping season known as the *postrera* or second season. Daily rainfall of 100 to 200 mm is not uncommon during this period. Rainfall is erratic, although still significant, during the latter part of the second season (January-March). In addition, the soil profile usually contains enough water for crops and natural vegetation to resist droughts of 4 to 6 weeks with little negative consequence. By April, the rains drop off, ushering in a short, relatively dry period known as *verano* or summer that runs through to the end of May.

The average annual temperature at sea level is about 26 ºC, ranging from 28 ºC (in May) to 24 ºC (in January). Evapotranspiration remains moderate during the rainy season (about 3 to 4 mm d^{-1}), increasing slightly during the dry season (5 mm d^{-1}).

Soil types vary by elevation and specific location but include Ultic Hapludalts, Typic Dystropepts, Typic Hapludults, Tropohumults and Tropohudults (Triomphe 1996), derived mainly from hard metamorphic rock originating in the Paleozoic era. Most of the hillside soils are relatively deep (typically 60-80 cm), mildly acidic (pH around 6.0) and have good levels of exchangeable bases to a depth of 60 cm or more, usually from 10 to more than 20 cmols (+) kg^{-1}. The major constraint for cultivation hence is not poor soils but very steep slopes and the susceptibility to erosion.

Human Environment

Settlement and Access to Land

Northern Honduras was still, in the 1990s, an active agricultural frontier and one of the main destinations for rural migrants originating from southern and western Honduras (Humphries 1994; Buckles et al. 1998). The availability of land drew displaced populations, helping reduce social unrest relative to other countries in Central America where striking inequalities in land distribution also existed. As a result of immigration, the population of northern Honduras grew at an annual rate of 4.2% between 1970 and 1990, compared to a 3.4% annual rate during this same period for the nation as a whole. During this period, population density in northern Honduras increased from 35 to 57 inhabitants per km^2, with a relatively high concentration of urbanites, primarily in La Ceiba, the third largest city in Honduras.

New migrants to northern Honduras found few opportunities on the fertile coastal plain, which hosts the plantations of the large transnational companies. Consequently, most of them had to settle on the hillsides and upper slopes of the Nombre de Dios mountain range. The lower hillsides were settled mainly during the 1970s and early 1980s and now have a relatively stable population living in about 110 small towns and hamlets. By contrast, the upper slopes of the Nombre de Dios range, generally unsuitable for agriculture, continue to be an active frontier (Humphries 1994). Even though, until the late 1960s, land could be claimed simply by clearing the forest cover and registering for squatters' rights, a decade-old change in agrarian law has allowed the change of squatters' rights into formal land ownership.

While northern Honduras is still actively receiving migrants, it is also a region with high rates of emigration to booming industrial cities outside the region (such as San Pedro Sula and its series of *maquiladoras*, or foreign-owned assembly plants) and to the USA. Virtually every household in some hillside communities has permanently had at least one member in the USA for the past 10-15 years. This emigration contributes to greatly improved livelihoods, because it provides cash for food, health care, improved housing, education and for establishing small businesses.

Even though hillside settlement is relatively recent, the distribution of land was already moderately concentrated by the early 1990s. Farm survey data (Buckles et al. 1992) indicated that 58% of the land in the hillside zone was owned by only 17% of the landholders, in holdings of more than 20 ha. For most of these households, ranching was the central component of their production strategy. By contrast, 46% of the landholders owned less than 10% of the total land area, in holdings of 5 ha or less.

The concentration of landownership is an important but not absolute limitation to farming. In the early 1990s, surveys indicated that some 21% of households did not own farmland, yet they were farming the land of others (either within family or outside) in exchange for cash, labour or a share of the harvest. Medium- and large-scale landowners dominated the formal land rental market by renting out land to small-scale and landless farmers, either for establishment of new pastures or for the re-establishment of degraded pastures, a process documented throughout Central America (Buckles et al. 1992; 1998). Three-quarters of all households surveyed in 1992 rented some land, typically a hectare or so, for maize and other annual crops.

Typology of Households

Significant differentiation of households evolved in the hillsides that can be related to the distribution of land and other resources (Table 1) (Buckles et al. 1998).

Ranchers represent 15% of the households. They own cattle and pigs, pastures and fallow land. Crop production is typically diversified, including above-average areas in maize, common beans (*Phaseolus vulgaris* L.) and rice (*Oryza sativa* L.). Annual crops are frequently grown for market. Ranchers have enough financial resources to establish small businesses and thus avoid low-paying, off-farm employment. They also most probably have the resources needed for investing in land-conserving and productivity-enhancing technologies.

Diversified farmers, also some 15% of the households, have fewer cattle than ranchers but more pigs, a less land-intensive form of livestock production. Nevertheless, they own enough land to grow a wide variety of annual crops. They also control some pasture and fallow. They typically engage in relatively stable and better paying forms of off-farm employment, such as factory work and teaching. This income in turn provides them with opportunities to accumulate land and livestock.

Table 1. Selected characteristics of the various kinds of household groups in northern Honduras (*Source*: Adapted from Buckles et al. 1998).

Criteria	Ranchers	Diversified farmers	Medium-scale farmers	Small-scale farmers	Subsistence workers
% of households	15.0	15.0	24.0	22.0	24.0
Farm size (ha)	32.0	12.3	7.7	5.1	2.0
Land owned (%)	96.0	94.0	83.0	58.0	11.0
Cropland (ha)	7.3	3.5	2.8	3.2	1.8
Pastures (ha)	15.1	4.5	0.5	0.0	0.2
Cattle (heads)	19.2	3.2	0.2	0.0	0.0
Farm work (%)	5.3	10.5	20.0	89.3	80.0
Off-farm wage (%)	0.0	31.6	3.3	0.0	20.0

Medium-scale farmers (24% of the households) own land, including sizable fallow areas, on which they grow small quantities of maize, rice and beans but normally only for subsistence. Some land may be under permanent tree crops. Land resources are too limited, however, for livestock production. Many avoid low-paying day work but most opt to complement farming with logging.

Small-scale farmers (22% of the households) own some cropland, mainly maize and beans for subsistence, but very little of it is under pasture or in fallow. Access to land through rental markets and to day work and self-employment (e.g. crafts and petty trade) are very important for them. Dependency on off-farm employment implies in turn that little labour would be available within the household to invest in land management practices, except perhaps using crop residues as mulch (no competition for residue use as small-scale farmers do not own livestock).

Subsistence workers frequently engage in low-paying, off-farm employment as day workers, their primary source of income. Crop production, often on rented land, is focused exclusively on maize, the main subsistence crop in the region. As with small-scale farmers, land resources are too limited to permit crop diversification and owning livestock is beyond their means.

Institutional Actors

Unlike the many other GMCC systems described in this volume, the one in northern Honduras has been basically established as a consequence of actions and decisions by farmers alone in response to the extremely rapidly changing circumstances prevailing in the region. Research and extension services have never been active in the hillsides and were virtually non-existent by the late 1980s-early 1990s. Whatever research or studies were conducted, they were rather extractive in essence, as they focused on learning from farmers' experience with the *Mucuna* system. Some development projects have conducted work relevant to the *Mucuna* system in the region. An externally funded forestry development project,

Proyecto de Desarrollo del Bosque Latifoliado (PDBL), has promoted since 1990 the diffusion of the *Mucuna* system in a number of hillside villages as a way of reducing deforestation (Szaraz and Irias 1993). But it merely proposed diffusing the existing system to new areas, without introducing modifications to farmers' original practice. In 1994, another project, focused on participatory technology development, started operating in several villages in the region (Humphries 1994). Despite the problems that farmers faced with the *Mucuna* system (see below), when they were given a chance to choose priority issues for technology development, they did not select improving the system as their highest priority. They focused instead on cocoa (*Theobroma cacao* L.) production, an option that at the time seemed to be the easiest way of improving their income.

CONVENTIONAL HILLSIDE MAIZE CROPPING SYSTEMS

Maize, beans and, to a lesser extent, upland rice are the most important annual crops grown on the hillsides of northern Honduras, accounting for 97% of the cropped area. Maize, grown on 34 000 ha in the Atlantida department alone (Barreto 1999) either sole- or intercropped with beans, can be cultivated during the first and second season. Common bean, usually planted as a sole crop or in relay after maize on very steep hillsides, to facilitate drainage and minimize web blight (*Thanatephorus cucumeris* [A.B. Frank] Donk), can be planted three times a year in the hillside zone—February, May and October (Humphries 1994). Yields of beans are low on average, between 0.4 and 0.8 t ha^{-1}. Moisture requirements limit upland rice to the first season only. Most hillside farmers also grow cassava (*Manihot esculenta* Crantz.), plantain (*Musa x paradisiaca* L. [pro sp.]), cocoa, coffee (*Coffea arabica* L.) and various citrus fruit trees (*Citrus* spp. L.) in small areas, typically on the house compound. A few farmers engage in the production of Tabasco chilli peppers (*Capsicum frutescens* L.) during the first season, one of the few non-subsistence crops grown on the hillsides.

Maize and other annual crops have been traditionally grown using techniques of no-till shifting cultivation, characteristic of the humid tropics of Mesoamerica. Although maize cultivation in most of Honduras is restricted to the main rainy season, climatic conditions in northern Honduras allow a second-season maize cycle during the *postrera*. Interestingly, most households cultivate both first- and second-season maize as a way of ensuring that enough will be available for consumption and/or sale throughout the year. Overall, surveys conducted by Buckles et al. (1992) and Sain et al. (1994) indicated that first-season maize yields were low and typically varied between 1 and 1.5 t.ha^{-1}, compared to the national average of 1.5 t.ha^{-1}. Second-season maize yields in the order of

1.5 to 2.5 $t.ha^{-1}$ were common, however, with considerably fewer labour costs and lower climatic risk compared to the first season.

First-Season Maize

For first-season maize, land preparation is undertaken between March and May and consists of the manual slashing of the fallow vegetation, followed by its drying and burning. One quarter of the farmers also apply a contact herbicide (2-4D or Paraquat) prior to planting. Planting time is at the onset of the first rains, typically by early June. Farmers use a dibble stick to plant three to five maize seeds per hole and establish populations ranging from 30 000 to 44 000 plants ha^{-1}. Two-thirds of the farmers use local maize varieties and the remainder use open-pollinated varieties. To date, hybrid maize is practically unknown in hillside maize production.

Maize is weeded twice, once manually at about 30 to 35 days after planting and the second time with herbicides (mostly Paraquat and 2-4D) applied via back-sprayers some 40-45 days after sowing. While herbicides are common, most farmers do not use fertilizer with first-season maize because of its high cost and production risk. When applied, rates are very low (average 30 kg N ha^{-1}). First-season maize plants are bent over below the ear after they reach physiological maturity to avoid lodging, facilitate drying and harvesting (ear insertion height on local cultivars is frequently more than 2 m) and protect them from bird damage and ear rot.

Second-Season Maize

For second-season maize, land preparation is usually initiated in November and most fields are planted by December or early January. Farmers do not burn prior to planting but rather leave the residues on the fields, which helps conserve soil moisture during the dry period. Weed pressure is less severe because of lower rainfall, hence weed control is less intensive (at least before the advent of *Rottboellia cochinchinensis* [Lour.] Clayton [itchgrass]; see below). Less than half (44%) of the farmers apply small amounts of nitrogen fertilizer to second-season maize. The application of fertilizer in this season is less risky than in the first season and potentially more profitable. Second-season maize reaches physiological maturity between April and June.

As in all fallow-based systems, the key to sustainable shifting cultivation in northern Honduras is related to the ratio of the length of the cropping to fallow periods. With an average of two cycles, cropping periods for first-season maize range from as few as one cycle (a common occurrence) to as many as seven. Fallow periods preceding a first-season maize range from 1 to 15 years, with an average of 4.2 years, slightly below the minimum period (5 years) needed to establish what is considered locally as a mature fallow. These figures, while highly variable, suggest that the cropping intensity in bush-fallow rotations in northern Honduras is, on average, as intensive as it can be in a sustainable manner. This is one of the main reasons why the *Mucuna*-maize system, which we will now describe in detail, represents such a valuable alternative for intensifying maize production in the region.

MANAGEMENT OF THE *MUCUNA*-MAIZE SYSTEM

Farmers establish *Mucuna* as a sole fallow crop during the humid first season. The mature stand is slashed in December with a machete and the second-season maize is planted into the layer of decomposing leaves and vines (Figure 3). The field is not burned prior to planting maize nor is the legume incorporated into the soil. *Mucuna* reseeds itself spontaneously from the unharvested seeds during the second-season maize cycle, and takes over the maize field around harvest time (April to June), using maize stalks as tutors. From that moment to the next slashing in December, no other field operations are performed, leaving the field under a short-term *Mucuna* fallow (Figure 3).

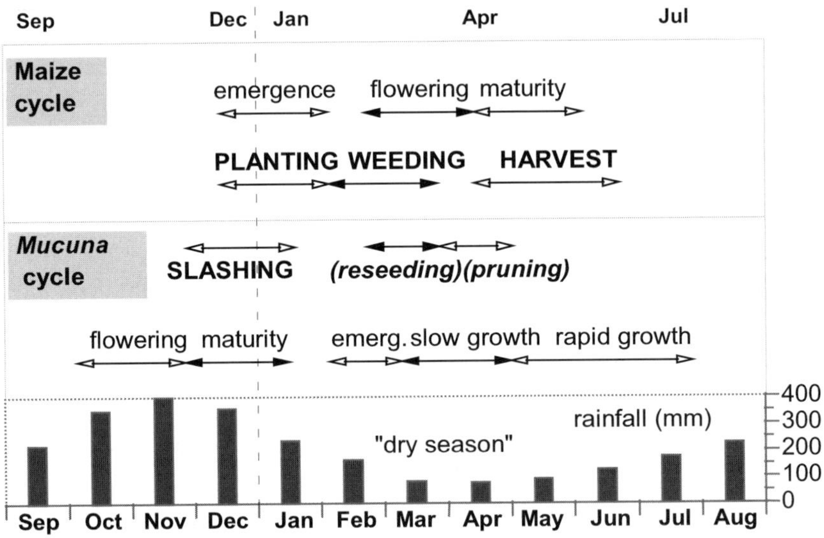

Figure 3. Simplified calendar of the maize-*Mucuna* rotation (*Source*: Triomphe 1996).

Initial Establishment

Mucuna seed usually consists of a mixture of three seed types: black, mottled and white, corresponding to three distinct types of *Mucuna* growing together in any given *Mucuna* field. It is planted just after weeding, which takes place about 40 to 60 days after maize planting. Farmers use a dibble-stick to plant 2 to 3 seeds per hole, in holes 1 to 2 m apart between the rows of second-season maize. This represents from 10 to 15 kg ha^{-1} of seed, which is typically collected from established *Mucuna* fields, as there are no *Mucuna* seed markets. An alternative method is to broadcast *Mucuna* seed but this is considered less effective at ensuring even establishment. A few farmers reported that they establish *Mucuna* directly after clearing a fallow field. Overall, labour costs associated with initial establishment of the *Mucuna* field are minimal.

Annual Re-Establishment

Once *Mucuna* is established, farmers can typically rely on natural reseeding, provided that the slashing of the crop is done after viable pods are produced in sufficient quantities. The pods left unharvested burst open when dry, ejecting seeds over the field reasonably evenly.

The resilience of a *Mucuna* field is remarkable. Farmers in San Francisco de Saco have relied on natural reseeding for more than 15 years without ever replanting their *aboneras* from new seed. Even after a marked failure of the *Mucuna* cycle during the winter of 1993-94, followed by extremely unfavourable conditions for seed germination, seed production from the sparse *Mucuna* stand reaching maturity the following year was so plentiful that most farmers did not need to replant them.

Although rarely necessary, some farmers toss *Mucuna* pods into the field at slashing time to ensure uniform stands, while others replant *Mucuna* later in the season in bare spots. The seed used for replanting is harvested usually from plants growing on trees or rocks, where seed production is high.

Reliance on natural reseeding of the *Mucuna* field allows farmers to maintain the crop permanently at no direct cost. The practice does have other less favourable management implications, however. *Mucuna* plants germinating in the maize field may grow so vigorously early in the season that farmers are obliged to prune the emerging plants. More problematic is the situation created by aggressive weed colonization (e.g. itchgrass) of patches devoid of *Mucuna* plants, when farmers do not complement natural reseeding with replanting. Itchgrass control implies significant extra weeding costs and when not done properly opens the way to the weed displacing *Mucuna*. This has occurred in some areas where itchgrass has become an important pest (Munguia 1992) to the point where profitability of growing a maize crop is jeopardized (Neill and Lee 2001).

Slashing

Slashing is the main management practice for *abonera* fields. A wide range of slashing dates is used but all farmers are careful to slash *Mucuna* only after it has produced viable pods. Once this is assured, the actual timing is influenced by the availability of labour and by considerations of risk of drought later in the season. Slashing involves cutting the pliant *Mucuna* cover with a machete while, with the help of a wooden hook, pulling *Mucuna* vines up from the ground or rocks. Farmers do not try to cut *Mucuna* finely because this would increase labour time and could destroy the pods needed for natural reseeding. However, some farmers insist that the slashed *Mucuna* material must be evenly spread on the field surface to ensure adequate soil cover and uniform maize growth. Slashing of *Mucuna* requires far less labour than that of a conventional woody fallow, about 10 d ha^{-1} for an *abonera* versus about 18 d ha^{-1} needed to clear a field fallowed for 4 to 5 years. During years with proliferation of rats (a cyclical pest, apparently not restricted to *Mucuna* fields), teams of three to five people may be formed to slash the *abonera* gradually, corner the rats and then kill them. Efficient rat control during slashing can significantly reduce the loss of maize seed and seedlings later on and farmers claim it is entertaining.

Planting

Most farmers prefer to plant maize as soon as possible after they have slashed *Mucuna*, thereby avoiding some of the competition provided by actively growing weeds. Many farmers proceed in tandem with slashing and planting. One or two days of slashing are followed by planting before continuing further. In practice, the interval ranges from a few days to a few weeks, depending mainly on the farmer's ability to mobilize labour.

Farmers dibble-stick maize seeds through the mulch and into the soil. Planting densities and seed type vary among farmers. The most common strategy involves planting 3-4 seeds per hole, in rows 80-100 cm apart, with an intra-row spacing of 50-80 cm. Seeds are frequently treated against a variety of insect predators (particularly ants), using an array of home recipes or strong pesticides such as Malathion. Sometimes farmers use pre-germinated seed to hasten emergence and provide young maize seedlings with a competitive edge against weeds. As with first-season maize, local genotypes (*Olotillo*, *Tuza morada*, *Raque*) reproduced on the farm are preferred, although improved open-pollinated germplasm is not uncommon.

Weeding

Weeding is a key practice for determining the fate of both maize and *Mucuna*. It keeps weeds from diverting nutrients and light from the growing maize crop and it creates a window for the successful natural reseeding of *Mucuna*.

Weeding strategies in *aboneras* are similar to those used in shifting cultivation systems and are equally varied. One exception is that farmers using chemical control in *aboneras* are careful not to apply 2-4D, or apply it very cautiously, because it can easily kill the emerging *Mucuna*. Also, manual weed control in *Mucuna* plots requires significantly less labour than in non-*Mucuna* plots (25 to >50% less according to farmer estimates). *Mucuna* gradually eliminates most weed species over the years, especially broadleaves, by preventing their germination, by outcompeting those that do emerge or possibly by allelopathic action. Infrequently, *Mucuna* itself can behave as a weed in some years when it competes with maize plants; labour involved in controlling it is minimal, however (<1-2 person-days ha^{-1} for pruning). According to farmers, weeds in a *Mucuna* system are also rooted much more superficially, owing to the presence of the mulch layer. Furthermore, the topsoil is looser and wetter, making it easier to pull out weed plants during a manual weeding.

An exception to this optimistic picture occurs when itchgrass has firmly established itself in an *abonera* field (see above), in which case labour costs associated with weeding may become relatively high once again (Neill and Lee 2001).

Fertilization

By and large, farmers feel that the *Mucuna* mulch provides enough nutrients to satisfy maize nutritional requirements. Farmers describe with delight the 'bluing' of maize planted in established *aboneras* as a proof of good plant health, a status generally confirmed by foliar analysis, fully justifying the local Spanish name, *abonera*. In some villages, however, it is a rather common practice for farmers at times (particularly in young *Mucuna* fields) to use small doses of urea in *aboneras* (25 to 50 kg ha^{-1}), surface-applied to maize from 40 to 60 days after planting. The effects of this fertilization on maize yields remain unclear (see below).

Harvest

Most farmers harvest their crop almost immediately after it reaches maturity, typically between mid-April and early June, to capture the best possible price and to avoid the summer rains of June-July. If maize plants are bent, the growth rate of *Mucuna* increases sharply, because it benefits from better light interception. This luxurious *Mucuna* growth can make harvesting maize more tedious because one literally has to fight the *Mucuna* to get at the maize ears.

Maize is the only harvested output in the *abonera* system (stover is left entirely in place). Because it is both the staple in their diet and a major source of income, farmers judge the performance of the *abonera* mainly through its ability to ensure a good maize yield, a criterion more important to them than the sustainability of the system.

Beyond Harvest

After harvest, the *abonera* is returned to the *Mucuna* summer fallow for a full 6 months, until it is time to slash again. A few weeks after harvest, *Mucuna* has usually managed to pull down all standing maize stalks and has achieved full canopy closure. *Mucuna* fields are not grazed or used for any other purpose during this time. Surprisingly perhaps, farmers seem quite unaware of the dormant opportunity lying in front of them to extract good quality forage from the *Mucuna* system at basically no other cost than labour. Inadequate access to information on this usage may be the reason behind it, an issue addressed in our concluding remarks.

Variability

Overall, farmers throughout the region closely follow similar management practices with respect to the *abonera* system, indicative of a common origin for the system and of a relatively uniform regional environment. For example, farmers do not use *Mucuna* in rotation or in association with crops other than maize.

There are, however, a number of minor differences among farmers, from field to field and from year to year, with respect to *Mucuna* management. The timing of *Mucuna* slashing and maize planting varies, as does the choice and timing of weed control operations or the planting densities of the maize crop. Also, as noted earlier, reseeding of *Mucuna* may be entirely spontaneous or completed by manual reseeding. These differences seem to take place in response to specific local conditions, such as timing of *Mucuna* maturity, weed pressure or intensity of rainfall during pod formation or at the time of slashing. Production constraints at the household level may also influence practices for which labour or cash availability is critical, such as hiring of wage labour or purchase of herbicides. Consequences of these seemingly minor adjustments on the productivity and sustainability of the system may be far reaching, as will be discussed later.

Before concluding this section, it must be noted that few of the soil conservation practices typically recommended for hillsides, such as contour planting or terraces, are used in conjunction with the *Mucuna* system. The only exception is the use of *Gliricidia sepium* (Jacq.) Walp. in some villages as a live fence around *Mucuna* fields and pastures. However, this seems to be related more than anything else to the potential of this fast-growing tree to provide posts for fencing nearby pastures.

Seed Availability and Exchange

The diffusion of GMCC systems often depends closely on seed availability. While this is the case also for the *Mucuna* system, the issue of seed availability in Honduras takes a very specific turn because, in the established *Mucuna* fields, natural reseeding is the major way of maintaining the *Mucuna* stand over the years (see above). Only newcomers rely on getting seed from others, mainly by borrowing small amounts of seed from their neighbours to establish their *Mucuna* fields.

Dependency on natural reseeding, added to the fact that *Mucuna* seed is not used for food or feed, makes it understandable that no established seed market for *Mucuna* exists in the region. At best, non-governmental organizations or development projects have made modest purchases of *Mucuna* seed but this has been sporadic and limited to very few communities or farmers.

This situation contrasts with what is happening in other parts of Honduras where seed markets are common for cover crops such as *Lablab purpureus* (L.) Sweet or *Phaseolus coccineus* L. (Solomon and Flores 1994; CIDICCO 1997).

PRODUCTIVITY AND THE MAIN IMPACTS OF THE *MUCUNA* SYSTEM

The productivity of the *Mucuna* system is approached from several angles. From an agro-ecological viewpoint the productivity of the *Mucuna* system can be analyzed using two complementary aspects: (1) the active accumulation of biomass and nutrients during the *Mucuna* growth cycle and (2) the subsequent decomposition of the *Mucuna* mulch and the associated release of nutrients during the maize cycle. Both aspects are summarized in the next section, using results from observational on-farm studies conducted between 1992 and 1994 by Triomphe (1996; also reported in Buckles et al. 1998). A section examining land productivity under the *abonera* system, relative to the conventional one, follows. The remaining sections examine the impacts of the system upon different aspects of the agro-ecological system.

Mucuna Biomass and Seed Production

Based on sampling in numerous fields from four communities in three successive years (1992, 1993 and 1994), the levels of total aboveground biomass at slashing fell in a relatively narrow range of 10 - 12.5 t ha^{-1} on a dry-matter basis (Table 2). The largest differences occurred among fields within the same year and community. For example, in one community, individual field minimums dropped to less than 7 t ha^{-1}, whereas maximums exceeded 15 t ha^{-1}. Pods were the most variable contributors to the total biomass—from as little as 6% of the total biomass (equalling less than 500 kg ha^{-1}) to as much as 24% (equalling over 2 t ha^{-1}) within a given field or site.

The overall stability of the biomass production probably stems from a combination of factors. These include the existence of a semi-permanent litter component throughout the year, averaging 60% of the total dry weight (Table 2). Also, the sheer length of the *Mucuna* cycle (8 months minimum) seemingly allows the *Mucuna*-weed stand to compensate over time for any temporary reductions in growth.

Table 2. Aboveground biomass and N content in *Mucuna* fields at slashing time in northern Honduras (*Source*: Adapted from Triomphe 1996).

Aboveground biomass and N content[a]	Sites				
	San Francisco 1992 (n = 44)	San Francisco 1993 (n = 32)	Las Mangas 1993 (n = 29)	Rio Cuero 1993 (n = 21)	Piedras 1993 (n = 19)
Total biomass (t DM ha^{-1}):					
Minimum	6.1	7.0	8.8	8.6	8.6
Maximum	15.9	16.3	13.9	14.5	16.2
Mean	10.8	12.4	11.0	10.7	11.3
s	2.3	2.1	1.4	1.6	1.9
% of total biomass:					
Green biomass	39.0	11.0	10.0	15.0	15.0
Pods		6.0	24.0	7.0	13.0
Vines		14.0	22.0	18.0	19.0
Litter	61.0	69.0	45.0	60.0	53.0
N content (kg ha^{-1}):					
Mean	263.0	316.0	278.0	272.0	310.0
s	75.0	67.0	41.0	52.0	60.0

[a.] DM = dry matter; Green biomass = leafy material and tender vines; pods = pods and immature seeds, vines = old stems, partly lignified and possibly about to start rotting and litter = dead material, including freshly shed leaves. Pods and vines were combined at San Francisco 1992.

As regards nutrients, this abundant biomass provides ample potential supplies of nitrogen, almost 300 kg ha^{-1} on average (Table 2), the actual level depending mainly on biomass production. In the Triomphe (1996) study, the accumulation of other key nutrients for crop growth was also impressive as calcium (140 kg ha^{-1} on average), potassium (100 kg ha^{-1}) and even phosphorus (15-20 kg ha^{-1}) were found at high levels.

Maize Yield

As is often the case in on-farm conditions, average maize yield levels varied markedly by community (Table 3), from about 2.5 t ha^{-1} in a community with poor soils to a high of 4.5 t ha^{-1} in a community with high-fertility soils.

Table 3. Average maize yields (t ha^{-1}) as a function of the duration of the *Mucuna* rotation in northern Honduras (*Source*: Adapted from Triomphe 1996).[a]

Year	Site	No *Mucuna*	With *Mucuna*					
			1–2 years	3–4 years	5–7 years	8–10 years	> 10 years	Mean
1992-93	San Francisco	1.9 b	2.2 b	3.7 a	3.0 ab	3.5 a	3.6 a	3.3
1992-93	Las Mangas	2.5 b	4.2 a	4.2 a	4.9 a	(4.4)	—	4.5
1993-94	San Francisco	2.0 b	3.3 ab	3.7 a	2.7 ab	3.6 a	3.4 a	3.5
1993-94	Las Mangas	1.4 b	1.8 ab	3.1 a	3.2 a	3.9 a	3.1 a	3.1
1993-94	Piedras	—	—	2.2 b	1.6 b	2.8 ab	3.0 a	2.5

[a] Means in a row followed by the same letter are not different according to Tukey's test at the 10% family probability level. The value in parenthesis contains only one observation.

Certain trends, however, appear to be valid across the region. In particular, maize yields in the presence of *Mucuna* were almost double those obtained by using the traditional bush-fallow rotation for second-season maize. Once the *Mucuna* rotation is well-established (usually after 3 years), maize yields seemed to remain fairly constant over time, with no apparent tendency for yields to decline despite continuous cultivation of the same field for at least 15 years (the oldest documented use of the system in the region).

Based on a regression approach, yields in the system were shown to depend significantly on years in *Mucuna* rotation and on density of maize stand. On average, every additional year in *Mucuna* yielded an extra 50 to 170 kg ha^{-1} of maize, whereas every additional 5000 plants harvested yielded between 250 and 500 kg ha^{-1} of maize. In addition, maize yields had a tendency to be more stable and maize cultivation posed less risk in the older *Mucuna* fields than what was observed for first- or second-season maize production in the bush-fallow system. Most of the added stability can undoubtedly be attributed to the positive effect the mulch layer has in improving water balance, especially during drier-than-usual years. This stability is noteworthy for what is, after all, a rain-fed environment.

System Response to the Application of N Fertilizer

The two main N sources for a growing maize crop in the *Mucuna* system are the biomass fractions of decomposing *Mucuna* (i.e. the litter) and the soil organic nitrogen, both of which release N in a gradual fashion upon mineralization (Triomphe 1996). Nitrogen fertilizer can constitute a third source for those farmers willing and able to invest in such an external input (see above).

While no clear trend was apparent in the response of a maize crop to the variable levels of N present in the aboveground biomass (except perhaps when biomass levels drop below 8 t ha^{-1}), there was a response to moderate rates (50 kg ha^{-1}) of N application in the form of urea. However, this response was found to be highly variable in research conducted during several cycles and sites. For example, only three of the 10 fields analysed in 1993-94 showed a statistically significant response to N. Reasons for this situation were not entirely apparent. A first factor was suspected to be related to environmental conditions. A highly significant response to N was observed during the unusually dry cycle of 1993-94, which can presumably be related to the slower-than-usual decomposition of the surface mulch. An important related factor was the amount of time a field had been in the *Mucuna* system. Older *Mucuna* fields, presenting the highest maize yields, highest maize nitrogen status and highest soil organic N levels, did not respond much to N fertilizer, whereas younger *Mucuna* fields responded markedly.

In contrast, there was no response to P applied at planting in the experiments conducted, which indicates satisfactory mobilization and availability of P.

Long-term Impact on Agriculture and Natural Resources

Continuous use of the *Mucuna* system allows a number of cumulative effects to take place, which gradually transform the soil profile. Here again, the net impact is in fact a balance between processes that tend to deplete stocks of nutrients and decrease soil fertility (particularly the repeated exportations via crop harvest) and those that tend to replenish stocks and increase soil fertility (such as N_2 fixation).

Based on a chronosequence scheme (Triomphe 1996), a number of tendencies were detected by comparing old *Mucuna* fields (10-15 years of continuous use of the *Mucuna*-maize rotation) with young *Mucuna* fields and with fields without *Mucuna*. In the old *Mucuna* fields, soil organic matter was shown to accumulate in the first few centimetres of the soil profile, at or very near the soil surface, in response to the humification of the litter. The organic matter increases reached 40-50% after a decade of *Mucuna* use (Figure 4).

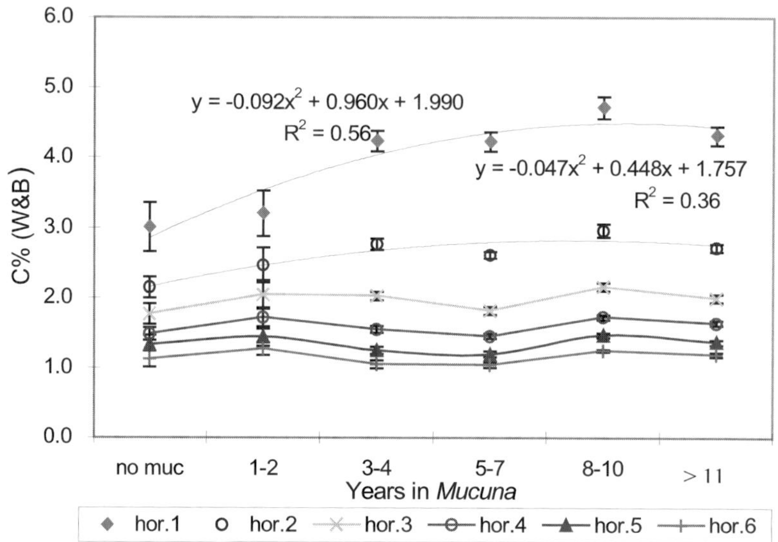

Figure 4. Changes in the distribution of organic carbon by 2.5 cm increments in the first 15 cm of soil profile over time in San Francisco de Saco, northern Honduras (From Triomphe [1996]).

Water infiltration doubled over 15 years, with a concomitant decrease in run-off rates. Total porosity also increased markedly. Despite a theoretical possibility that nitrogen imbalances might lead to soil acidification, measurements indicated that soil pH remained stable, while exchangeable bases, such as Ca and Mg, tended to increase throughout the soil profile (Tables 4 and 5). At the same time, available forms of most other nutrients remained at stable levels, despite significant yearly exports through maize harvests.

The above observations tend to confirm the existence of a relatively efficient nutrient cycling in the *Mucuna* system. In the studies conducted, soil biological life appeared to prosper as indicated by the proliferation of earthworms, insects and fungi at the litter/soil interface, while the quasi-absence of serious pests or soil-borne pathogens also points towards a healthy functioning of the soil profile. Last, but not least, erosion was visibly marginal in most *Mucuna* fields because of the permanent cover provided by the *Mucuna* biomass (live or dead). *Mucuna* use may, however, contribute to an increase in localized landslides under steep slopes in case of heavy rains, as a consequence of its ability to store water and to loosen up the topmost portion of the soil profile.

Table 4. Soil pH at different depths in soils from fields with different duration of *Mucuna* rotation in northern Honduras (*Source*: Buckles and Triomphe 1999).

Years in *Mucuna*	pH in water, 0–10 cm		pH in water, 30–60 cm	
	San Francisco	Las Mangas	San Francisco[a]	Las Mangas
0	6.05	6.51	5.91 ab	6.20
1–2	5.94	6.68	5.70 b	6.50
3–4	5.82	6.50	5.76 ab	6.36
5–7	6.01	6.33	6.02 a	6.32
8–10	5.96	6.44	5.95 ab	6.27
>10	6.06	6.62	6.09 a	6.38

[a] Means in a row followed by the same letter are not different according to Tukey's test at the 10% family probability level.

Table 5. Exchangeable Ca and available P at 0-10 cm depth in soils from fields with different duration of *Mucuna* rotation in northern Honduras (*Source*: Buckles and Triomphe 1999).

Years in *Mucuna*	Exchangeable Ca (cmol kg^{-1}), 0–10 cm		Extractable P (mg kg^{-1}), 0–10 cm	
	San Francisco[a]	Las Mangas	San Francisco[a]	Las Mangas
0	8.7 ab	18.6	0.4 b	7.6
1–2	6.9 b	20.6	0.8 ab	8.8
3–4	8.8 ab	19.4	1.8 ab	9.0
5–7	10.7 a	18.9	1.7 ab	4.7
8–10	11.1 a	20.9	2.0 ab	7.8
>10	12.0 a	23.9	2.4 a	22.8

[a] Means in a row followed by the same letter are not different according to Tukey's test at the 10% family probability level.

In short, in the observational study conducted, all long-term agronomic indicators associated with the sustainability of the *Mucuna* system seemed positive or at least satisfactory. Perhaps the clearest sign of technical sustainability from a farmer's perspective is that maize yields were as high as or higher than in old, continuously cropped, *Mucuna* fields compared to young ones and on average about double those obtained in fields not planted to *Mucuna*.

These favourable consequences of the agronomic sustainability of the *Mucuna* system at field level have had several consequences regionally. They include a relative stabilization of frontier agriculture because farmers using the *Mucuna* system did not need to open new land every year, as well as an increase in the sale and rental value of land under the *Mucuna* system. Also, there has been a clear shift away from summer maize cultivation towards winter maize cultivation (Sain et al. 1994;

Buckles et al. 1998), even though summer maize was not abandoned altogether because of concerns of food security and post-harvest losses.

Profitability of the *Mucuna* System

Sain et al. (1994) and Buckles et al. (1998) have compared the *Mucuna* system to the traditional fallow-maize system over a 6-year period (Tables 6 and 7). They estimated that the profitability per unit of land and labour was greatly superior in the *Mucuna* system. However, this advantage became apparent only after the first 2 years (considered the investment period) since the introduction of *Mucuna*, implying that farmers investing in the system made use of a planning horizon of at least 2 years. The authors concluded that, as the *Mucuna* system had diffused very rapidly in the Atlantic littoral, it was probably the superior return to labour that had triggered adoption, something consistent with farmers' own evaluation of the system.

Table 6. Annual flow of average net returns per unit of land in the *abonera* and bush-fallow systems (*Source*: Buckles et al. 1998).

Year[a]	Average net return (US$ ha^{-1})		
	Abonera	Bush fallow	Incremental flow
1	97.85	119.92	-22.08
2	89.30	135.54	-46.24
3	192.79	0.00	192.79
4	192.79	0.00	192.79
5	192.79	0.00	192.79
6	192.79	137.87	54.92
NPV (10%)	734.60	328.75	405.84
NPV (30%)	487.80	261.32	226.48
NPV (100%)	232.87	192.00	40.87
NPV (150%)	183.66	175.55	8.11

[a] NPV, net present value; values in parentheses are discount rates used.

Table 7. Annual requirements of family labour and summary results of the simulation of the net present value of net returns per unit of family labour (*Source*: Buckles et al. 1998).

Year	Person-day ha^{-1}		Average net return (US$ person-day ha^{-1})		
	Abonera	Bush fallow	*Abonera*	Bush fallow	Incremental flow
1	54.5	49.5	3.75	4.38	-0.63
2	19.0	45.5	6.65	4.93	1.72
3	19.0	0.0	12.10	0.00	12.10
4	19.0	0.0	12.10	0.00	12.10
5	19.0	0.0	12.10	0.00	12.10
6	19.0	27.0	12.10	7.06	5.04
Total	149.5	122.0	--	--	--

Humphries (1994) compared the costs and benefits of the *Mucuna* system, the slash-and-mulch winter maize system without *Mucuna* and the slash-and-burn summer maize system. The study was conducted over one cycle and in one community. In her analysis, the *Mucuna* system provided net profits that were 52% higher than the winter slash-and-mulch system, whereas the summer maize cultivation was returning a small loss. The profitability of bean production (US$300-400 ha^{-1}) was estimated to be three (summer beans) to four times (winter beans) greater than that of the maize-*Mucuna* system (US$100 ha^{-1}). However, risks of crop failure and environmental degradation (particularly during the summer) were very high, especially for summer beans. When a similar analysis was run using data that Triomphe (1996) obtained in higher yielding sites, the profitability of *Mucuna* production was shown to be equivalent to that of bean production. Note, however, that these benefits can be derived only from healthy fields not affected by itchgrass. If the weed is well established, returns may drop sharply, pushing many farmers to abandon their maize-*Mucuna* fields altogether (Neill and Lee 2001).

In a further analysis, Humphries (1994) estimated that a farmer with only three milking cows could realize as high annual profits as a typical hillside producer of maize and beans, with less effort and risk but with more land because of the low productivity of pastures. The selling of Tabasco chilli peppers to a nearby factory was by far the best income-generating enterprise (US$2000 ha^{-1}, or seven times the profits of summer beans and 20 times those of the *Mucuna* system) but capital costs and risk of crop failure were extremely high. Buckles et al. (1998) estimated that a farmer managing a herd of 10 cows can generate an income 10 times higher than a day labourer working 200 days during the year, with far less risk and physical effort than are required for the *Mucuna* system. For his part, Flores (1993), comparing the *Mucuna* system to a mechanized fertilizer-based system on flat cooperative land, concluded that although the latter provided farmers with 18% higher net profit per hectare, the return per unit of capital invested was 30% higher for the *Mucuna* system. He also observed that the way expenses were incurred in the two systems was radically different. In the *Mucuna* system, 52% of the cost went back to local farmers in the form of wage labour, whereas in the mechanized system, 71% of the expenses ended up paying for inputs and services bought from outside.

In summary, a number of comments can be made:

- Undoubtedly, the *Mucuna* system—provided that farmers know how to maintain it healthy (no infestation by itchgrass)—is by far the most profitable way of producing maize in the hillsides, from the viewpoint of return to labour, land and cash expenditures. Furthermore, this source of income is both relatively stable and sustainable over time.
- Profitability of bean production, either summer or winter, can be greatly superior to that of the maize produced in the *Mucuna* system,

particularly if maize yields remain moderate (2 t ha^{-1} or less); however, production and environmental risks in bean production are greater.

- If farmers can access enough land and/or capital to buy and maintain even a few cows and have an easy way to market milk or cheese, the income from raising livestock is equal or frequently superior to that from annual cropping, at a lower cost in labour and at small risk.
- If capital is freely available and farmers are willing and capable of taking big risks, high-value crops such as Tabasco chilli peppers can produce incomes an order of magnitude higher than those derived from maize or beans. It is unclear, however, what number of farmers it would take to saturate this type of niche market.

While comparisons among existing cropping systems at the local level (which has been the main focus here) provide critical insights, one needs also to consider the opportunity costs for investing in off-farm employment. In that respect, it is hard for any agricultural production system to provide incomes comparable to those of a migrant worker in the USA, explaining the continued out-migration of many small-scale farmers in the region.

USE AND ADOPTION BY FARMERS

Mucuna Use in Mesoamerica

While *Mucuna* has become such an integral part of the landscape and the culture of the northern Honduras hillsides, it has not always been so: in fact, its presence in the region is relatively recent (Buckles 1995; Buckles et al. 1998).

The United Fruit Company probably introduced the plant into Mesoamerica in the 1920s as a forage crop. Elderly banana (*Musa* sp. L.) plantation workers in Morales and Puerto Barrios, Guatemala, report that *Mucuna* was grown in maize by plantation workers on company land and grazed by mules used to transport bananas from the plantations to the railway depots (Buckles 1995).

Use of *Mucuna* as a forage crop by the banana companies faded as tractors replaced mules during the 1930s but started to spread outside the banana plantations. The Ketchi, originally from the densely populated highland area of Verapaz, were employed on banana plantations in Guatemala and may have become familiar with *Mucuna* on these estates. Carter (1969, cited by Buckles 1995) reported that Ketchi, migrating to the lowland valley of Polochic in the Department of Izabel, Guatemala, had planted *Mucuna* in rotation with second-season maize since their arrival in the 1950s. Commercial farmers, also settling in the valley during the 1950s, used *Mucuna* as a dual-purpose soil improver for maize and forage crops for cattle. Today, while its use has declined owing to diversion of land to pasture for cattle (Buckles 1995), *Mucuna* is still used by Ketchi in the valley of Polochic, the northern coastal mountains near Livingstone, the Petén and border areas in Belize. Indigenous farmers also have used the crop since at least the 1950s in the Mexican states of Chiapas, Oaxaca, Tabasco and Veracruz, by Mames of south-western Chiapas and by Nahuas in southern Veracruz (Buckles et al. 1998).

It is not known how *Mucuna* spread among these populations. Migration patterns and trade links among indigenous people in the region may have played a role. Today, *Mucuna* seed produced in the Guatemalan lowlands is marketed as a coffee substitute among indigenous groups in the highlands linked culturally to the Mames of Chiapas. Throughout southern Mexico, *Mucuna* is known as 'Nescafé', a food use that may also have stimulated diffusion of the seed, if not the cover crop management practices as well.

Introduction of *Mucuna* to Northern Honduras

Possibly, two Guatemalan brothers who settled in Planes de Hicaque near Tela introduced *Mucuna* into northern Honduras during the early 1970s. However, this early diffusion resulted in very low levels of adoption (a few percent of all farmers) during most of the 1970s.

One Honduran brother-in-law is credited with introducing the seed into San Francisco de Saco, also one of the earliest sites of *Mucuna* use in northern Honduras, where it grew wild, 'unnoticed' for a number of years (CIDICCO 1997). A few farmers in the community observed the crop's ability to control weeds and improve maize yields in fields it dominated, thereby rediscovering the rotation practice of the Ketchi and others. Don José Maria Ayala in San Francisco de Saco explained that this happened because smallholder farmers were pushed away from the fertile coastal land into the less favourable hillsides and hence were forced to look for ways of intensifying their maize production. This explanation is very much in line with the concept of induced innovation (Hayami and Ruttan 1985). Farmers in other regions have made similar accounts of local rediscovery of the use of *Mucuna* as a soil-improving cover crop (Buckles 1995).

Toward Massive Adoption: 1970 to 1990

By the early 1980s, annual increases in adoption rates rose sharply and remained high (around 5% y^{-1}) until the end of the decade, when they clearly levelled off. By the early 1990s, adoption had reached about 65% of the hillside farmers (Figure 5). Certain interesting features characterize this adoption:

- Despite the increased popularity of the *Mucuna*-second-season maize system, a notable proportion of farmers needs to keep on growing first-season maize in areas of their farms not planted to *Mucuna*, as a means of ensuring food security for the household.
- As mentioned earlier, the *abonera* system was developed and diffused without the intervention of formal extension services or formal incentive programmes. For the most part, farmers learned of the technology from family members or other farmers living in the same community or nearby. The borrowing of *Mucuna* seed (if not direct harvesting) appeared to have been the main vehicle for diffusion, with new adopters copying the management by observing the practices used by prior adopters.
- Finally, even as the adoption process was advancing at full speed, an inverse movement, abandoning the *abonera* system, was already becoming apparent. Almost 20% of the farmers surveyed by Buckles et al. (1992) had abandoned *Mucuna* cultivation, mainly because of change in land use and tenure status of the field.

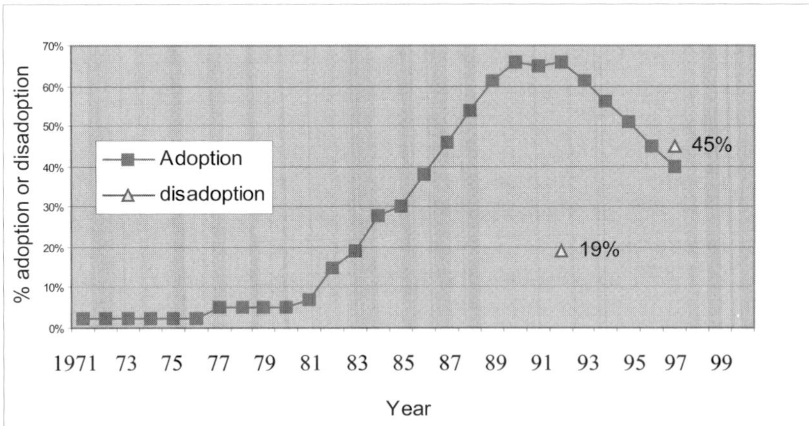

Figure 5. Dynamics of adoption and disadoption of the *Mucuna* system by hillside farmers of northern Honduras, 1971 to 1997 (*Source*: Adapted from Buckles et al. 1998 with data from Neill and Lee 2001).

In the same survey (Buckles et al. 1992), farmers ranked their key **reasons** in adopting *Mucuna* (Table 8). Foremost among these were the productivity boost caused by *Mucuna*, the ease of land preparation and weed control after *Mucuna* fallow (i.e. savings in labour) and the improved moisture status of the second-season maize, an effect linked to a lower risk of winter crop failure. Interestingly, erosion control arrived at the bottom of the list of advantages identified by farmers. Finally, the higher price farmers obtain from selling winter maize was not mentioned, even though it is an implicit advantage for any system with a second-season maize crop.

Table 8. Farmers' evaluation of the advantages of the *Mucuna*-maize system in northern Honduras.

Reasons for using *Mucuna* system	Percentage of farmers	
	First selection	Second selection
1. Fertilizer effect	39	20
2. Ease of land preparation	23	28
3. Moisture conservation	22	25
4. Weed control	8	24
5. Erosion control	8	3
Improves land productivity (1, 3, 5)	69	48
Improves labour productivity (2, 4)	31	52

Triomphe (1996) argued that adoption was facilitated by the fact that the *Mucuna* system was a **multi-purpose innovation** that **simultaneously** solved key constraints for small-scale hillside agriculture. He speculated further that the control farmers exerted over the technology and the extremely nice fit of the system with respect to natural ecological controls and processes might have contributed greatly to its success.

Although most farmers were able to adopt *Mucuna*, not everyone did and adoption **intensity** was not similar for all types of farmers described earlier (Table 9). As is common, subsistence workers accessing land primarily through rental agreements had the lowest adoption rate. In contrast, ranchers and medium-scale farmers (two groups relatively free of land constraints) adopted at a very high rate. With diversified and small-scale farmers, adoption rates were intermediate. Generalizing, it seems clear that security of access to land, whether through ownership or squatters' rights, was a fundamental condition for investing in the *Mucuna* system. However, the lack of land ownership is not an absolute limitation, as *abonera* land-rental markets have developed throughout the region.

Table 9. Adoption of the *Mucuna* system by farm-size class and household group in northern Honduras (*Source*: Adapted from Buckles et al. 1998).

Farm size (ha)	Percentage of households		Household group	Percentage of households	
	With *Mucuna*	Without *Mucuna*		With *Mucuna*	Without *Mucuna*
Landless[a]	33.3	66.7	Subsistence workers	36.7	63.3
1–2 ha	56	44	Small-scale farmers	64.3	35.7
2–5 ha	76	24	Medium-scale farmers	76.7	23.3
5–10 ha	71	29	Diversified ranchers	68.4	31.6
> 10 ha	86	14	Ranchers	84.2	15.8
All farm sizes	64.3	35.7	All household groups	64.3	35.7

[a] Use by means of land rental.

Sharp Decrease in the Use of the *Abonera* System: The 1990s

While adoption had been extremely rapid in the 1980s, it revealed itself to be very fragile. In surveys conducted barely 5 years after the surveys of Buckles et al. (1992), Neill and Lee (2001) reported a discontinuation of the system by about 45% of the farmers, that is, an annual discontinuation of about 4.5% between 1992 and 1997, an intensity similar to that of the adoption process in the 1980s (Figure 5).

Neill and Lee (2001) reported diverse reasons for the discontinuation (Table 10). Of the agronomic factors, one of the most influential stemmed from the negative impact of itchgrass on labour costs and productivity of maize cultivation (see above). Many of those who had discontinued also had a limited understanding of certain key agronomic aspects of the *Mucuna* system, particularly of the need to maintain the *abonera* by annual reseeding to avoid leaving room for itchgrass. From the economic standpoint, *Mucuna* use seemed closely associated with the dependency of farmers on maize cultivation. The more maize that farmers grew, the less off-farm income they received, the more likely they were to continue using *Mucuna*. Unsurprisingly, infrastructure development had a lot to do with this situation, as farmers with better access to road and markets were less likely to cultivate maize and *Mucuna*. Contrary to common observations in other regions, neither land tenure security, shifting land markets or the rise of extensive cattle raising appeared to exert any direct influence on the decision to discontinue *Mucuna* use.

Table 10. Profiles of adopters, abandoners and non-adopters of the *Mucuna* system in northern Honduras in 1997 (*Source*: Adapted from Neill and Lee 2001).

Characteristic	Adopters	Abandoners	Non-adopters
Farm size (ha)	13.2	12.6	3.8
Gross cropped area (ha)	5.2	4.3	2.9
Gross maize area (ha)	4.0	3.2	2.4
Dry-season maize profit (L ha^{-1})[a]	2065	1014	1258
Share of farm income of total income (%)	82	71	62
Good access to roads and markets (%)	41	59	72
Reseeds *abonera* (%)	49	21	---

[a] 15 L=$US1 (average rate for 1998).

CRITICAL FACTORS AND PROSPECTS FOR THE FUTURE

Based on evidence available up to 1998, the following factors appear critical for future use and performance of the *Mucuna* system. They are discussed in two related sections: those that impact the agronomic and economic performance of the system and those that impact its adoption.

Improving the Performance of the *Mucuna* System

While there are few concrete difficulties for successful management of the *Mucuna*-maize rotation, its short-term productivity and long-term sustainability are associated with a judicious use of, and control over, natural reseeding. The correct dates for slashing of the *Mucuna* crop and manual reseeding of the gaps that may develop in the *Mucuna* stand are both critical, especially if itchgrass is present.

Ensuring better stands of the maize crop and making more systematic use of improved germplasm (Fournier 2000) also seem important. Average current maize yields (3 t ha^{-1}) could probably be increased two-fold relatively easily by increasing densities from the present 30 000 plants ha^{-1} at harvest to 50 000 or 60 000 plants ha^{-1}. While these high densities should be broadly compatible with the high fertility status and adequate availability of water and nutrients of *abonera* fields, they would require substituting current landraces for improved, shorter maize varieties as the former would probably be affected by severe lodging beyond 40 000 plants ha^{-1}. In addition, applications of small doses of N fertilizer as side dressing may have to become more systematic, to complement the somewhat uncertain N-supplying capacity of the *Mucuna* mulch (Triomphe 1996; Fournier 2000). Undoubtedly, this would imply a new dependency on commercial maize seed and fertilizer suppliers. It could also result in the disappearance of valuable landraces, unless participatory plant breeding schemes allowing the conservation and even enhancement of local genetic diversity were established.

As long as the profitability of the system relies solely on the maize harvest, it may never bring to the region's small-scale farmers anything more than a solid contribution to household food security. Were *Mucuna* to find uses as forage, feed or food, however, the profitability of the system and its ability to generate income could be greatly enhanced. Because of increasing demand for fodder, the use of *Mucuna* biomass as forage should be explored and transformed from a negative factor, interfering with the maintenance of soil cover and with *Mucuna* re-establishment, to a way of making the system more attractive. Perhaps an even better alternative would be to promote the utilization of *Mucuna* seed as animal feed or even human food. This, however, may depend on gender issues within the household (women are usually the ones responsible for feeding livestock). It may also depend on the potential for *Mucuna* detoxification (Flores et al. 2002).

Whatever can be done on these aspects, however, diversification of crop production and income sources at the farm level should not be limited to what the *Mucuna* system has to offer. A household deriving its only or main source of income from the sale of surplus maize or *Mucuna* forage or seed cannot strive for much more than ensuring that its most basic needs are fulfilled. It could be argued that perhaps one of the 'best' roles for the *Mucuna* system may be to achieve food security for hillside farmers and community members in an environmentally-friendly, low-risk, labour-saving way on a relatively small acreage. On the remainder of the farm, other sustainable systems could be established to contribute to household income generation more efficiently than the current or modified *Mucuna* system. In addition, off-farm activities could provide yet another avenue for household income. In other words, diversification and astute exploitation of market niches and/or comparative advantages of hillside environments may be better than a blind reliance in the (limited) possibilities of the *Mucuna* system.

The ability to tackle the factors mentioned above is strongly dependent on the existence, or the lack, of a solid research and development capacity that would strengthen farmers' own capacities to adapt to and innovate in a rapidly changing environment. Improving the *Mucuna* system, optimizing forage or seed production and diversifying at the farm level are all topics on which farmers could benefit tremendously from outside help.

Factors Affecting Future Adoption

At the more macro level, adoption seems to depend significantly on access to land, whether through direct ownership or via land rental. This access must be possible both during summer and winter maize production since most farmers do not appear to trust one single harvest per year as a reliable way to meet their food security needs. The conversion of land to pasture or other crops (see below) will possibly affect the use of the *abonera* system by those small-scale farmers who used to rely on rental markets for producing their first-season maize while maintaining their own land in the *Mucuna*-winter maize system. These farmers may have to convert some of their *aboneras* into the maize-fallow system.

Another factor shaping adoption is the future price of winter maize. At present, winter maize has a higher price in the region but future trends depend on the fate of grain trade policies in the whole of Central America. Undoubtedly, a powerful feature for adoption has been and certainly will be the attractiveness and feasibility of other alternatives that small-scale farmers may have for using their land and labour. Whereas, in the past, the *Mucuna* system was attractive compared to traditional shifting cultivation systems, it is now competing in a number of villages against dual-purpose livestock raising, African Palm (*Elaeis guineensis* Jacq.) production, and other high-value alternatives.

Rapid growth in regional demand for milk and beef products has driven the conversion of cropland to pastures, especially in the lowland communities. Already, in the early 1990s, some of the poorest farmers had started to sell their land to better-off farmers or ranchers and had moved further uphill (Humphries 1994). The establishment of permanent pastures on the best hillside land is likely to reduce the availability of fallow land, putting so much pressure on the remaining bush-fallow cropping systems that land degradation is likely to take place. Furthermore, higher rental costs that result from the reduced supply of fallow land may undermine the already precarious access to land by the poor and threaten the viability of their livelihood strategies.

While the former discussion is partly speculative, the final outcome will primarily depend on the pace of economic development in the region, which itself will shape the pace at which cities and agro-industries, among others, will grow. The prospects for off-farm migration (whether to work in *maquiladores* on the North Coast or to migrate to the USA) will also most certainly continue to play a significant influence on the decisions of small-scale farmers regarding the intensity and nature of their farming activities. Additional impacting forces will include future policies on land tenure and inputs, as well as labour prices. Finally, a critical factor is whether or not incentives will be created for favouring good landscape husbandry and soil conservation by farmers in the hillsides. One can only hope that in the aftermath of the devastating Hurricane Mitch in 1998, the benefits to society at large associated with the use of environmentally friendly cropping systems, as documented by World Neighbours (1999), will become evident to policy-makers and taxpayers.

CONCLUSIONS

Many useful lessons can undoubtedly be derived from this overview of the maize-*Mucuna* system in northern Honduras. First, it has been shown that both spontaneous adoption and disadoption of the system occurred at a massive scale within less than two decades. The great agronomic advantages of the system are undisputably soil conservation, improvement of soil fertility, efficient nutrient cycling and water conservation. But, similarly undisputable, is its inherent economic fragility under rapidly changing external circumstances. Even if improvements are made to the system (see above), it is impossible to continue believing, as perhaps many did in the early 1990s, that *Mucuna* systems constitute a miracle solution to the dilemma of sustainable hillside management in the tropics, even in the wet lowland tropics. Other options have to be actively developed and, while the *Mucuna* system may retain an important role for local food security in the future, it will probably never be nearly as prominent as it was a decade ago.

Furthermore, there is no doubt that much of the characteristics and trajectory of the system are relatively unique to the context of northern Honduras. The agro-ecological and socio-economic niches of the system appear to be small and very difficult to find elsewhere. At best, similar conditions can be found in limited areas of Central America, such as in Siuna province of Nicaragua and the Polochic Valley of Guatemala (Barreto 1999).

Last, but not least, it is remarkable that, for the most part, formal agricultural institutions were never capable and perhaps simply never willing to accompany farmers in their experiences with the *Mucuna* system. They were not present to help them adopt the system and they did not have any ready solutions to contribute to help them overcome its limitations as they became apparent. This provides us much to think about on the limits to spontaneous adoption and the role outsiders may have in strengthening local knowledge and capacity for innovation.

REFERENCES

Barreto H (1999) The *Mucuna*-maize system. A report submitted to Centro Internacional de Información sobre Cultivos de Cobertura (CIDICCO), Tegucigalpa, Honduras. (Mimeo)

Buckles D (1995) Velvet bean: A 'new' plant with a history. Econ. Bot. 49: 13-25

Buckles D, Triomphe, B (1999) Adoption of *Mucuna* in the farming systems of northern Honduras. Agrofor. Syst. 47: 67-91

Buckles D, Ponce I, Sain G, Medina G (1992) Tierra cobarde se vuelve valiente. Uso y difusión del fríjol de abono (*Mucuna deeringianum*) en las laderas del litoral atlántico de Honduras. Centro Internacional de Mejoramiento de Maíz y Trigo (CIMMYT), Mexico. 36 p.

Buckles D, Triomphe B, Sain G (1998) Cover crops in hillside agriculture. Farmers' innovation with *Mucuna*. International Development Research Centre (IDRC) and Centro Internacional de Mejoramiento de Maíz y Trigo (CIMMYT), Ottawa, Canada. (Also available in Spanish). 218 p.

CIDICCO – Centro Internacional de Información sobre Cultivos de Cobertura (1997) Lo que el fríjol nos ha enseñado. Video produced by CIDICCO, Tegucigalpa, Honduras.

Flores M (1993) ¿Tienen razón los agricultores de usar el fríjol abono? *In:* Buckles D (Ed.) Gorras y sombreros: Caminos hacia la colaboración entre técnicos y campesinos. Centro Internacional de Mejoramiento de Maíz y Trigo (CIMMYT), Mexico. p. 33-40.

Flores M, Eilittä M, Myhrman R, Carew L, Carsky R (Eds.) (2002) *Mucuna* as a food and feed: Current uses and the way forward. Proceedings of a workshop, 26-29 April 2000, Tegucigalpa, Honduras. International Cover Crops Clearinghouse (CIDICCO), Center for Cover Crops Information and Seed Exchange in Africa (CIEPCA) and Judson College, Tegucigalpa, Honduras. 413 p. www.cidicco.hn.

Fournier J (2000) Technological gap for maize (*Zea mays* L.) cultivation in the Polochic watershed of Guatemala. Ph.D Thesis, Swiss Federal Institute of Technology, Zurich, Switzerland, 118 p.

Hargreaves G H (1980) Monthly precipitation probabilities for moisture availability for Honduras. Utah State University, Logan, Utah. 63 p.

Hayami Y, Ruttan V (1985) Toward a theory of technical and institutional change. *In*: Hayami Y, Ruttan V (Eds.) Agricultural development. An international perspective. Johns Hopkins University Press, Baltimore, USA. p. 73-114.

Humphries S (1994) Land use in humid tropical hillsides of Central America: A case study of migrant farmers in the Atlantic littoral area of northern Honduras. Annual Report 1993-1994 Centro Internacional de Agricultura Tropical (CIAT) Hillsides Program, Cali, Colombia. p. 153-216.

Munguia A (1992) Distribución de la caminadora (*Rottboellia cohinchinensis*) en la región el litoral Atlántico de Honduras. Agronomic Engineer Thesis, Centro Universitario Regional del Litoral Atlántico (CURLA) - Universidad Nacional Autónoma de Honduras (UNAH), La Ceiba, Honduras.

Neill S, Lee D (2001) Explaining the adoption and disadoption of sustainable agriculture: The case of cover crops in northern Honduras. Econ. Dev. Cult. Change. 49: 793-820.

Saín G, Ponce I, Borbón E (1994) Profitability of the abonera system practiced by farmers on the Atlantic Coast of Honduras. *In:* Thurston H D, Smith M, Abawi G, Kearl S, (Eds.) Tapado. Slash/mulch: How farmers use it, and what researchers know about it. Centro Agronómico Tropical de Investigación y Enseñanza (CATIE) and Cornell International Institute for Food, Agriculture and Development (CIIFAD), Ithaca, New York, USA. p. 273-282.

Solomon T, Flores M (1994) Intercropping corn and fríjol chinapopo (*Phaseolus coccineus*). Centro Internacional de Información sobre Cultivos de Cobertura, Honduras (CIDICCO), Tegucigalpa, Honduras.

Szaraz,G, Irías D (1993) Development of the Honduran tropical moist forest: Experiences in integrated management areas. For. Chron. 69: 672-679.

Triomphe B (1996) Seasonal nitrogen dynamics and long-term changes in soil properties under the *Mucuna*/maize cropping system on the hillsides of northern Honduras. Ph.D Thesis, Cornell University, Ithaca, New York, USA. 217 p.

World Neighbours (1999) Changing course. Video. World Neighbours, Oklahoma, USA.

Chapter 4

Cultivating Maize with *Mucuna* in the Los Tuxtlas Region of South-eastern Veracruz, Mexico

M. Eilittä[1], F. L. Arteaga[2], M. Diaz[3], C. Guerrero[2], B. Herrera[2], G. Narvaez[3], L. Paré[2], G. C. A. Robles[4] and B. Triomphe[5]

[1] *Rockefeller Foundation-funded Post-doctoral Fellow*
[2] *Cooperativo Mok-Cinti, Mexico*
[3] *Universidad de Chapingo, Mexico*
[4] *Desarrollo Comunitario de Los Tuxtlas (DECOTUX), Mexico*
[5] *Centre de Coopération Internationale en Recherche Agronomique pour le Développement, France - Centro Internacional de Mejoramiento de Maíz y Trigo, Mexico (CIRAD-CIMMYT)*

SUMMARY

Los Tuxtlas, a region of volcanic hills in southern Mexico, has high rainfall (1200-2500 and up to 4000 mm per annum) but soil fertility has decreased 70-75% in the past 40 years. Maize (*Zea mays* L.) is the subsistence crop in the region and is typically cultivated in sole-cropped conditions, with or without a short fallow, and with residue burning. Yields of summer maize (planted in June and averaging 1.2 t ha^{-1}) and the relatively less important winter maize (planted in October-November and averaging 0.5 t ha^{-1}) are low, seemingly mostly caused by poor soil fertility. Two Mexican non-governmental organizations, Proyecto Sierra de Santa Marta and Desarrollo Comunitario de Los Tuxtlas (Community Development of Los Tuxtlas), introduced *Mucuna pruriens* (L.) DC. to the region's farmers as a means of improving soil fertility. This introduction was achieved by conducting campaigns through farmer extension workers (*promotores*) and adaptive on-farm research (1992-2000). Typically, *Mucuna* was intercropped between summer maize rows in August (40-70 days after summer maize planting). It was then either slashed in late October and winter maize planted in the dead mulch (summer-winter maize system) or it was left to continue growing and no winter maize planting took place (summer maize system). Before the following summer maize harvest, all crop residues, including those of *Mucuna*, were typically burned. Research efforts (and some of traditional

M. Eilittä et al. (eds.), Green Manure/Cover Crop Systems of Smallholder Farmers, 99-127.

use) in the region also focused on a rotational system where *Mucuna* was sole-cropped during summer and was followed by winter maize. On-farm assessments revealed that *Mucuna* could grow well in the region (particularly if grown during the high rainfall summer season as in the rotational system) and exert a strong, positive impact on winter maize yield (particularly in the rotational system). Allocating the main maize cropping season, summer, to *Mucuna* (as is done in the rotational system) is questionable. Positive impacts on summer maize yield are either not found or are inconsistent.

Cost-benefit studies reveal that maize production systems with and without *Mucuna* have an extremely low profitability because of low maize prices and yields, and the lack of multiple uses and consequently of markets for *Mucuna* seed. While some studies on farmer perceptions of the *Mucuna* technology have been conducted, no formal adoption studies have been made in the region. *Mucuna* use seemingly varies by community and has greatly fluctuated in response to the various extension efforts. The future of *Mucuna* in the region depends on a number of factors, not least of them the general economic policies of Mexico. Development and policy, as well as a research agenda, are outlined for improved adoption and positive impact of the *Mucuna* technology in the region.

INTRODUCTION

Overview

South-eastern Veracruz, Mexico, is a microcosm of the drastic and recent changes in the environment and farming systems that have taken place throughout tropical regions. The region was once a renowned centre of biological diversity and known throughout Mexico for its beautiful forests. However, most of these have been cut due to cattle-ranching and population growth and a great many of its animal species have been driven almost to extinction (Ramirez 1991; Barrera-Bassols et al. 1993; Ramirez et al. 1997). For the region's subsistence farmers, the recent changes have translated into increased land pressure and lower soil fertility, causing decreased yields of maize (*Zea mays* L.), their staple food, and severely curtailing the cultivation of beans (*Phaseolus vulgaris* L.), another important crop.

In looking for solutions to such problems, two local non-governmental organizations (NGOs; Proyecto Sierra de Santa Marta [PSSM] and Desarrollo Comunitario de Los Tuxtlas [DECOTUX]) conducted extension and research efforts on *Mucuna pruriens* (L.) DC. in the region in the 1990s. Most commonly, farmers in the region relay-cropped *Mucuna* in their summer maize. The experiences of the region's farmers and the two organizations indicate that, while *Mucuna* has the potential to increase productivity and sustainability of maize-based cropping systems in the region, further work is needed on *Mucuna*-based systems both in the region and in general to solve certain field-level bottlenecks.

CONTEXT OF *MUCUNA* WORK IN THE REGION

The Biophysical Environment

The Los Tuxtlas region consists of volcanic hills rising from the coastal plateau by the Gulf of Mexico. This mainly rural region encompasses an area of about 78 by 40 km in the southern part of Veracruz state at lat 18°10'N to 18°45'N, long 92°42'W to 95°27'W (Figure 1). The land is undulating to mountainous, with altitudes ranging from 200 to about 1700 m. The average annual temperature is between 22 and 26 °C, with highest extremes in May and lowest extremes in January (Dirzo et al. 1997). The total annual rainfall varies from 1200 to over 4000 mm. The rains typically start in June and are highest between July and September; rainfall is greatly reduced between January and May. Average annual rainfall in the regional town of Catemaco, located close to the efforts of *Mucuna* diffusion, is 1890 mm (Figure 2). The region's soils have not been studied in detail; they are variable and consist of Andisols, Alfisols, Ultisols, Vertisols and Entisols (Tasistro 1994). As already mentioned, the region is renowned for its geographic and species diversity, although in recent decades severe deforestation, attributable to cattle ranching and to endemic population growth and in-migration, has caused general deterioration of natural resources, including agricultural soils. In the past 40 years, soil fertility has been estimated to have decreased 70-75% (Tasistro 1994 citing Uresti et al. 1992) and deficiencies of N and P are common.

Degradation of natural resources has affected most of the region's inhabitants, in that often only patchy secondary growth remains and maize fields and/or pastures dominate the landscape. Few forest products other than firewood are still available to many communities.

The Human Environment

The region has been the site of major maize-based civilizations since Olmec times; maize has been cultivated locally for over 4000 years (Chevalier and Buckles 1995). Today, the region is inhabited both by Spanish-speaking *mestizos* (of mixed indigenous and European origin), who populate most of the area, and by indigenous people of Nahua and

Popoluca origin, who are concentrated primarily in the Sierra de Santa Marta area. Most of the region's inhabitants are devout Catholics; in addition, small Protestant groups can be found virtually in every village.

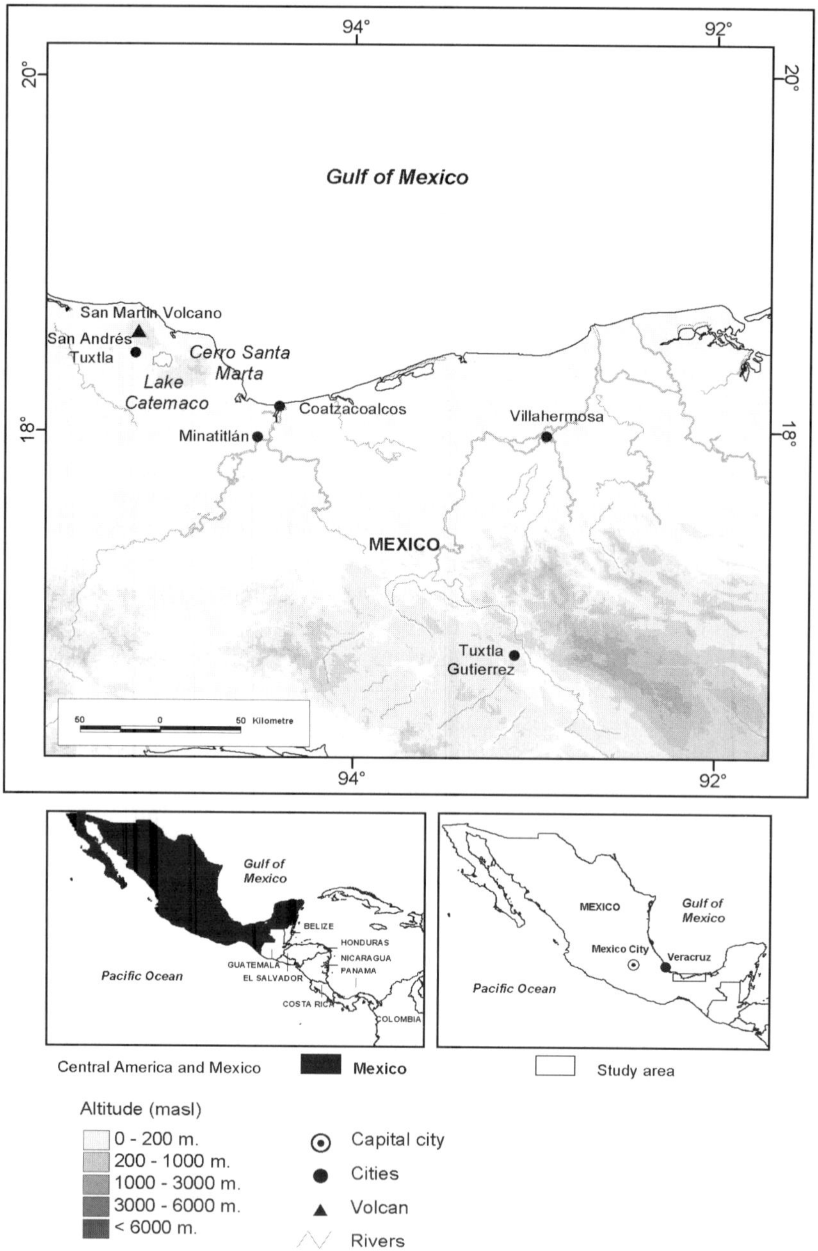

Figure 1. Location of the study area in southern Veracruz, Mexico.

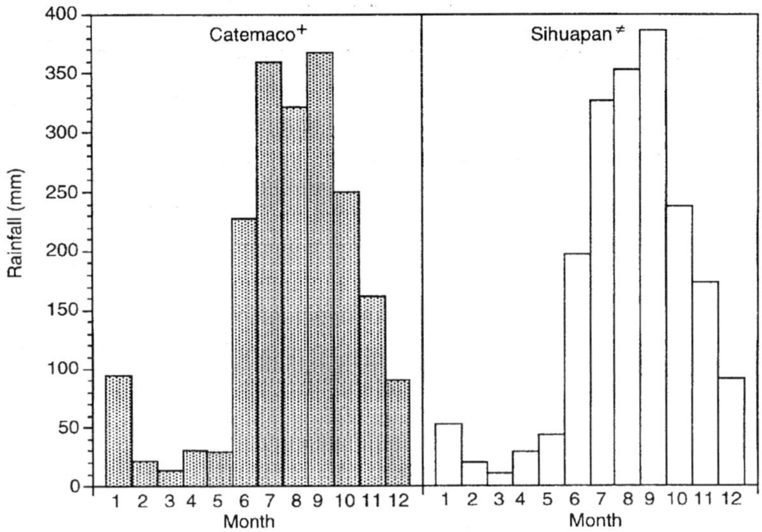

\+ Based on calculated average of monthly rainfall measurements between 1/1987 and 4/1997 obtained from the local weather station. No data was available for the following months: 1/1998, 9/1990, 1/1992, 4/1995 4/1996.

≠ Based on calculated average of monthly rainfall measurements between 1/1990 and 3/1997 obtained from the local weather station.

Figure 2. Average monthly rainfall in Catemaco and Sihuapan, Veracruz, Mexico, 1987 to 1997.

The region's rural people typically live in tightly clustered villages, which usually have electricity and potable water and are served by roads that are accessible throughout the year. However, poverty is widespread. Only a proportion of inhabitants have completed primary school (50% in the Sierra de Santa Marta area) and health problems afflicting the intestinal tract and skin are common (PSSM n.d.).

The traditional landowning patterns were based largely on the institution of communal land holding, the *ejido*. However, in recent years, land titles have been offered to some *ejido* members, leaving others without secure access to land. The share of households owning land varies greatly by community, ranging from 20% (in the San Andres area and in the large municipal centres of the Sierra de Santa Marta area) to a great majority.

Farming forms the base of subsistence for most of the rural people; fishing predominates on the sea coast and around Lake Catemaco. Maize is the subsistence crop of the region and is cultivated typically in sole-cropped conditions, with or without short fallows. Minor food crops include beans (mainly *Phaseolus vulgaris* L.), groundnuts (*Arachis hypogaea* L.), chayote (*Sechium edule* [Jacq.] Sw.), cassava (*Manihot esculenta* Crantz) and various fruits. Cash crops of varying importance include coffee (*Coffea* L.), while cattle ranching and/or tobacco

(*Nicotiana tabacum* L.; grown typically by large companies) dominate agricultural production in some parts of the region.

Maize cultivation dictates labour availability, and labour constraints are typically experienced in May-August (summer maize planting and weedings), late October-early November (winter maize planting) and February-March (harvest). In the coffee-growing communities, the coffee harvest, starting in November, constitutes an additional labour bottleneck period. Within the household, men do most of the fieldwork, while women are responsible for the housework and cooking, including the laborious preparation of tortillas (local staple made from maize flour into thin unleavened bread). However, women also work in the fields, especially at the busiest work times.

Challenges to Maize Farming

The traditional Mayan *milpa* cultivation system has all but disappeared (Stuart 1978). It commonly consisted of maize together with 10 to 20 other crops grown during the year's cycle. Loss of this system and the decreased area planted in beans serves to exaggerate the importance of maize, which for many is the only crop. In other respects, maize farming continues to employ traditional practices including residue burning, use of traditional maize varieties planted widely spaced, reliance on labour inputs and the utilization of simple tools, such as the machete and the dibble stick.

In the areas where *Mucuna* has been adopted, rainfall typically varies between 1500 mm and 2500 mm, enabling the cultivation of two maize cycles: *maíz temporal* (hereafter referred to as summer maize) and *maíz tapachole* (hereafter winter maize). **Summer maize** is the main crop because of the more favourable environmental conditions. It is typically planted in June, weeded in June through August, 'doubled' (i.e. bending of the stem just below the ear to facilitate drying and prevent wind and bird damage) in late September and harvested the following January-March. Typically, 1 to 3 ha of summer maize is planted. Summer maize yields in most of the region are low, averaging about 1.2 t ha^{-1} (Buckles and Erenstein 1996; Eilittä 1998). **Winter maize** is typically relay-intercropped in the rows between mature summer maize in late October or early November. Average areas cultivated are small, at 0.3 to 0.6 ha per family, and average yields are very low, at just 500 kg ha^{-1} (Buckles and Erenstein 1996) because of low and erratic rainfall, high winds and greater damage by birds as compared to summer cultivation. In some parts of the region, better winter maize yields are possible because of higher residual moisture.

Low maize yields have been linked to lower soil fertility caused by shortened or absent fallows, as well as to increased weed and—a particular concern of the region's farmers—pest and disease pressures

(Buckles and Erenstein 1996; Soule 1997). In the early 1990s, 60% of smallholders used fertilizers (Soule 1997) but by the mid-1990s many credit programs had been discontinued and fertilizer prices had increased, resulting in very low fertilizer use in the late 1990s. In contrast, the use of herbicides, typically Paraquat, has continued to be widespread.

Socio-Political Context

The work of the PSSM and DECOTUX responded not only to this biophysical and human context but also to the socio-political context of Mexico in the 1990s. By that time, the rapid economic and social changes resulting from Mexico's market liberalization, the creation of the North American Free Trade Association (NAFTA), and economic turmoil created increasingly difficult conditions for the region's smallholders. The price of maize, their main crop, had been very low (typically 1.0 to 1.5 pesos kg^{-1}; US$1 = 6-9 Mexican pesos during this period). The region's smallholder farmers received little institutional support in terms of extension, research or credit. Prices of all goods, including inputs, had increased greatly, especially since the December 1994 devaluation. Sources for off-farm income are generally limited to occasional farm employment but recently out-migration to work at the *frontera*, the USA border, has increased.

The NGOs saw green manure/cover crop (GMCC) systems as a means of intensifying the region's agriculture in an environmentally friendly way, in contrast to the mono-cropping systems with high external inputs that the Mexican government had often promoted. These mono-cropping systems were considered poorly suited to the area's environment and farming systems and helped cause food insecurity, poor competitiveness of smallholder agriculture and health hazards because of herbicide and pesticide use. Instead, the two NGOs consciously sought guidance from the traditional *milpa* system where increasing soil fertility through organic inputs was seen as a necessary stepping-stone. The low cost and relatively low labour requirements of GMCC systems, *Mucuna*'s popularity in the early 1990s and its traditional use within the region made it the choice GMCC.

The political ideologies of those active in the NGOs led to their reliance on farmer-to-farmer diffusion and farmer experimentation. Their main strategies originated from perceptions that Mexico had long discriminated against the small-scale farmer, the *campesino*, and that the national and state governments were corrupt and had policies that were paternalistic. The NGOs strove instead to allow the smallholder farmer to be the subject, not the object, of development and a partner together with the organizations.

The two NGOs were not alone in their efforts. By the late 1980s and early 1990s in southern Mexico, and more widely throughout

Mesoamerica, diverse organizations had grown increasingly interested in GMCCs and farmer-to-farmer diffusion. In south-eastern Mexico, many NGOs and academic institutions working in sustainable agriculture were collaborating in a network funded by the Rockefeller Foundation (Arteaga et al. 1997), while the International Maize and Wheat Improvement Centre (CIMMYT, its Spanish acronym) was active in GMCC research in both Mexico and Central America (Buckles and Barreto 1996; Buckles and Erenstein 1996). This interest was partly fuelled by the 'discovery' of traditional GMCC systems throughout Mexico (including in some communities of the Los Tuxtlas region) and Central America, and the exchanges of information between various tropical regions. Many researchers worked on GMCC systems and exchanged experiences between Central America and Mexico but Daniel Buckles, formerly of CIMMYT and a founding member of the PSSM, was particularly influential in *Mucuna* efforts in Veracruz.

RESEARCH AND EXTENSION EFFORTS ON *MUCUNA* IN THE REGION

The efforts on *Mucuna* in the region included both research, in the form of student projects, and extension (Buckles and Arteaga 1993; Buckles et al. 1994; Buckles and Perales 1995; Buckles and Erenstein 1996; Arteaga 1997; Robles and Alemida 1998). A brief overview of these efforts is given below.

A Chronology of the Work

The PSSM, an NGO established in 1990[1], was the prime mover behind the *Mucuna* diffusion efforts in the region. Based in Jalapa, the state capital, and with some technical staff living on site in the region, the organization worked in agriculture, forestry, horticulture and fisheries through seven to eight researchers and a number of administrative and technical staff. Across its programs, the PSSM's primary focus was research in the social sciences but it also had a strong developmental component, typically working with 10 farmer extension workers (*promotores*). The PSSM was mainly active in the Sierra de Santa Marta region and in the area west of Catemaco (Figure 1).

Another NGO, DECOTUX, loosely affiliated with a political party then in opposition, initiated *Mucuna* extension campaigns in 1995 in the San Andres Tuxtla area (Robles and Almeida 1998). The organization is far smaller than the PSSM and during its *Mucuna* extension work employed two to three technical staff, typically including one agronomist. In their work on GMCC systems, both organizations have actively cooperated with other institutions, such as with CIMMYT, with the

Rockefeller Foundation-funded RED (a network of organizations working in sustainable agriculture) and on two occasions with government agencies. Moreover, the two projects have cooperated closely with one another through farmer-to-farmer training and joint field visits.

Efforts of the PSSM

The PSSM's *Mucuna* work was conducted in several phases. Over the course of the 1990s, the PSSM's emphasis shifted from research to extension and back to research, as outlined below.

Preliminary On-farm Research Conducted by the PSSM and CIMMYT (1991-93)

Beginning 1 year prior to the extension efforts, two researchers involved with the PSSM and CIMMYT led participatory farmer trials comparing traditional maize cultivation without *Mucuna* to several maize-*Mucuna* systems (Buckles and Perales 1995). Most farmers chose to experiment with a relay-cropping system. The investigators concluded that short-term yield increases from *Mucuna* might be expected only in winter maize. Farmers familiar with *Mucuna* in the region evidently agreed that *Mucuna* is 'best for winter maize' (Buckles and Perales 1995).

Large-Scale Extension Conducted by the PSSM (1992-93)

Extension efforts started in April 1992 when the research results of the previous year were presented to the farmers, and seven farmers—later hired as village-level farmer-extension workers—were asked to hold meetings in their villages. Typically, village meetings consisted of presentations on *Mucuna* cultivation and free distribution of its seed. Between 1992 and 1993, 2250 people in over 45 villages were reached through the meetings; of these people, 1164 accepted *Mucuna* seed (Buckles and Perales 1995; Arteaga 1997). The farmer-extension workers periodically visited nearly one-fourth of those who planted *Mucuna* to collect information on *Mucuna* and maize growth. Over 5 tons of the harvested *Mucuna* seed was purchased from the farmers in the region (Buckles and Perales 1995). This work was accompanied by a CIMMYT-affiliated Ph.D. research project on farmer expectations of benefits from *Mucuna* use (Soule 1997).

Large-Scale Extension Conducted with State Government Collaboration (1994)

A common focus on soil conservation brought the PSSM and the state government of Veracruz together to promote *Gliricidia sepium* (Jacq.)

Walp. hedgerows and *Mucuna* intercropping in a large regional effort involving 56 farmer groups and 1457 farmers that received seed and incentives. The PSSM discontinued the collaboration after only 1 year, feeling that the state government bureaucracy prevented effective work, that training of farmer extension workers was inadequate and that the late paying of many incentives caused dissatisfaction. Small quantities of *Mucuna* seed were purchased from the farmers.

Return to PSSM Extension (1995-)

In 1995, the PSSM continued the extension work alone, reaching 461 farmers. That year, another CIMMYT-affiliated Ph.D. project, an agronomic assessment of the farmer-managed maize-*Mucuna* systems, was begun and continued until 1997 (Eilittä 1998).

On-farm Research and Farmer Experimentation (1996-)

In 1996, *Mucuna* work was refocused on on-farm research that addressed certain field-level bottlenecks. Research was concentrated on:

(1) Identification of varieties which would survive the dry season and thus synchronize better with the nutritional requirements of summer maize (PSSM);
(2) Feeding trials with pigs; and
(3) Management techniques for winter maize grown after sole-cropped summer *Mucuna*.

Between 1996 and 1998, the RED network conducted an evaluation of GMCCs in two to three communities of the region (Arteaga et al. 1997).

Collaboration with the National Government's Ministry of Environment, Natural Resources and Fisheries (1998-99)

Concerned over the quickly deteriorating soil quality, the Secretaria de Medioambiente, Recursos Naturales y Pesca (SEMANARP) bought GMCC seed from smallholder farmers in five states to distribute to other farmers across southern Mexico. In all, 55 PSSM-affiliated farmers sold 26 000 kg of *Mucuna* seed at an extremely attractive price, with the NGOs acting as intermediary. Despite delays in payments, the project was successfully concluded.

Post-1999 Development

In 1999, the agricultural development work separated from the PSSM and a farmer cooperative, Mok-Cinti, was initiated to continue it. Several of the technicians who had formerly been PSSM staff members began work with the cooperative. Work on *Mucuna*, however, has decreased in importance and by the year 2001 no extension efforts on *Mucuna* were ongoing in the villages where the PSSM was active.

Although the PSSM's *Mucuna* efforts have been accompanied by a relatively substantial body of research, much of it in association with CIMMYT, the research projects have been relatively short-term and have therefore lacked continuity.

DECOTUX Work with *Mucuna*

DECOTUX is a younger organization than the PSSM and its work with *Mucuna* has been of shorter duration. In 1995, after conducting initial surveys, DECOTUX began an extension program in which small amounts of seed were distributed in community meetings. In the first year, 20 communities were included but, feeling overextended, DECOTUX reduced the number of communities to 10 in 1996. At that time, it also began a program of farmer experimentation, focusing on uses of *Mucuna* for high-value crops such as chilli pepper (*Capsicum annuum* L.). The organization also participated in SEMANARP's seed-buying effort in 1998-99 and 64 farmers affiliated with the project sold 33 000 kg of *Mucuna* seed during the year.

Diffusion Strategies

The PSSM and DECOTUX operated through a strategy of farmer-to-farmer diffusion, wherein the community-based farmer-extension worker (*promotor*) plays the central role and coordinates efforts in the community, acts as a liaison to the project, disseminates information and distributes seed to farmers in his/her region. The *promotores* continually received training through visits to other communities and regions and conducted periodic meetings among themselves and with the NGO technicians.

Many visitors to the region have been impressed by the self-confidence, enthusiasm and relative autonomy of the *promotores* from the projects. Several of the more experienced of these *promotores* became trainers themselves. Within the PSSM only, they received a part-time salary for their efforts. The community did not choose the *promotores* in either project; rather, the projects selected certain farmers. It should be noted that this model has not been free of problems. On the contrary, in communities often divided both politically and by religion, the role and work of the *promotores* has been difficult. Training the farmer extension workers in both technologies and communicative skills was found essential.

STRUCTURE OF THE MAIZE-*MUCUNA* SYSTEMS IN THE REGION

Three different maize-*Mucuna* systems can be distinguished in the region: the traditional rotational system (summer *Mucuna* followed by winter maize cultivation) and the two relay-intercropping systems (one with summer and winter maize cultivation and another with summer maize cultivation only).

As mentioned earlier, *Mucuna* has been grown for several decades in a few communities in the municipalities of Mecayapan and Soteapan in the Sierra de Santa Marta region. Most of this cultivation had been in the form of a **traditional rotational system**, where summer cultivation of sole-cropped *Mucuna* was followed by its slashing in October-November, after which winter maize was planted (Sohn López-Forment 2000).

However, most farmers who began using *Mucuna* as a result of the extension efforts chose to employ one of **two relay-intercropping systems** in which *Mucuna* is planted in an established summer maize crop. In these systems, the recommended date of planting *Mucuna* was at 40 days after planting maize but most farmers planted it later, up to 70 days after maize planting. Almost always, planting took place in August, when maize is already close to or at flowering and often strongly shades the developing *Mucuna* until maize 'doubling' in late September. Thereafter, *Mucuna* growth is often rapid provided that rainfall is adequate. Farmers typically planted *Mucuna* only in a small part of the area in summer maize (generally not exceeding 0.3 to 0.5 ha) and often rotated the area in *Mucuna* within a field.

Two strategies, in effect different systems, were utilized to manage *Mucuna* after late October:

(1) **In the summer-winter maize system**, both summer and winter maize were grown. The August-intercropped *Mucuna* was slashed in late October-early November. Summer maize was either harvested at that time or, more commonly, left in the field. *Mucuna* was then slashed around the summer maize plants and winter maize planted in the rows between mature, drying summer maize. In the summer following, maize was planted in June, after field slashing and often burning the dead mulch (including *Mucuna*). In this system, *Mucuna* did not produce mature seed.

(2) **In the summer maize system**, only summer maize was produced. The August-intercropped *Mucuna* was allowed to mature (typically in December) and senesce (typically in January) in the field. Summer maize was typically harvested in January-February and *Mucuna* seed, if harvested, was collected in March-April. The following summer maize was planted in June, after field slashing and often burning all mulch, including that of *Mucuna*.

The choice of the system depended partly on the environment and partly on farmer preference. In some typically higher altitude areas of the region, no winter maize can be cultivated and therefore only the summer maize system was practiced. At the other extreme, in certain locales winter maize is as important as summer maize and therefore the summer-winter maize system was always used. However, in most areas where the extension efforts took place, winter maize is a minor crop and planted in a smaller area than summer maize. Therefore the farmers had the option of slashing August-planted *Mucuna* in late October to early November or allowing it to mature.

Farmer practices have varied greatly, particularly in planting dates (40-90 days after planting maize) and densities of *Mucuna* (Eilittä 1998). Farmers have also investigated ways to solve problems that have been encountered with *Mucuna* use. For example, farmers started slashing *Mucuna* prior to maturity or used herbicides to control aggressive *Mucuna* self-reseeding, which has been especially problematic in fields that have not been burnt and where *Mucuna* has been cultivated for several years. At times, farmers also allowed *Mucuna* to self-reseed at the beginning of the summer growing season and only supplemented the density by planting. Finally, some farmers established seed plots to replenish *Mucuna* seed supply. Interestingly, farmers in the area often point out that they plant *Mucuna* in the parts of the field with lowest fertility (Soule 1997; Eilittä 1998).

PERFORMANCE OF THE MAIZE-*MUCUNA* SYSTEMS

Mucuna can grow very well in the region and exert a strong, positive impact on maize yield. Yet certain biophysical factors and suboptimal farmer management evidently constrain both *Mucuna* biomass productivity and *Mucuna*'s impact on maize yield. Perhaps more importantly, however, the economic productivity of the local maize-*Mucuna* systems is poor because of the low maize prices and the lack of markets for *Mucuna* seed. While most of those exposed to *Mucuna* during the extension efforts chose to intercrop it with summer maize, the following assessments of the systems' performance also include the rotational system as a comparison.

No research has been conducted on *Mucuna*'s impact on soil quality, soil erosion, or weed incidence in the region, all of which are important topics. Presumably, these impacts are more beneficial in the traditional rotational system with its higher *Mucuna* biomass production and good soil cover during heaviest rains (August-October).

Mucuna Productivity

In many ways, the growing conditions under the traditional rotational system are ideal for rapid biomass production because *Mucuna* is sole-cropped during the rainy summer season. In researcher-managed, on-farm trials (where management mimicked farmer conditions), *Mucuna* grown sole-cropped during the summer season produced up to 11.0 t of dry matter (DM) per hectare (leaf-stem-mulch component) and average biomasses of 4.9 to 7.9 t DM ha^{-1} have been reported for various trials (Eilittä et al. 2003a; 2003b). This biomass production is similar to that in northern Honduras, where farmers have spontaneously adopted *Mucuna* (Triomphe and Sain, Chapter 3, this volume).

In the relay-intercropped, farmer-managed fields, *Mucuna* growth was reported to be poorer and highly variable. If slashed in late October or early November (summer-winter maize system), maximum *Mucuna* biomass at slashing time was relatively low because of the very short growing season (from August to late October) and the shading of maize until maize 'doubling'. Agronomic monitoring of farmer fields demonstrated that average leaf-stem biomass in the summer-winter maize system was less than 1 t ha^{-1} in 1995 and 1.43 t ha^{-1} in 1996 (Eilittä 1998). An additional 1.80 t ha^{-1} of mulch (including *Mucuna*, maize and weed residues, measured only in 1996) was reported. In this summer-winter maize system, no mature seed was produced because *Mucuna* is slashed early.

In the summer maize system, where *Mucuna* matures until its January senescence, *Mucuna* biomass was typically greater. In farmer-managed fields, *Mucuna* leaf-stem biomass averaged 1.32 t ha^{-1} in 1995 and 1.01 t ha^{-1} in 1996, while the dead mulch fraction averaged 2.70 t ha^{-1} in 1996; this included dead mulch of no-*Mucuna* origin (Eilittä 1998). Such higher production is caused by a longer growing season, because *Mucuna* growth continues into January. Clearly, however, *Mucuna*'s growth during the dry season was poorer because of advancing maturity, low rainfall and often high winds, and much of the dried mulch is lost during the following dry season.

On-farm trials mimicking farmer conditions compared the two relay-cropped systems to the rotational system and supported these findings (Table 1). Notably, quite good seed production of *Mucuna* was possible in the intercropped conditions, affirming the later findings of the SEMANARP project (Narváez Carvajal 1999). In addition, the quantity of N and P in the biomass fractions was quantified (Table 2), with results mirroring those of the biomass.

Table 1. *Mucuna* biomass (t DM ha^{-1}) in maize-*Mucuna* systems in Los Tuxtlas, Mexico, 1996-97. Data are means across four fields (*Source*: Eilittä et al. 2003a).

Maize-*Mucuna* system	Leaf	Stem	Dead mulch	Pod[a]	Weed	Leaf+stem +dead mulch
1. Summer-winter maize[b]	0.46	0.50	1.79	0.02	1.24	2.75
2. Summer maize[c]	0.31	0.53	4.23	1.40	1.41	5.06
3. Rotational[b]	0.57	1.64	5.13	0.39	1.01	7.34
Contrasts and *P*-values:						
System 1 vs. 2	0.034	0.770	0.003	<0.001	0.236	0.005
System 2 vs. 3	0.001	<0.001	0.189	<0.001	0.068	0.004

[a] Included immature pods.
[b] *Mucuna* biomass sampled at slashing of *Mucuna*, typically in early November.
[c] *Mucuna* biomass sampled at maturity of *Mucuna*, typically in January.

Table 2. Nitrogen and phosphorus content (kg ha^{-1}) of the weed and *Mucuna* leaf + stem + mulch biomass fractions in maize-*Mucuna* systems in Los Tuxtlas, Mexico, 1996-97. Data are means across four fields (*Source*: Eilittä et al. 2003a).

Maize-*Mucuna* system	Nitrogen		Phosphorus	
	Weed	Leaf+stem+ dead mulch	Weed	Leaf+stem+ dead mulch
1. Summer-winter maize system[a]	28	50	1.8	3.4
2. Summer maize system[b]	28	101	1.8	5.5
3. Rotational[a]	22	147	1.3	9.4
Contrasts and *P* values:				
System 1 vs. 2	0.748	<0.001	0.878	<0.001
System 1 vs. 3	0.162	0.007	0.067	0.001

[a] *Mucuna* biomass sampled at slashing of *Mucuna*, typically in early November.
[b] *Mucuna* biomass sampled at maturity of *Mucuna*, typically in January.

In all systems, soil fertility and management had an impact on *Mucuna*'s growth in farmers' fields. The impact of soil fertility can be seen within communities (biomass production in some fields was clearly less because of poor fertility) and between communities (*Mucuna* productivity was seemingly favoured by higher average fertility in some communities). Management factors that limited *Mucuna* productivity across farmer-practiced systems included low *Mucuna* planting density and late planting date (Eilittä 1998). The effect of choice of maize-*Mucuna* system on *Mucuna* biomass production combined several effects – that of management (sole vs. intercropping, density) and climate (production during wet vs. dry time, winds).

In summary, potential biomass production under the rotational conditions in the region is relatively high but studies have found that intercropping and suboptimal farmer management greatly reduce it.

Moreover, the high variability in biomass production in the farmer-practiced systems sets Los Tuxtlas apart from the spontaneously adopted maize-*Mucuna* system in Honduras, where biomass variability is relatively low (Triomphe 1996). Elsewhere, strong variability in GMCC productivity has been cited as one of the reasons explaining the poor adoption of green manure crops (Becker et al. 1995; Drechsel et al. 1996).

Mucuna's Impact on Maize Yield

Mucuna's impact on maize yield has been studied only during the first years of use, because of the relative newness of the technology and the short-term nature of the various research projects conducted. Research to date indicates the following:

- In the summer-winter maize system, where *Mucuna* is slashed just prior to winter maize planting, winter maize yield is clearly increased in the presence of the slashed *Mucuna*. However, there is no carry-over effect on the following summer maize (even without residue burning, a common farmer practice). Despite the low *Mucuna* biomass at the late October/early November slashing time, the winter maize yield increases averaged 50% in on-farm trials (Table 3), indicating effective synchronization of nutrient release of *Mucuna* and nutrient uptake by maize. However, because of low winter maize yields in most of the region, such high **relative** yield increases amounted only to a few hundred kilograms in **absolute** terms and may therefore not compensate for the labour involved in *Mucuna* planting and slashing.
- In the summer maize system, where no winter maize is planted and *Mucuna* is allowed to senesce in the field, no clear positive yield impacts on the following year's summer maize have been detected in the first years of Mucuna cultivation. This lack of response in summer maize, documented both in an on-farm trial (Table 3) and in agronomic monitoring of 38 (in 1995-96) and 26 (1996-97) farmer-managed fields (Eilittä 1998), is particularly worrisome because summer is the main season for maize production. Several factors may contribute to such low or absent impacts: the recency of *Mucuna* use in the area, the practice of burning residues prior to summer maize cultivation (most farmers, including those cultivating *Mucuna*, continue to burn their fields), the relatively low nutrient content of *Mucuna* and the large potential for mulch and nutrient losses because of the rapid decomposition of *Mucuna* and the long, windy, dry season. It should be noted, however, that preliminary research results from an inter-institutional evaluation project in the region seem to indicate that increased summer maize yields may be expected from the inclusion of *Mucuna* (G. Narvaez, unpublished data). Field

observations support the conclusion that such impacts are neither consistent nor strong.

- In a rotational system that resembles the traditional maize-*Mucuna* system in the region, strong and constant positive impacts on winter maize yield are also evident when sole-cropped, summer season *Mucuna* is followed by winter maize planting. Such yield increases varied between 50 to 120% in on-farm trials (Eilittä et al. 2003a; Eilittä et al. 2003b).

Table 3. Maize yield (t ha^{-1}) in maize-*Mucuna* systems in Los Tuxtlas, Veracruz, Mexico. Data are means across four fields (*Source*: Eilittä et al. 2003a).

Maize-*Mucuna* system	Second season 1995-96	First season 1996	Second season 1996-97	First season 1997
1. Summer-winter maize[a]	0.41	1.00	0.74	0.94
2. Summer maize[b]	None[c]	1.09	None[c]	1.00
3. Rotational[d]	[e]	None[c]	1.11	None[c]
4. No *Mucuna* (summer-winter maize)	0.26	0.96	0.51	0.81
Contrasts and *P*-values:				
1 vs. 4	0.005	0.752	0.036	0.731
2 vs. 4	-	0.312	-	0.321
1 vs. 3	-	-	0.002	-
1 vs. 2	-	0.752	0.002	0.509

[a] First- and second-season maize with first-season *Mucuna*.
[b] First-season maize with first- and second-season *Mucuna*.
[c] Maize not produced during this season.
[d] Second-season maize following first-season *Mucuna*.
[e] Treatment included in second year only.

While the two systems with winter maize production clearly benefited from including *Mucuna*, the long-term sustainability of the summer-winter maize system is doubtful because of its pattern of intensive double cropping with relatively small additions of *Mucuna* mulch. Moreover, winter maize production is a risky enterprise in most of the region because of low and erratic rainfall and high winds. It is therefore uncertain if farmers in the region would sacrifice summer season maize production (as they must in the rotational system) in order to benefit winter maize production. However, the rotational system may have potential as a short-season fallow from which some additional maize production can be obtained.

Costs and Benefits of the Maize-*Mucuna* Systems

Researchers' Views

Several studies have examined the costs and benefits associated with the local maize-*Mucuna* cultivation systems (Buckles and Erenstein 1996; Arteaga et al. 1997; Soule 1997). Estimates of costs and benefits are difficult to make, especially since long-term yield impacts of *Mucuna* systems have not been documented.

The studies agree that current maize production systems with and without *Mucuna* have an extremely low profitability and typically negative returns if labour is valued at its opportunity cost, that is, at local wages. The subsistence orientation of maize production and low valuation of family labour explain why maize continues to be produced in the region. Similarly, the studies agree that *Mucuna* systems require more labour than the conventional maize production systems (Buckles and Erenstein 1996; Soule 1997). In Soule's survey, farmers estimated that a maize-*Mucuna* intercrop demanded 74-79 labour days per hectare, in comparison to 65 days for maize with no *Mucuna*. Buckles and Erenstein (1996) estimated increased labour use of 24-29 days in the upland zone and 11 days in the lowland zone; these calculations also include utilization of commercial fertilizer, conservation tillage and contoured hedgerows. It should be noted, however, that the distribution of labour is different in the maize-*Mucuna* system. Perhaps most importantly, time spent slashing fields prior to summer maize planting—an extremely demanding task conducted at the hottest time of the year—is decreased because of *Mucuna*'s ability to suppress tough woody weeds. Farmers also reported decreased time spent on weeding. On the other hand, *Mucuna* can self-reseed to a great extent in some fields, necessitating increased time spent on weeding.

Buckles and Erenstein (1996) found that the profitability of a maize–*Mucuna* system with added fertilizer and other resource-conserving technologies (including residue management) might eventually become higher than that of the system without *Mucuna*. Still, the maize–*Mucuna* systems were unprofitable during the first years, an observation that in the authors' view called for the use of incentives prior to this period.

Clearly, the low profitability is mainly caused by a combination of the very low selling price of maize (at 1–1.5 pesos kg^{-1}, or 16-24 US cents), the low maize yields and the lack of multiple uses and consequent absence of markets for *Mucuna* seed. Although a few farmers affiliated with the two projects have been able to increase their profits through the sale of *Mucuna* seed, those markets have been sporadic. In fact, in the absence of secure seed markets and strong positive yield impacts, direct subsidies may be the only way to pay off the labour investments for *Mucuna* utilization.

Farmer Expectations and Views

Although no extensive study exists on farmer perceptions on *Mucuna*, farmer viewpoints have been solicited on a number of occasions. Interestingly, Soule (1997) in August of 1993 surveyed farmers to assess their expectations of benefits from incorporating *Mucuna* in their maize system. At that time, most *Mucuna* farmers in the study had not yet harvested one maize crop following a previous year's *Mucuna* cultivation. Farmers with less access to fertilizer and herbicide expected higher profits from including *Mucuna* in their maize system. Such an expectation may indicate that poor farmers stand to gain more from the technology (Soule 1997) or, alternatively, that poorer farmers simply had higher hopes of having found a beneficial technology.

The Pachuca Group of the RED network assessed the sustainability of the maize-*Mucuna* systems in south-eastern Mexico, including an evaluation by a farmer group with participants from two communities of Los Tuxtlas (Arteaga et al. 1997). Farmers identified certain indicators of sustainability and then qualitatively evaluated *Mucuna*'s performance against those indicators. Results show that *Mucuna* was found to have a positive overall impact but that such an impact was not always constant:

- Although no overall trend of the effect of *Mucuna* on maize yield was found in the region, in the two Los Tuxtlas communities included in the study farmers reported a positive impact on maize.
- Fewer edible plants were associated with the maize-*Mucuna* systems, because *Mucuna*'s rapid and luxuriant growth prevents intercropping of other food crops in maize, such as beans.
- *Mucuna*'s positive impact on soil quality was found to be clear and taking place in the short term, except in soils recently taken into cultivation.
- Farmers reported that *Mucuna* reduced labour because of decreased time spent in slashing and decreased weeding time.
- *Mucuna* was found to be best adapted to areas with medium rainfall, because dry conditions do not favour vegetative growth and *Mucuna* seed spoils in humid conditions.
- Farmers judged that *Mucuna* is relatively well suited to a number of soil types.
- *Mucuna* use was found to reduce the need for fertilizers and herbicides, even allowing farmers with little land to abandon herbicide use completely.

In the community meetings organized by the Pachuca Group, farmers were also asked to name the advantages and disadvantages of *Mucuna* cultivation (Table 4). Easier slashing because of *Mucuna* was a particularly important advantage for the farmers, while increased labour for weeding (if *Mucuna* itself becomes a weed) and increased presence of rats and snakes were some of the disadvantages mentioned.

Table 4. Advantages and disadvantages of *Mucuna* in maize systems as cited by adopting (A) and non-adopting (N) farmers in two communities of Los Tuxtlas, Mexico (*Source*: Arteaga et al. 1997)[a].

Advantages and disadvantages	Santa Rosa		Chuniapan de Arriba	
	A	N	A	N
Advantages:				
Fertilizes the soil	x	x	x	x
Controls weeds	x	x	x	
Conserves humidity	x		x	
Reduces erosion	x		x	
Reduces input costs	x	x	x	
Controls disease/pests	x		x	
Softens hard soil	x			
Makes slashing easier	x			
Helps winter maize development	x			
Increases yield			x	
Disadvantages:				
More rats	x	x	x	
Mucuna cannot be controlled and lodges maize	x		x	
Maize grain deteriorates	x			
Herbicide use burns *Mucuna*		x		
No seed when needed		x		
Beans cannot be planted		x		x
Maize planting is more difficult because of mulch		x		
Seed cannot be sold				x
Mucuna seed banks formed in the soil	x			
Impact of mulch on pests not known				x
Too little adoption to evaluate well		x		

[a] Farmer views were solicited in 1996, i.e. 4 years after the initiation of *Mucuna* cultivation in Santa Rosa but only slightly over 1 year after the initiation of *Mucuna* cultivation in Chuniapan de Arriba. Not surprisingly, farmers in Santa Rosa, particularly non-adopters, knew more about the crop.

MUCUNA USE AND ADOPTION

As noted earlier, no formal adoption studies have been made in the region. Several additional factors make it difficult to accurately assess *Mucuna* adoption there. First, *Mucuna* has been introduced relatively recently in most of the region. Its use has not stabilized but rather fluctuates, at least partly in response to the activities of the various projects. Moreover, *Mucuna* use varies greatly even between the communities where the two NGOs have been working, making it difficult to conduct an assessment at the regional level.

In the following, information on *Mucuna* use and adoption has been assembled from a number of sources, including the authors' observations and impressions.

Trends in *Mucuna* Use (1992-2000)

Before 1992, when the PSSM began *Mucuna* extension in the area, about 150 farmers used *Mucuna* in a traditional system, the result of a spontaneous diffusion process dating to the 1960s-70s. These traditional *Mucuna* users were mainly located in the municipalities of San Pedro Soteapan and Mecayapan, in the region of Sierra de Santa Marta (Buckles and Perales 1995).

As mentioned in the historical overview above, the two NGOs in their initial years of *Mucuna* work reached a large number of farmers annually. It is not always known how many of these planted *Mucuna*, or how many of those who planted in one year continued planting in successive years. In two communities where the PSSM works, Santa Rosa and Venustiano Carranza, 78% of those who planted in 1992 were documented to have continued planting in 1993. Based on these rates, an estimated 723 farmers planted *Mucuna* in all communities where the PSSM worked in 1993 (Arteaga 1997). *Mucuna* cultivation varies greatly from one community to another. For example, it has been widely adopted in some of the communities where DECOTUX works, while in others no adoption has occurred. In one of the communities with most *Mucuna* adoption, Chuniapan de Arriba, an evaluation in 1996 estimated that 53 of the 272 farmers with land rights utilized *Mucuna*. In contrast, in 1995 when Eilittä (1998) began an agronomic assessment in the region, *Mucuna* use was declining and, in several of the communities studied, further declined between 1995 and 1997.

The program of SEMANARP, which bought seed from farmers affiliated with the PSSM and DECOTUX in 1998-99, naturally encouraged *Mucuna* cultivation in the region but such purchases were bound to have only a short-term effect.

To summarize, it can be said that, overall, clear increases in the number of *Mucuna* users between 1992 and 1994 were followed by disadoption and decreases in their number. The reasons for this decrease have not been studied formally but several likely factors are discussed in the following section.

Profile of *Mucuna* Farmers

The extension campaigns of 1992-93 purposely targeted the small- and medium-sized subsistence farmers who make up most of the region's rural population. Other important sectors of the population, including cattle ranchers, wealthier farmers and farmers with no land, were therefore not explicitly made targets of the efforts and *Mucuna* continued to be cultivated almost entirely by small- to-medium-sized subsistence maize farmers. Little utilization of *Mucuna* for fodder or grain production has developed.

Some differentiation occurs in adoption between and within communities. The following patterns have been observed:

- Increased interest in *Mucuna* among farmers with secure access to land: Farmers who rely, for example, on a family member lending them a piece of land are not inclined to plant *Mucuna*, because they may not hold the same piece of land the following year. The impact of the recent privatization of landholding on *Mucuna* adoption is still unclear. Those who gained secure landholdings may be more interested in its cultivation.
- Increased interest in *Mucuna* among farmers who attach greater importance to maize production: Conversely, less interest has been shown in *Mucuna* in communities where cattle ranching or coffee cultivation, with their competing labour demands and alternative income possibilities, is important.
- Increased interest in *Mucuna* in areas with clear soil fertility decline: In areas where soil fertility is still relatively good because of low population pressure and adequate fallow periods, less interest has been shown in *Mucuna*.

In summary, *Mucuna* seems to be most attractive to smallholder maize producers farming degraded lands who have few other livelihood options. As discussed earlier, such producers have had to contend with increased input prices and low maize prices, which themselves may act as an impetus for adopting *Mucuna* for at least subsistence maize production.

PROSPECTS FOR *MUCUNA* AND OTHER GMCCs IN THE REGION

Strong forces, both external and internal, are shaping the future of smallholder farming in the Los Tuxtlas region of south-eastern Veracruz, and factors such as NAFTA, the liberalization of the economy, and deterioration in environmental quality, at least in the short term, are creating difficult conditions for smallholders. One encouraging development was the creation of a Federal Reserve—Reserva de la

Biosfera de Los Tuxtlas—in November 1998, which aims to protect some of the region's remaining forests. The reserve program is supposed to increase funding for organizations that work on creating sustainable livelihoods for the region's poor. Clearly, prospects for *Mucuna* and other GMCC systems in the region will be strongly tied with these often contradictory and unpredictable factors, and with how they will play out in the future. Other factors, which can be impacted either within the development or research spheres, are discussed below.

Development and Policy Agenda

Favourable prospects for *Mucuna* in the region would appear to depend on future developments in the following areas.

Diffusion Efforts that Focus on Key Constraints

While the deteriorating soil quality in the region is important, smallholder farmers face a number of constraints and seem to consider pests and diseases as a more important problem than soil fertility. The increased adoption and positive impact of *Mucuna* in the region may therefore depend as much on work to improve the pest and disease situation as on direct work on *Mucuna*-related issues themselves.

Development of the Use of *Mucuna* as a Food and Feed

If *Mucuna* were to find increased utilization as a food and feed, it would lead to improved nutrition, added value through improved production of livestock and development of secure markets for *Mucuna* seed. Two observations support this. In the region, relatively large quantities of good-quality seed can be produced and, because of the low profitability of farming, farmers are constantly searching for alternative and cash crops in their farming systems. Soil conservation, often stressed by technicians, is a secondary goal for farmers for whom immediate benefits are more important. Research on the use of *Mucuna* for food and feed needs to largely precede the development efforts (see below).

Development of Secure Land Tenure

Access to land varies greatly by community. For example, in the San Andres area and in the large municipal centres of the Santa Marta region, only 20% of smallholder farmers own land or have permanent access to land; while in other communities, most farmers are either landowners or have relatively stable access to land. Farmers without secure access to land are less likely to invest their time in resource-conserving technologies.

Policies Promoting Sustainable Agriculture

Several policies of the Mexican government have not effectively promoted sustainable agriculture in the region. For example, the government's Programa de Apoyos Directos al Campo (PROCAMPO), although providing farmers with cash (meant for purchasing of inputs), in fact encourages the cultivation of unprofitable maize in large areas. The extension efforts of national and state government organizations in GMCCs have typically been short lived. In contrast, all government programs should include active promotion of conservation measures.

Development of Positive Social Organizations

At present, the region's smallholders are relatively isolated and neglected and only have been able to passively respond to social, economic and biophysical changes. This situation could be improved by building cooperative organizations in which farmers would envision, experiment and utilize appropriate technologies, aided by technicians working for the farmers and by other collaborating farmers in the region and in other regions.

Research Agenda

Realization of the potential regional contributions from *Mucuna* and other GMCCs will depend partly on the success of future research. A number of bottlenecks to *Mucuna*'s further use and impact in the region can only be solved through research. Priorities include the following areas of research.

Mucuna and Other GMCCs as Food and Feed

As discussed earlier, some studies involving *Mucuna* and pigs have already been conducted in the region but concerns over *Mucuna*'s L-dopa content and lack of guidelines for processing limit its adoption. *Mucuna*'s extensive utilization as a feed in the southern USA and southern Africa in the first half of the twentieth century indicates its potential as a feed crop (Eilittä and Sollenberger 2001), while traditional use as a minor food crop in several tropical countries (Osei-Bonsu et al. 1995; Gilbert 2002) may indicate that it could be developed into a food crop at a larger scale. Further food and feed studies would need to evaluate various components of the GMCC biomass from a number of perspectives, including palatability, weight gain, processing needs and perishability of the products. Regarding L-dopa, the potential of breeding and biotechnology to lower L-dopa content should be considered. Several of these issues are the focus of an ongoing inter-institutional research project, Increasing

Mucuna's Potential as a Food and Feed, which is led by the Center for Cover Crops Information and Seed Exchange in Africa (CIEPCA)-International Institute for Tropical Agriculture (IITA).

Diversification of GMCCs

Studies that contribute to the local diversification of GMCCs are necessary not only because of *Mucuna*'s limited multiple uses but also because of the inherent dangers (e.g. increased occurrence of pests and diseases as well as saturated markets) in relying on only one species. Research should focus on both traditional edible species as well as introduced legume species.

Assessment of Nutrient Dynamics in GMCC Systems

The lack of a clear, strong impact on summer maize in the region is disconcerting and calls attention to the relatively limited general knowledge of nutrient dynamics in GMCC systems. The following issues should receive particular attention:

(1) Avenues of nutrient loss in field conditions during the dry season, including N volatilization and mulch loss: In areas with a strong dry season, monthly mulch losses of 1 t ha^{-1} have been observed (Carsky et al. 1998) and field observations support high rates of mulch loss also in the conditions of Veracruz. Additionally, volatilization losses of surface-applied mulch have been documented to be high in laboratory conditions (Costa et al. 1990). Little work has been conducted in field conditions of the combined effect of the various phenomena impacting the availability of nutrients to the maize crop.

(2) Dynamics of mulch formation and its impact on soil nutrient status, temperature and moisture in intercropped systems as compared to sole-cropped *Mucuna* systems: Clearly, mulch formation is very different in the sole-cropped rotational system, where *Mucuna* starts shedding leaves already 2 months after planting (because of self-shading) and forms a thick mulch layer soon after. In comparison, because of better light conditions, the intercropped *Mucuna* loses few leaves before its senescence. A study comparing nutrient release from such contrasting mulch layers has not been conducted.

Combining GMCC Use with Fertilizers

GMCC research worldwide is indicating potentially high returns if organic and inorganic fertilizers are combined. Locally, the potential of such applications should be investigated, with special focus on increasing the regional winter maize production in the rotational system.

Designing Cropping Systems with Lower Pest and Disease Incidence

Farmers in the region report that pest and disease incidence has greatly increased in the past decades. Changes in cropping systems may lead to lower pressure and be affordable to the farmers.

Evaluating the Processes for Generation and Diffusion of Technology

We know relatively little of what constitute particularly effective and efficient methods to promote resource-conserving technologies or integrate research and diffusion efforts, but recent studies have started focusing on the role of, for example, temporary seed markets during GMCC diffusion (Douthwaite et al. 2002). Learning from past and ongoing efforts, both in Veracruz and beyond, is therefore needed.

CONCLUSION

Biophysically, *Mucuna* clearly is well adapted to the conditions of the Los Tuxtlas region of Veracruz, Mexico. It can produce large amounts of biomass and exert positive impact on maize yield. Highest biomass production and relative yield impacts were found for a system where only winter maize is produced, which is problematic, given the low yields and riskiness of winter maize. More importantly, improvement of the maize-based agriculture is generally difficult because of the low maize prices and general unprofitability of agriculture in the region, factors that are strongly tied to economic policies. Incorporation of *Mucuna* in these systems is difficult because it lacks uses as a food and feed as well as other, stable markets for its seeds. While the cultivation of *Mucuna* in the region during the past decade has waxed and waned as a response to efforts of various organizations, the long-term future of *Mucuna* cultivation in the region is uncertain and will be strongly tied to the economic policies of Mexico.

NOTES

1. The PSSM was established as a joint program of the Instituto de Investigaciones Sociales (the Institute of Social Research) of the Universidad Nacional Autónoma de México (UNAM, the National Autonomous University of Mexico); Carleton University; and International Development Research Centre (IDRC).
2. On-farm trials covered only the first 3 years of *Mucuna* cultivation, while the agronomic monitoring was conducted in fields where *Mucuna* had been cultivated for 1 to 4 years.

REFERENCES

Arteaga L. (1997) Campaña de extensión de abonos verdes: 1992-1994. *In*: Paré L, Velázquez H E. (Eds.) Gestión de recursos naturales y opciones agroecolócas para la Sierra de Santa Marta, Veracruz. Instituto de Investigaciones Socíales, Universidad Nacional Autónoma de México, Mexico. p. 87-96.

Arteaga L, Carranza T, Eilittä M, González M, Guerrero C, Guevara F, Herrera B, López A, Martinez F, Mendoza A, Narváez G, Puentes R, Reyes H, Robles C, Sohn I, Triomphe B (1997) El uso de sistemas de cultivo con plantas de cobertura en algunas comunidades del Sureste Mexicano: Contexto, resultados y lecciones aprendidas. Paper presented in a workshop on Green Manure/Cover Crops Systems for Smallholders in Tropical and Subtropical Regions, 6-12 April 1997, Chapeco, Brazil.

Barrera-Bassols N, López Binquist C, Palma Gayeb R (1993) Vacas, pastos y bosques en Veracruz: 1950-1990. *In*: Barrera N, Rodriguez N (Eds.) Desarrollo y medio ambiente en Veracruz: Impactos económicos, ecológicos y culturales de la ganaderia en Veracruz. Fundación Friedrich Ebert, Xalapa, Veracruz, Mexico. p. 35-71.

Becker M, Ladha J K, Ali M (1995). Green manure technology: Potential, usage, and limitations. A case study for lowland rice. Plant Soil 174:181-194.

Buckles D, Arteaga L (1993) Extensión campesino a campesino de los abonos verdes en la Sierra de Santa Marta, Veracruz, México. *In*: Buckles D (Ed.) Gorras y sombreros: Caminos hacia la colaboración entre técnicos y campesinos. Centro Internacional de Mejoramiento de Maíz y Trigo (CIMMYT), Mexico. p. 51-59.

Buckles D, Barreto H (1996) Intensificíon de sistemas de agricultura tropical mediante leguminosas de cobertura: Un marco conceptual. CIMMYT Natural Resources Group (NRG) Document No. 96-06. Centro Internacional de Mejoramiento de Maíz y Trigo (CIMMYT), Mexico. 13 p.

Buckles D. Erenstein O (1996) Intensifying maize-based cropping systems in the Sierra de Santa Marta, Veracruz. CIMMYT Natural Resources Group (NRG) Document No. 96-07. Centro Internacional de Mejoramiento de Maíz y Trigo (CIMMYT), Mexico. 55 p.

Buckles D, Perales H (1995) Farmer-based experimentation with velvet bean: Innovation within tradition. Centro Internacional de Mejoramiento de Maíz y Trigo (CIMMYT), Mexico. Internal document. 22 p.

Buckles D, Arteaga L Soule M (1994) Extensión campesino a campesino de los abonos verdes en la Sierra de Santa Marta, Veracruz, México. *In*: Memoria: Taller sobre las políticas para una agricultura sustentable en la Sierra de los Tuxtlas y Santa Marta, Veracruz. Centro Internacional de Mejoramiento de Maíz y Trigo (CIMMYT), Mexico. p. 139-154.

Carsky R J, Tarawali S A, Becker M, Chikoye D, Tian G, Sanginga N (1998) *Mucuna* – Herbaceous cover legume with potential for multiple uses. Resource and Crop Management Research Monograph No. 25. International Institute of Tropical Agriculture (IITA), Ibadan, Nigeria. 52 p.

Chevalier J M, Buckles D (1995) A land without gods: Process theory, maldevelopment, and the Mexican Nahuas. Zed Books, London. 374 p.

Costa F J S A, Bouldin D R, Suhet A R (1990) Evaluation of N recovery from *Mucuna* placed on the surface or incorporated in a Brazilian Oxisol. Plant Soil 124:91-96.

Dirzo R, González S E, Vogt R C (1997) Introducción general. *In*: González E, Dirzo R, Vogt R C (Eds.) Historia natural de los Tuxtla. Universidad Nacional Autónoma de México, Mexico. p. 3-6.

Douthwaite B, Manyong V M, Keatinge J D H, Chianu J (2002) The adoption of alley farming and *Mucuna*: Lessons for research, development and extension. Agrofor. Syst. 56(3):193-202.

Drechsel P, Steiner K G, Hagedorn F (1996) A review on the potential of improved fallows and green manure in Rwanda. Agrofor. Syst. 33:109-136.

Eilittä M (1998) On-farm assessment of maize-*Mucuna* systems in the Los Tuxtlas region of south-eastern Veracruz, Mexico. Ph.D. thesis, University of Florida, Gainesville, Florida, USA. 224 p.

Eilittä M, Sollenberger L E (2001) The many uses of *Mucuna*: Velvetbean in the southern United States in the early 20th century. *In*: Flores B, Eilittä M, Myhrman R, Carew L B, Carsky R J (Eds.) Food and feed from *Mucuna*: Current uses and the way forward. Centro Internacional de Información sobre Cultivos de Cobertura (CIDICCO), Center for Cover Crops Information and Seed Exchange in Africa (CIEPCA) and World Hunger Research Center, Judson College, Tegucigalpa, Honduras. p. 73-100.

Eilittä M, Sollenberger L E, Littell R C, Harrington L W (2003a). On-farm experiments with maize-*Mucuna* systems in the Los Tuxtlas region of Veracruz, S. Mexico: I. *Mucuna* biomass and maize grain yield. Exp. Agric. 39(1):5-17.

Eilittä M, Sollenberger L E, Littell R C, Harrington L W (2003b) On-farm experiments with maize-*Mucuna* systems in the Los Tuxtlas region of Veracruz, S. Mexico: II. *Mucuna* variety evaluation and subsequent maize yield. Exp. Agric. 39(1): 19-27.

Gilbert R A (2002) *Mucuna pruriens* in Malawi: A promising legume with a troubled history. *In*: Flores M, Eilittä M, Carew L, Carsky R J, Myhrman R (Eds.) Food and feed from *Mucuna*: Current uses and the way forward. Proceedings of a workshop held 26-29 April 2000 in Tegucigalpa, Honduras. Centro Internacional de Información sobre Cultivos de Cobertura (CIDICCO) – Center for Cover Crops and Seed Exchange in Africa (CIEPCA) – International Institute of Tropical Agriculture (IITA)-World Hunger Research Center, Tegucigalpa, Honduras. p. 48-59.

Narváez G (1999) Produccíon de semilla para abono verde y cobertura en el Sureste de México: Informe de actividades julio de 1998-abril de 1999, síntesis. Available at the Centro Regional Sur de Chapingo, Oaxaca de Juárez, Mexico. Unpublished report. 17 p.

Osei-Bonsu P, Buckles D, Soza F R, Asibuo J Y (1995). Traditional food uses of *Mucuna pruriens* and *Canavalia ensiformis* in Ghana. ILEIA Newsl. 12(2):30-31.

PSSM – Proyecto Sierra de Santa Marta (n.d.) La Reserva Especial de la Biosfera Sierra de Santa Marta, Veracruz: Diagnosticos perspectives. Unpublished mss. PSSM, Jalapa, Mexico. 77 p + 12 p (appendices).

Ramirez R F (1991) Importancia ecologica de la Sierra de Santa Marta. Proyecto Sierra de Santa Marta, Jalapa, Mexico. 21 p.

Ramirez R F, Gutiérrez R, Paré L (1997) Características de la Sierra de Santa Marta. *In*: Paré L, Velázquez H E (Eds.) Gestión de recursos naturales y opciones agroecolócas para la Sierra de Santa Marta, Veracruz. Instituto de Investigaciones Sociales, Universidad Nacional Autónoma de México, Mexico. p. 17-37.

Robles G C A, Alemeida M E (1998) Experimentación campesina y tecnología sustentable en Los Tuxtlas. Red de Gestión de Recursos Naturales –Rockefeller Foundation, Mexico. 125 p.

Sohn López-Forment I M (2000) Ecological and socio-cultural dynamics of traditional and legume-based *Milpa* agriculture in southeast Mexico. Ph.D. thesis, The Ohio State University, Ohio, USA. 328 p.

Soule M J (1997) Farmer assessment of velvet bean as a green manure in Veracruz: Experimentation and expected profits. CIMMYT Natural Resources Group (NRG) Paper No. 97-02. Centro Internacional de Mejoramiento de Maíz y Trigo (CIMMYT), Mexico. 24 p.

Stuart J (1978) Subsistence ecology of the Isthmus Nahuat Indians of southern Veracruz, Mexico. Ph.D. Thesis, University of California, USA. 408 p.

Tasistro A S (1994) La labranza de conservacion para la Sierra de Los Tuxtlas y Santa Marta. *In*: Memoria: Taller sobre las políticas para una agricultura sustentable en la Sierra de los Tuxtlas y Santa Marta, Veracruz. Centro Internacional de Mejoramiento de Maíz y Trigo (CIMMYT), Mexico. p. 89-122.

Triomphe B (1996) Seasonal nitrogen dynamics and long-term changes in soil properties under the *Mucuna*/maize cropping system on the hillsides of Northern Honduras. Ph.D. thesis, Cornell University, Ithaca, USA. 217 p.

Uresti G J, Uribe S G, Fransisco N (1992) Erosión en terrenos agrícolas de ladera – cuatro estudios de caso en San Andrés Tuxtla, Veracruz. *In*: Sims B G (Ed.) Ingeneria apropriada para el pequeño productor de ladera en México, Honduras, y Nicaragua. Santiago Tuxtla, Veracruz, Mexico. p. 76-85.

Chapter 5

Improved Fallows in Eastern Zambia

J. Peterson[1], F. R. Kwesiga[2], S. Franzel[2], D. Phiri[3] and C. Gladwin[4]

[1]*Consultant*
[2] *International Centre for Research in Agroforestry (ICRAF), Kenya*
[3]*World Vision*
[4]*University of Florida, USA*

SUMMARY

In Zambia, the production of the main staple crop, maize (*Zea mays* L.), has fallen steadily over the past few years due to the removal of fertilizer subsidies, the collapse of agricultural credit programs and the occurrence of periodic droughts. In 1992, the International Centre for Research in Agroforestry (ICRAF) and the Zambian Ministry of Agriculture, Food and Fisheries initiated on-farm research in eastern Zambia, a hilly region with an average rainfall of 960 mm, to develop soil fertility systems that are easily accessible to smallholder farmers. Together with a number of national and international partners, they have developed systems with *Sesbania sesban* (L.) Merr., *Tephrosia vogelii* Hook. f., *Gliricidia sepium* (Jacq.) Walp. and pigeon pea (*Cajanus cajan* [L.] Millsp.) where maize yields following 2-year improved fallows approach those of fully fertilized fields. The fallow species are either planted in a nursery (*Sesbania* and *Gliricidia*) or directly in the field after the beginning of the rains. In the first year, they can be intercropped with maize or sunflower (*Helianthus annuus* L.). They are weeded during the first year and a fire-break is constructed around them. At the end of the second year, the trees are cut and the field is replanted to maize or, occasionally, to cotton (*Gossypium hirsutum* L.), groundnuts (*Arachis hypogaea* L.) or sunflower. The principal product removed from the improved fallow system is the main crop but others include firewood and building materials, tree seeds and at times *Tephrosia* leaves for use as a pesticide. After 2-year *Sesbania*, average maize yields are 3000 to 5000 kg ha^{-1}, and after 2-year *Tephrosia* or pigeon-pea fallows, they are 2000 to 3000 kg ha^{-1}, in comparison with 3000 to 5000 kg ha^{-1} for fully fertilized, and less than 1000 kg ha^{-1} for unfertilized, continuously cropped maize.

M. Eilittä et al. (eds.), Green Manure/Cover Crop Systems of Smallholder Farmers, 129-152.

These yields decrease over time and, typically after 2-3 years of maize cultivation, another improved fallow needs to be planted. Over a 5-year period (2 years fallow and 3 years cultivation), improved fallows require 11% less labour than unfertilized maize and 32% less labour than fertilized maize. Overall returns per hectare are highest for fertilized maize and lowest for unfertilized maize. However, returns to labour are highest for improved fallows and lowest for unfertilized maize. The main problems include insects (*Mesoplatys* beetles on *Sesbania*, termites), grazing by livestock (pigeon pea), the additional labour required for establishing a fallow (especially *Sesbania*) and the long period required before benefits accrue. An estimated 20 000 farmers have planted improved fallows. A regional network of farmers, non-governmental organizations, research organizations and government extension personnel promotes the system. It utilizes farmer participatory methods and attempts to ensure the availability of a wide range of options for farmers as regards species, planting methods and period of fallow.

Adoption data from ICRAF, collected since 1996, indicate that 50-70% of those farmers who tested improved fallows planted a second one and that women were initiating an increasing number of on-farm trials. Future prospects for the continued expansion of the system in Eastern Province are reasonably good. Factors with particular impact on future adoption include fertilizer prices, existence of other income-generating opportunities, access to livestock and seed availability.

INTRODUCTION

Zambia has one of the highest population growth rates in the world and was once the most urbanized country in sub-Saharan Africa (Celis et al. 1991). With the removal of fertilizer subsidies, the collapse of agricultural credit programs and the occurrence of periodic droughts, production of the main staple crop, maize (*Zea mays* L.), has fallen steadily over the past 5 years. As a result of falling maize production and labour shortages exacerbated by the HIV/AIDS (Human Immunodeficiency Virus/Acquired Immunodeficiency Syndrome) epidemic in the region, the country's food security has deteriorated (FEWS 1999).

To combat maize production constraints, the International Centre for Research in Agroforestry (ICRAF) and the Zambian Ministry of Agriculture, Food and Fisheries (MAFF) have undertaken a research program to develop alternative soil fertility systems that will allow small-scale farmers without access to credit or cash to improve their maize production. Using nitrogen-fixing tree species including *Sesbania sesban* (L.) Merr., *Tephrosia vogelii* Hook. f., *Gliricidia sepium* (Jacq.) Walp. and pigeon pea (*Cajanus cajan* [L.] Millsp.) in improved fallow systems, maize yields following 2-year improved fallows approach those of fully fertilized fields (Kwesiga and Beniest 1998; Kwesiga et al. 1999). Following these research efforts, local non-governmental organizations (NGOs) and MAFF extension agents have demonstrated the technology to farmers in the Province and over 20 000 farmers have planted improved fallows.

Improved fallows build upon traditional fallow practices by incorporating fast-growing, nitrogen-fixing vines, shrubs and trees into traditional fallows to speed up the natural soil regeneration process. Whereas traditional fallows can require up to 20 years to improve the soil, improved fallows can significantly improve soil fertility in just 2 years, although this time can be longer in areas where rainfall and other factors limit the growth of the selected tree species. Improved fallows also have secondary benefits, such as the provision of firewood, medicine, fodder, erosion control, weed suppression and shade (Kwesiga and Beniest 1998).

This case study documents the development and extension of the improved fallow systems in eastern Zambia.

THE ENVIRONMENT IN EASTERN ZAMBIA

Biophysical Environment

The Eastern Province of Zambia is located between lat 10-15°S and long 30-33°E and covers an area of 70 000 km^2, or about 9% of the total land area of Zambia (Figure 1). It borders Malawi to the east and Mozambique to the south and is characterized by gentle to moderate slopes intermingled with hills, ridges and minor escarpments. Eastern Zambia retains large tracts of natural forest cover, consisting mainly of miombo (*Brachystegia* Benth.-*Julbernardia* Pellegr.) and *mopane* (*Colophospermum mopane* J. Kirk ex Benth) woodland and is divided into a high altitude plateau area (900-1500 masl) and the Luangwa River valley (300-600 masl). In addition to the Luangwa River system, there are many seasonally waterlogged low-lying areas, or *dambos*, which are exploited locally for dry season gardens and improved fallow nurseries.

Eastern Province has three distinct seasons: a warm wet season from November to April (the main agricultural season), a cool dry season from May to August (the gardening season) and a hot dry season in September and October. Rainfall averages 960 mm per year, with most of the rain (85%) falling from December to March. The driest month (August) experiences no rainfall, whereas the wettest month (December) averages 231 mm. However, over the past 10 years, Eastern Province has experienced several droughts, with average rainfalls of less than 600 mm per year. Average daily minimum and maximum temperatures vary from 18 to 31 °C during the hottest month (October) and from 9 to 23 °C during the coldest month (July).

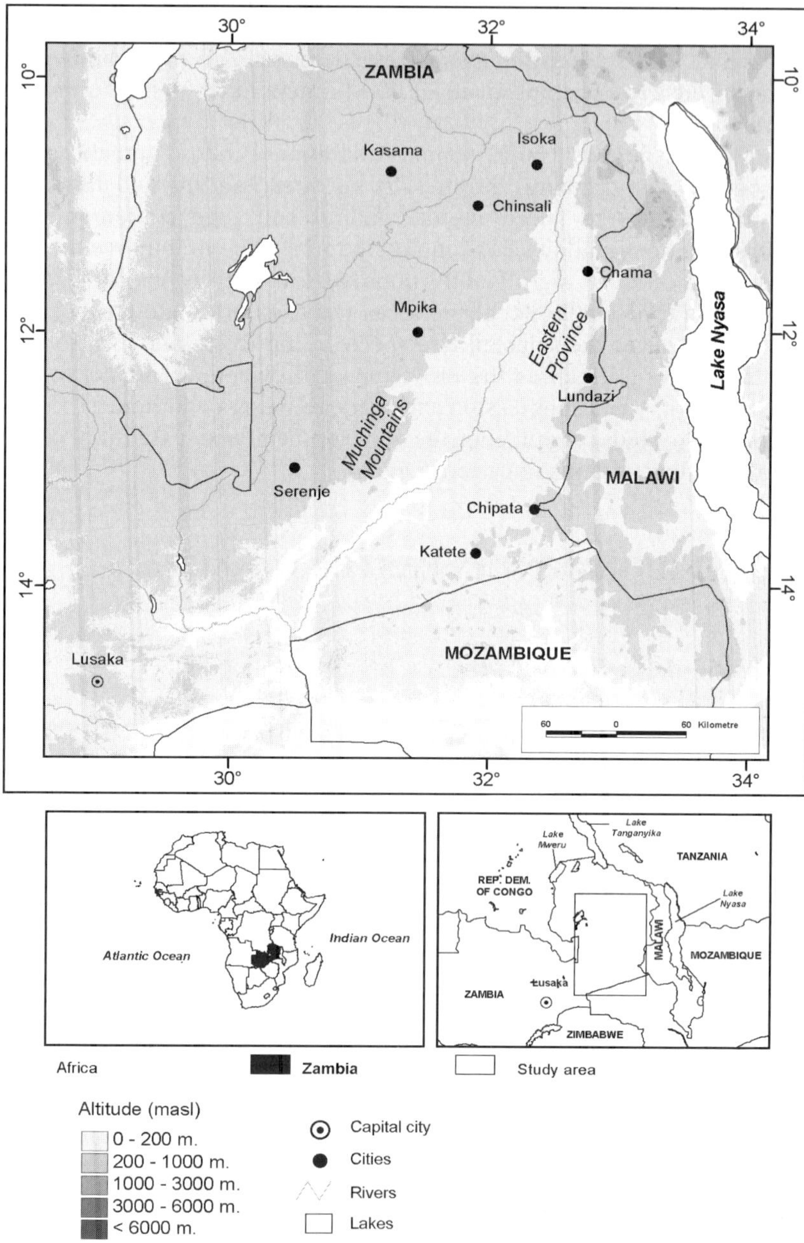

Figure 1. Location of study area, Eastern Province, Zambia.

Soils in Eastern Province are a combination of medium-textured, alluvial soils in the valley and sandvelt soils on the plateau. The most common types are loamy sands or sands (Acrisols; Food and Agriculture Organization-United Nations Educational, Scientific and Cultural

Organization [FAO-UNESCO] Classification, Alfisols and Luvisols). These soils are moderately leached, well drained and relatively fertile, but their water- and nutrient-holding capacity is low and they suffer from excessive leaching during the rainy season. Around Chipata, red clays and red-brown loams (ferric Luvisols; FAO-UNESCO classification) are common, whereas shallow, gravelly soils (Lithosols) are found on hillsides and escarpments. The clays in the area are non-sticky, making them relatively easy to hoe or plough. Soil pH levels vary considerably in the region, from 4.6 to 6. Organic carbon levels on farmers' fields vary from 0.27 to 2.99%. Grey-brown loamy sands or sands as well as hydromorphic Gleysols are found in *dambo* areas or in other poorly drained sites where organic carbon and soil pH are highly variable.

Socio-economic Environment

Currently, the population of the Province is estimated at 1 225 000 (Kupanda 1996) and is largely rural (91%). The population density has doubled since 1969 but is still low at 14.1 persons per km^2. The average annual population growth rate from 1980 to 1990 was 4.0%.

In the early 1990s, there were about 93.4 males per 100 females in Eastern Province and women headed 20% of the households (CSO 1992). The number of males per household has increased in the past decade, as men have returned from the copper mines due to decreased copper prices and privatization of the mines.

A number of different ethnic groups live in Eastern Province, including Chewa, Ngoni, Tumbuka, Nsenga, Lala, Yao and Kunda. However, the main ethnic groups include the Chewa, who are matrilocal, and the Ngoni, who are patrilineal. Chewa and Ngoni have also settled in neighbouring Malawi and Mozambique.

HIV/AIDS (Human Immunodeficiency Virus/Acquired Immunodeficiency Syndrome) is one of the country's leading health problems. In mid-1993, 36% of expectant mothers attending pre-natal clinics at the University Teaching Hospital in Lusaka tested HIV positive. The United Nations currently estimates that 1.1 million Zambians will die of AIDS by the year 2005. The disease is having a marked impact on households throughout the country. By the end of 1993, 40% of households in Zambia had one or more orphans in their care, half of whom were orphaned as a result of HIV/AIDS. Many grandparents are now providing care to their dying children and are raising their grandchildren. The impact of AIDS on food production is also anticipated to be significant (ZARD 1996).

Eastern Province has about 3.8 million ha of arable land (of 6.9 million total ha) but only 35% of it is currently utilized. Land in agricultural production has been continuously cultivated for an average of 12 years, although most farmers (about three-fifths) have access to land that has never been cleared in their lifetime. Seventy percent of the farmers in the region have fallow fields, which average 1.8 ha in size and have been fallowed for an average of 5 years (Peterson et al. 1999).

There are two types of land tenure systems in Zambia. State land, constituting 4% of Eastern Province, is controlled by the government and is often used by townships or for resettlement schemes. Originally land appropriated by the British colonial government, it is typically fertile. Rights to traditional land, under the control of village headmen, are typically given to the family that first clears the land, who can then transfer its user rights to another person (also through inheritance) without consulting the chief. However, traditional land cannot be bought or sold. Few people have title deeds to their land, although most enjoy relatively secure land tenure (Celis et al. 1991). A woman can gain access to land only if a close male relative (or husband) will give her a part of his own land, which can create problems for women in female-headed households. Despite traditional restrictions in women's land tenure status, tree tenure status is unrestricted and secure.

Eastern Province is divided into eight administrative districts, where a District Agricultural Coordinator (DACO) coordinates agricultural activities. Each district includes a number of camps (a cluster of villages) staffed by an agricultural camp officer (extension agent). Technical specialists are employed at the district level. In addition to government extension services, a number of donor-funded initiatives in Eastern Province are providing technical assistance to farmers.

The main source of farm labour is the family. Hired labour is often used during critical periods, such as cultivating, planting, weeding and harvesting. Some farmers also exchange their labour for maize once their maize stocks are exhausted. Labour availability from December to January is a major constraint to increased crop production. The main form of credit currently available for small-scale farmers comes from companies that buy cash crops.

Agricultural Production

Main agricultural products include maize, groundnuts (*Arachis hypogaea* L.), cotton (*Gossypium hirsutum* L.), tobacco (*Nicotiana tabacum* L.), sunflower (*Helianthus annuus* L.) and livestock; food processing is the major industry in the Province. Before the introduction of mineral fertilizers and hybrid maize varieties in the 1970s, farmers in Eastern Province practiced mound cultivation of maize, groundnuts and squash (*Cucurbita* sp. L.). They farmed land for 3-5 years and then left it fallow for up to 50 years (Raussen 1997). However, during the 1970s and 1980s, farmers in the region had access to hybrid maize seeds, mineral fertilizers, agricultural credit, technical advice and government-operated agricultural marketing boards. As a result, they also adopted row cropping systems and began to continuously crop their fields (Celis et al. 1991).

In the past 15 years, most agricultural credit programs have been discontinued and credit for hybrid seeds and mineral fertilizers is no longer widely available. Consequently, the use of hybrid seed and fertilizers has decreased and the area devoted to maize production has diminished. However, farmers continue to practice row production and to continuously cultivate their fields (Peterson 1999).

Hand cultivators begin clearing new land (*mphanje* land) in March and continue clearing it through September. Ox cultivators (less than half the farmers in the Province) prepare land in October/November. In October, the previous season's crop residues are placed in between existing furrows, or rows. Old furrows are then destroyed and new furrows are created incorporating the previous season's crop residues. A cropping system known as *galaousa*, meaning dry planting, typically involves maize intercropped with pumpkin (*Cucurbita* sp. L.) and cowpea (*Vigna unguiculata* [L.] Walp.). Planting is done in October, before the rains begin. Most weeding is done by hand and labour for weeding often determines how much land is cultivated.

Chicken, goats, cattle, pigs and sheep are common components of local farming systems. Most households (70%) keep at least a few chickens and 30% of the farmers in Eastern Province keep cattle (Ngugi 1988; Peterson et al. 1999). During the main cropping season, cattle are herded away from fields and goats and sheep are confined to the village. During the dry season, animals browse freely without herding except in certain areas designated by Chiefs.

Although women are usually responsible for collecting fuelwood, the felling of trees is a man's job. As firewood has become more difficult to collect (requiring walks of over 6 km and an ox cart for transport), it has become commercialized. In such areas, men now are collecting firewood for the family and for sale. Men are also responsible for building fences.

STRUCTURE AND FUNCTIONING OF THE SYSTEMS

Crops Involved and Main Practices

The main crops involved in the improved fallow system being adopted in eastern Zambia are nitrogen-fixing tree species (*Sesbania*, *Tephrosia*, *Gliricidia* and pigeon pea) and maize. Maize is the most important crop in the region from both a social and political standpoint. However, some farmers have followed their improved fallows with other crops, including cotton, groundnuts and sunflower (Kwesiga and Beniest 1998; Peterson 1999). More research is needed to identify the potential of improved fallows to increase agricultural production in other cropping systems, including cash crop systems.

Box 1 presents technical specifications for the improved fallow systems. For farmers using *Sesbania* or *Gliricidia*, the improved fallow cycle starts in September/October with planting of seeds in a nursery, either in plastic pots or directly into raised beds. Typically after 5-6 weeks, seedlings are transplanted to fields. *Tephrosia* and pigeon pea seeds are planted directly into fields as early in the rainy season as possible (usually in December or January).

***Box 1.* Technical specifications of improved fallows**

- The recommended *Tephrosia* planting patterns are 0.9 m × 0.75 m, or 14 763 trees ha^{-1}, either in a pure stand, or intercropped with maize. *Sesbania* is planted at 1 × 1 m. However, often ox-cultivators use 0.9 m between ridges and 0.5 m between plants. Hand-hoe cultivators use 0.75 m between ridges and 0.5 m between plants for *Tephrosia*. In the case of *Sesbania*, the between-ridges spacing is the same as indicated for *Tephrosia* but between-plants spacing is 1 m.
- Trees must be weeded the first year of production and a fire-break (a 2- to 4-m wide cleared strip) should be constructed around each plot early in the dry season (April - June).
- Tree survival rates of 60% or more are considered good and roughly 30% is required to achieve expected fertility benefits. The minimum number of trees is lower for *Sesbania*, which has a wider canopy, than for *Tephrosia*.
- At the end of the second year of fallow, the trees are cut and the field is replanted to maize. Some farmers fallow for longer periods.
- *Sesbania*, *Tephrosia* and pigeon pea die after they are cut, so there is no competition between the trees and crops following an improved fallow. *Gliricidia* resprouts and continues to grow in the field for over 10 years in a modified hedgerow intercropping system. This increases tree-crop competition but decreases establishment costs. The resprouts are removed and applied to the crop as green manure during the normal weeding process. As a result of this periodic application of the green manure, the fallow effect lasts longer in *Gliricidia* fallows than in *Sesbania*, *Tephrosia* or pigeon pea fallows.

All trees can be intercropped with maize or sunflower during the first year of establishment or they can be planted in pure stands. Research findings suggest that, in dry years, intercropping reduces tree survival and maize yields. A further disadvantage of intercropping is that livestock may trample the trees during the first dry season, when they enter fields to graze on maize stover. However, many farmers prefer intercropping because they can use their scarce land and labour to produce a maize crop and an improved fallow at the same time (Franzel et al. 2002).

Trees need to be weeded at least once during the first production cycle (when intercropped, weeding occurs naturally as part of the regular agricultural calendar) and a fire-break must be cleared to protect the trees in June and July, when farmers set fires to hunt for mice. Subsequently, trees are left to grow a second year, after which they are cut and the plot is replanted to a crop (usually maize).

Farmers have been involved in numerous modifications of the improved fallow system. Several of these modifications have to do with the planting phase. Whereas, in early on-station trials, tree species were planted in plastic pots in nurseries, farmers began testing bare root transplants as well as the direct seeding of several improved fallow species. They also have started dry planting of *Tephrosia* fallows (sowing the seeds before the onset of the rains as a way of dealing with the issue of labour demand during the peak period). Several farmers have planted *Sesbania* at weeding time into parts of fields where maize is performing poorly. Others have also experimented (with good success) in establishing fallows in uncultivated fields.

Farmers were also instrumental in the evolution of improved fallows to a partial intercrop system with maize in the first year of establishment. In areas with pest problems, they have begun to mix tree species in their fallow fields to reduce the risk and impact of insect attacks. Farmers have also experimented with other ways to diminish the impact of insects in their nurseries and to mix different crops into the improved fallow matrix.

It should be noted that farmers begin by planting relatively small proportions of their fields to improved fallows. For example, initial plantings in 1995-96 by females were only 0.03 ha, compared to 0.07 ha by males. Since the same percentage of males and females stated that they had obtained enough planting material, it appears that females wanted smaller plots, perhaps because they lacked land or were more risk adverse than men (Franzel et al. 2002). Keil (2001) found that farmers who had started planting improved fallows in 1996-97 or earlier, planted 0.23 ha to improved fallows in 2000-01, demonstrating their strong interest in the technology. The index of adoption, the annual area planted as a proportion of the greatest possible area the farmer could plant to improved fallow in a single year, was about 32%.[1] In general, farmers view improved fallows as one of several alternative methods to improve soil fertility and use other methods, such as mineral fertilizer or manure, on other plots.

In terms of tasks associated with the production and maintenance of improved fallows, both men and women manage nurseries, although men are responsible for creating a fenced area to protect the nursery and when pots are used, women often fill them with soil. Technically speaking, lowland areas traditionally used for nursery production belong to men but land borrowing is common. Men, women and especially children water the young seedlings and everyone participates in the transplantation process. When trees are direct-seeded into fields, men, women and children participate. The entire family weeds improved fallows. Men generally create and maintain fire-breaks. When the trees from an improved fallow are cut for firewood, women are generally involved. However, when trees are cut for building material, or simply to clear the fallow, men are often involved. In the production of the subsequent maize crop, all family members are involved.

Products and Impacts

The main product removed from the improved fallow system is maize, which is typically for home consumption, although some is exchanged for goods or labour. Maize stover is generally returned to the soil; often, cattle are also allowed into a harvested field to graze. Additional products removed from the system in some villages include firewood and building materials (especially from *Gliricidia* and *Sesbania* fallows), tree seeds (which are sold to ICRAF and to local extension programs) and fodder (all of the tree species are browsed by ruminants). In isolated cases, farmers are harvesting *Tephrosia* leaves to use as a pesticide. Other than tree seeds, the products of the system are not sold. Since it is not uncommon for village headmen to withdraw uncultivated land from people, some farmers are now planting fallow species in their fields not only for soil fertility restoration but also to secure the tenure of their land.

Average maize yields after 2-year *Sesbania* fallows range from 3000 to 5000 kg ha^{-1}. Maize yields after 2-year *Tephrosia* and pigeon pea fallows vary from 2000 to 3000 kg ha^{-1} (Kwesiga and Beniest 1998). Whereas fully fertilized maize yields an average of 3000 to 5000 kg ha^{-1} on farmers' fields, unfertilized, continuously cropped maize fields and traditional fallows average less than 1000 kg ha^{-1}. Thus, farmers planting improved fallows generally double or triple their yields over maize, which is typically unfertilized or under-fertilized. However, maize yields in improved fallow plots decrease over time and, eventually, after 4-5 years of production, the fields must be replanted with another fallow. In the case of *Gliricidia* fallows, the residual effects may last longer than this and since *Gliricidia* resprouts, there is no need to replant trees; just revert the land to a fallow. Maize yields obtained from various production systems on-station at Msekera, eastern Zambia, are compared in Table 1.

In one on-station trial, a field cultivated with maize and *Gliricidia* in a modified improved fallow/hedgerow system (where the trees are cut each year and allowed to regrow along with the maize) produced 2870 kg ha^{-1} of maize grain after 3 years of continuous cultivation. The yield was not significantly different from the yield of maize grain produced with fertilizer applied at recommended rates (250 kg ha^{-1} N). The improved fallow plots with *Gliricidia* produced twice as much maize grain as unfertilized plots (Zambia-ICRAF 1998). Interestingly, productivity of improved fallows is further increased by fertilizer application (Figure 2).

The productivity and yield effects of the different improved fallow species vary significantly depending on rainfall, soil type and pest attacks. Whereas *Sesbania* and *Gliricidia* generally outperform other tree species on station, *Tephrosia* and pigeon pea perform admirably in farmers' fields, depending on the location (Zambia-ICRAF, 1992-1998).

Table 1. Yields of maize grown after different production options (*Source*: Kwesiga and Beniest 1998).

Option	Maize yield (t ha^{-1})
2-year *Sesbania sesban* fallow	5.36
2-year *Tephrosia vogelii* fallow	3.18
2-year *S. macrantha* fallow	2.97
2-year *Cajanus cajan* fallow	2.78
1-year *S. sesban* fallow	3.43
1-year *T. vogelii* fallow	2.80
1-year *C. cajan* fallow	2.40
1-year *S. macrantha* fallow	2.07
Groundnut-maize rotation	1.87
Grass fallow	1.84
Maize without fertilizer	1.09
Maize with fertilizer[a]	3.96

[a] Fertilizer rate of 112 kg N ha^{-1}.

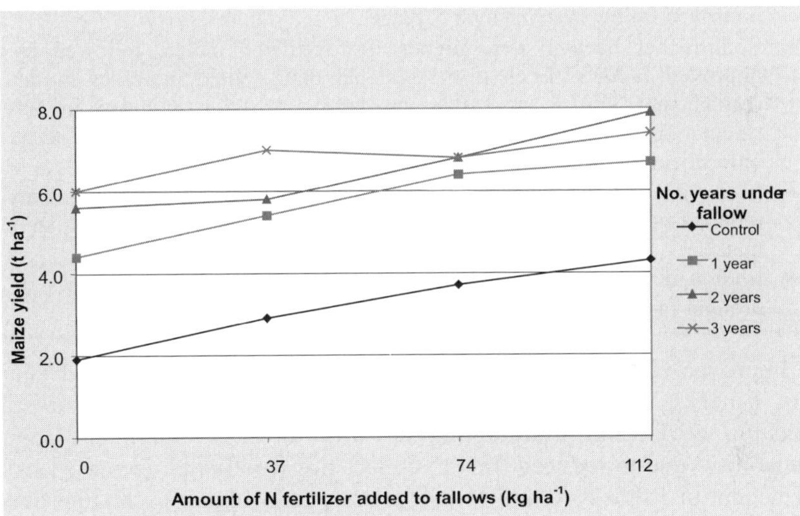

Figure 2. Interaction between fertilizer application and improved fallow system with *Gliricidia.*

Costs and Benefits Associated with the Systems

Costs associated with the systems include labour costs (for nursery production, weeding, maintenance of the fire-break and harvesting of the wood), land and time. Opportunity costs (the value of production lost due to the use of land and labour in the improved fallow) are also incurred.

Benefits of the systems include increased soil fertility, including improved soil organic matter and water use efficiency. This leads to improved maize production and increased household food security, decreased weeds (including parasitic weeds such as *Striga asiatica* [L.]

Kuntze), production of potential botanical pesticide products (from *Tephrosia*), of forage and browse (both from the trees species and from increased amount of maize stover) and of firewood and building materials, and decreased deforestation from firewood harvesting.

As mentioned above, maize yield increases following a 2-year improved fallow generally last for 2 to 3 years; seed sale is a short-term benefit for some farmers. Box 2 describes the profitability of improved fallows, based on farmer-managed on-farm trials (Franzel et al. 2002).

***Box 2.* Profitability of improved fallows (*Source*: Franzel et al. 2002)**

- Over a 5-year period (2 years in fallow and 3 years in cultivation), a 1-ha improved fallow required 11% less labour than 1 ha of unfertilized maize and 32% less labour than fertilized maize using the recommended rate.
- Improved fallows increase total maize production 87% above unfertilized maize over a 5-year period (even without any yield in the first and second year). Fully fertilized maize yields 2.5 times more maize than improved fallow systems over 5 years.
- Returns per hectare were highest for fertilized maize, followed by improved fallows. Even if the residual maize yield increases in the second and third years after the fallow are not included in the calculations, improved fallows were still more profitable than unfertilized maize.
- Improved fallows gave significantly higher returns to labour than unfertilized maize and slightly higher returns to labour than fertilized maize.
- Improved fallows are much less risky than fertilizer use in times of drought (a common phenomenon in Eastern Province).

Improved fallows have a number of other more intangible and long-term benefits, including reduction of erosion (due to improved soil structure, cover and trees acting as wind breaks), improved labour availability (due to reduced time to collect fire wood) and decreased risk. As evident in Table 2, returns to labour are clearly higher in the improved fallow system than in the continuous cropping system and are comparable to returns in continuously cropped but fertilized maize.

Table 2. Labour inputs and returns to labour in the improved fallow, continuous unfertilized maize and continuous fertilized maize systems[a] (*Source*: Franzel et al. 2002).

Cropping system	Labour input (days)	Maize yield (t ha^{-1})	Return to labour (US$ d^{-1})
Improved 2-year fallow	441	8.8	1.50
Continuous unfertilized maize	499	4.8	0.65
Continuous fertilized maize	645	21.9	1.45

[a] Data are for a 5-year period because the improved fallow system is assumed to have a 5-year cycle, 2 years of fallow followed by 3 years of maize cultivation.

Limitations of the Improved Fallow Systems

The main problems that farmers encounter with the technology are insects (*Mesoplatys* beetles on *Sesbania*, termites) and poor germination. Additional problems identified by farmers include fire, drought, livestock browsing and weeds (Kwesiga and Beniest 1998; Peterson 1999). Other drawbacks include the need to wait 2 years to receive the benefits of the technology, lack of land (especially for older widows), lack of time (especially when taking care of sick family members) and lack of strength to manage an improved fallow (also a problem for older widows and pregnant women). In some cases, farmers have been able to organize and share their nursery tasks amongst themselves to reduce the individual amount of labour required to develop and manage a nursery (Peterson 1999). All four tree species are subject to browsing damage during the dry season and none are fire resistant. Fire management may turn out to be the most serious difficulty associated with the system. Labour requirements, especially of transplanted trees, could also limit the ability of farmers to plant larger improved fallows (Peterson 1999; Franzel et al. 2002).

Each of the improved fallow species is associated with a unique set of problems:

- Several pests and diseases, particularly beetles, nematodes and termites, attack *Sesbania*. However, the degree to which they lower the quality and the impact of already established, improved fallows has not been measured. Insect or fungal attacks in the nursery have severely curtailed the ability of several farmers to plant the desired number of trees. In addition, *Sesbania* has not performed well in some parts of the region where shallow, sandy soils and nematodes are thought to affect production (Kwesiga and Beniest 1998).
- Livestock heavily browse pigeon pea in some areas and insects often attack the seeds and flowers. Seed quality has also been a problem for pigeon pea. As a result, in some areas, farmers have abandoned pigeon pea fallows. However, in areas without livestock populations, pigeon pea remains a viable improved fallow species.
- *Gliricidia* seeds are expensive to obtain and establishment by cuttings has not proven successful on station. There are few serious production or performance problems with *Gliricidia* fallows. However, the trees must be pruned in a timely fashion during the cropping season to avoid competition with adjacent crops. *Tephrosia* has fewer pest problems, possibly because of the insecticidal properties of the plant.

Seed Availability and Exchange

Seed supply may be a problem mainly during the early expansion stage, which requires special means for supply, such as contracted seed producers, or buying large quantities of seed from outside suppliers. Other problems may emerge after the expansion stage if, for example, a certain provenance of a species becomes susceptible to pests/diseases and needs to be substituted by a new one.

In eastern Zambia, farmers, research institutions and NGOs maintain the availability of improved fallow seeds. Although a few farmers have established seed banks, most farmers rely on natural regeneration of coppiced trees, border plantings and seeds produced in their own improved fallows, or those of neighbours, to plant additional fallows. In the early stages of technology dissemination, the Zambia-ICRAF team provided free seed to all interested parties. This seed was either purchased from farmers who participated in earlier on-farm trials, or produced on station or imported from other sites. In later dissemination efforts, the World Vision International Project is distributing seed on a loan basis, that is, farmers are required to pay back twice the amount they receive. Various partners in Zambia produce the seeds that local organizations provide. Seed is not currently sold among farmers or by private vendors. Often, during farmer trainings, demonstrations and farmer-to-farmer exchanges, farmers will exchange seeds informally amongst themselves.

Fortunately, *Sesbania*, *Tephrosia* and pigeon pea pose no great problems. They produce prolific seed early in improved fallows. For *Gliricidia*, the situation is worst because the normal management approach generally precludes seed production, it is a poor seed producer in many environments and its seed often has poor storability and very low viability (15% in some cases!).

In addition, seed quality may become a problem in the future. The problems are threefold. First, the introduction of a narrow genetic base may lead to reduced adaptive capacity to change with farmer requirements and environmental conditions, leading to lower sustainability of production. Second, multiple provenances of *Sesbania* and *Tephrosia* have been used without an understanding of the interaction between provenances, which in fact may be negative and lead to reduced sustainability. This necessitates that seeds are acquired from a known and well-documented provenance, and from a reliable source. Third, it is not clear whether seed production and distribution systems will evolve and how sustainable they would be. Whereas seed likely will be available in the main areas where improved fallows are planted, it is not clear how the germplasm will move to new areas.

USE AND ADOPTION BY FARMERS

Diffusion Efforts

Improved fallows are still a fairly recent introduction in the area. Although on-farm trials began as early as 1992, large-scale dissemination of the technology did not begin until 1996. On-farm trials with *Gliricidia* did not begin until 1997.

The improved fallow system has been actively promoted by an adaptive research and dissemination network of farmers, NGOs, research organizations and government extension personnel (Katanga et al. 2002). The main institutions involved in promoting improved fallows include:

- Farmers' clubs and cooperatives, women's clubs and individual farmers;
- Government of Zambia/Ministry of Agriculture, Food and Fisheries/Department of Field Services, the Department of Land Management, and the Research Branch;
- ICRAF, partially funded in the Province by the Swedish International Development Agency (SIDA) and the Canadian International Development Agency (CIDA);
- World Vision International, partially funded by the United States Agency for International Development (USAID);
- The University of Florida, funded by USAID;
- Soil Conservation and Agroforestry Extension Program, funded by the government of Zambia and SIDA;
- The Reformed Church of Zambia;
- Lutheran World Federation;
- The Finnish Volunteer Services;
- The Catholic Church of Zambia; and
- The Baptist Church of Zambia.

The network's members in the area meet twice yearly to review progress, discuss problems, plan training activities, arrange for seed distribution and develop training materials. Many local NGOs share seeds and extension materials and benefit from the critical mass of experienced improved fallow farmers that have developed in the region. Biannual planning and monitoring and evaluation meetings involving all stakeholders have been critical to the success of the extension campaign.

Network members are involved in both on-farm research and extension. The main diffusion strategies used so far include on-farm trials, farmer-to-farmer training and farmer field days, demonstration fields, the development of farmer trainers (lead farmers) and farmer field schools. Other methods used include rural radio, community theatre, songs, video, community meetings, on-station field days, formal training sessions (mobile courses and residential courses) and handing out of leaflets, brochures, newsletters, T-shirts and hats.

In several ways, the technology development and dissemination process for improved fallows has been particularly successful in Eastern Province. From the beginning (in 1992, after only 5 years of on-station testing), researchers took their trials on farm, listened to farmers' valuations of the technology, encouraged farmers to innovate and modified the technology based on farmers' recommendations and innovations. As the technology evolved and improved, more and more farmers were involved in the testing process, which created a critical mass of experienced farmers ready and waiting to teach new farmers about the technology. Another key element contributing to the success was the wide range of options available for farmers to try. These options included different species, planting methods and length of period of fallow. Some of these options originated with researchers, whereas farmers discovered others. Local facilitators helped spread information about the wide range of improved fallow options.

In addition, the ICRAF team included a technology dissemination specialist to test and adapt innovative extension pathways and processes that helped the improved fallow program to reach more farmers more effectively. The dissemination specialist worked with local NGOs and district officials to reach women's groups in the area. At all stages of the development and dissemination process, ICRAF staff worked with government officials and local extension agents, providing them with training in the technology and developing critical problem-solving skills.

Problems encountered in disseminating improved fallows include:

- Drought in the early years of technology development, which reduced tree nursery production and survival;
- Lack of farmer, scientist and extension agent experience with, and information on, improved fallow technologies and species;
- Low morale of government extension staff and lack of operational funds, transportation and material support for their activities; and
- Shortage of seed – particularly of *Gliricidia* and pigeon pea.

Patterns of and Factors Impacting Adoption

In 1996, the ICRAF on-farm research team began collecting adoption data. These initial adoption surveys indicated that about 50% of those farmers who tested improved fallows with ICRAF planted a second improved fallow on their own and that women were initiating an increasing number of on-farm trials (rising from 22% of the 113 trials during the 1994-95 season to 49% of the 657 trials in 1995). In a more recent survey by World Vision and the University of Florida (Peterson et al. 1999), 66% of farmers in the region were aware of improved fallow technologies, while 52% were aware of other green manure technologies and 94% were aware of crop rotations. In all, 7% had tried the technology. Of those who tried the technology, 23% adopted it, that is, planted a second improved fallow. Keil (2001) surveyed 100 farmers who had first planted improved fallows during 1996-97 or earlier. By 2000-01, 71% had adopted, 23% had abandoned the practice and the status of the remaining 6% could not be determined (Table 3).

Table 3. Categories of respondents regarding the planting of improved fallows (n = 100) (*Source*: Keil 2001).

Category	Percentage of farmers
Planting improved fallows regularly without outside support	49
Planting improved fallows regularly with outside support (seeds)	22
Stopped planting improved fallows	23
Uncertain status	6

Some differences occur in the adoption patterns of women and men (Donald Phiri and Steve Franzel, ICRAF internal communication, 1996; Franzel et al. 2002):

- Men tend to have bigger plots than women (649 m^2 vs. 326 m^2).
- Women tend to plant direct-seeded *Tephrosia* more than men do (41% vs. 24%).
- Women have a higher survival rate for *Sesbania* than men have (47% vs. 29% had survival rates above 75% of trees planted).
- More men than women had seen ICRAF improved fallow trials at Farmer Training Centres (39% vs. 25%).
- More women than men chose to intercrop their trees with maize (42% vs. 37%).

The main users of improved fallow systems today are smallholder farmers who have limited means available to them to improve their soil fertility. Most of these farmers use little or no fertilizer and their maize yields are quite low (averaging less than 1000 kg ha^{-1}). They are farmers who believe that low soil fertility is the main agricultural production constraint and believe that adequate maize production is not possible without the use of some sort of fertilizer (Peterson 1999; Peterson et al. 1999). Both testers and non-testers of the technology come from a wide range of age groups and ethnic groups and have differing education levels and life experiences. However, improved fallow testers have a tendency to be middle-aged (30-45 years old), relatively well-off, active club members, who work closely with their camp extension officers. They often have access to animal traction, cultivate larger land areas and have bigger households than other farmers. Women testers tend to be unmarried (either single, widowed, or divorced) members of female-headed households because they have more authority to make their own land and labour use decisions than women in male-headed households.

Although well-off farmers are more likely to test the technology, it is the poorest farmers (especially women in female-headed households) who are most likely to adopt it. When given the opportunity, both men and women appear to be adopting the technology in similar proportions (Peterson 1999; Franzel et al. 2002). In four villages where improved fallows had been widely disseminated, 32% of the male-headed and 23%

of the female-headed households had planted them, with no significant difference between the two. On the other hand, there was a strong association between wealth level and testing improved fallows, as 53% of the well off, 40% of the fairly well off, 22% of the poor and 16% of the very poor farmers tested improved fallows (Phiri et al. 2003). Keil (2001) found that among farmers who had tested the practice, the very poor, the poor and the fairly well off were more likely to adopt improved fallows than the well-off farmers. This tendency probably reflects the fact that well-off farmers can afford mineral fertilizer and, after testing improved fallows, many of them decided to use fertilizer instead.

People who are not using the technology tend to be very old or very young, poor, lack access to animal traction, cultivate smaller land areas and have smaller households. In some cases, they are new residents of an area, who have not yet established their farming patterns (Peterson 1999).

Generally, farmers who abandon improved fallows have experienced a change in their resources, which alters either their motivation or their ability to plant another improved fallow. For example, some farmers are suddenly able to acquire fertilizer from extended family members or from the sale of livestock and as a result are no longer interested in planting an improved fallow. They would rather use a faster, easier soil fertility amendment. For many farmers, having to wait 2 years to realize the benefits of the technology is a real constraint. Some farmers decide to use their land to produce cash crops, such as cotton (which comes with a fertilizer package) or groundnuts (which do not require fertilizer), rather than plant another improved fallow (Peterson 1999). Although original extension efforts concentrated on *Sesbania*, because of its easier establishment or labour requirements, more farmers now are planting *Tephrosia*. For example, in 2000-01, over 11 000 of the 14 395 improved fallows facilitated by the World Vision Project were planted to *Tephrosia* (Table 4).

Table 4. Improved fallows species planted by the farmers who were facilitated by World Vision Integrated Agroforestry Project during the 2000-01 cropping season (*Source*: World Vision Integrated Agroforestry Project data).

Men	Women	Farmer groups	Total no. of fallows
7077	4360	32	11469
1261	294	13	1698
511	457	4	972
202	48	6	256
9051	5159	55	14395

Farmers whose original improved fallows failed also appeared less likely to continue or to expand their use of the technology. Termites, fire, beetles, drought and livestock browsing were the most commonly mentioned causes of technology failure. Poor nursery management was also sometimes an important problem. However, in some cases, farmers who were not successful with their first improved fallows continued to cultivate improved fallows because, after visiting other farmers, they were convinced that the technology works. Other constraints to the continued practice of the technology include lack of seeds or seedlings to plant another improved fallow (this could also be due to lack of information on how to manage current fallows for seed production), poor health and lack of additional land (Peterson 1999).

THE WAY FORWARD

Critical Factors Affecting the Future of the System

Currently, farmers in eastern Zambia have the land, labour and extension support required to adopt improved fallows. In addition, they have few other viable options for enhancing soil fertility. The application of sufficient quantities of mineral fertilizers to maize is beyond the means of most small-scale farmers in the region and without fertilizer or large quantities of manure they cannot produce adequate quantities of maize. Thus, for many of these farmers, the benefits (particularly food) of adopting improved fallows exceed the costs (particularly labour and land).

In 1998, ICRAF project staff assessed the critical factors affecting the adoption potential of improved fallows in the area (Franzel et al. 2002). They evaluated the feasibility of the technology (the degree to which farmers are able to manage it on their own), profitability, degree of farmer interest and the institutional support available to assist the spread of improved fallows. They determined that the technology is feasible, because farmers (including poor women farmers) are able to plant and maintain improved fallows on their own, with survival rates exceeding 50%. They also determined that the technology is profitable. Using data from farmers' fields, the returns to labour and land for farmers cultivating improved fallows for 2 years and maize for 3 years were double those from continuously cultivated maize fields, and slightly higher than those from fully-fertilized maize fields. The degree of farmer interest was evaluated based on the vast increase in the number of farmers interested in trying the practice, their continued planting of improved fallows after initial testing and institutional support, which has been provided by the informal agroforestry network established in Eastern Province since 1996. Thus, the study concluded that prospects for a continued adoption and use of the system were good.

In addition, researchers from the University of Florida have conducted ethnographic studies to identify the critical factors that have motivated and constrained farmers in the region to test and adopt improved fallows. Their research indicates that poor soil fertility, high fertilizer prices, awareness of the technology and visual verification of its efficacy are motivating factors which encourage farmers (both men and women alike) to try improved fallows in the region. Willingness to wait 2 years for the benefits of the technology is a prerequisite to testing. Limited access to time and labour to manage the technology and lack of access to seeds, knowledge or land are constraints that limit the number of farmers who adopt the technology. Some farmers felt they would rather use other soil fertility techniques, such as mineral fertilizer or manure, than plant improved fallows (Peterson 1999; Peterson et al. 1999).

These findings, combined with the findings from ICRAF, indicate that future prospects for the continued expansion of the system in Eastern Province are reasonably good, provided that extension efforts continue and fertilizer prices remain at present levels. Currently, several new programs are involved in the widespread dissemination of the technology and by 2001-02 over 20 000 farmers in Eastern Province had planted improved fallows.

Factors with particular impact on the future adoption of improved fallows are discussed below.

Fertilizer Prices

If the price of mineral fertilizer increased, more farmers might be motivated to use improved fallows, perhaps not only on their maize fields but also on their cash crops and gardens. However, if the price of fertilizer decreased, many farmers who have not yet tried improved fallows might not be motivated to do so and many current practitioners would probably abandon the technology. Whereas farmers in male-headed households might switch to mineral fertilizers if prices decreased, women farmers probably would not, unless the decrease was significant - making the price of fertilizer per bag equal to the price of a chicken, a decrease of 75% (Peterson 1999). Interestingly, some farmers have indicated that, even if the price of fertilizer went down, they would continue to plant improved fallows to reduce risk ('because you never know what fertilizer prices will do in the future') and because the benefits of improved fallows last more than one season. In addition, some farmers do not like mineral fertilizers and believe they harm the soil and create dependence (Peterson 1999).

Income-generating Opportunities

If improved marketing opportunities for maize resulted in increased prices for the crop, farmers might be motivated to increase their maize production, which would require (and allow for) the increased use of mineral fertilizers. If additional cash crop markets opened up for which farmers were given fertilizer on credit, farmers might shift some land out of maize or require more of their land and labour to produce cash crops, either of which could result in decreased demand for improved fallows. In addition, increased income-generating opportunities could provide farmers with enough disposable income to purchase all the mineral fertilizer they need, in which case many farmers might abandon the practice.

Livestock Resources

Currently, few farmers have a sufficient number of livestock to effectively apply manure to their maize fields. If they could build up their livestock resources, they might be motivated to abandon improved fallows. However, several factors impede farmers from building up their herd: theft, animal diseases, as well as the cost of ox carts or ploughs (required to facilitate manure transport and to cultivate fields where manure has been applied) are major constraints to livestock production and the use of manure.

Seed Availability

Seeds for all species are available from local sources, at no cost to farmers, either from extension personnel or from other farmers. However, one of the main positive attributes of the improved fallow system is that little if any cash is required to adopt the technology. If seeds become a marketed product, it may become more difficult for farmers to adopt the system. The most sustainable solutions may be ones where seed is valued in ways that require no money, such as seed-lending schemes.

Research Needs

The Zambia-ICRAF team has embarked on a number of research efforts and surveys in order to understand and improve the technology. It has designed and participated in basic biophysical assessments, species screening trials, forage trials, impact assessments, wealth ranking exercises, adoption and household surveys, monitoring surveys and labour studies (see References). In addition, it has continuously consulted with farmers informally to evaluate their interest in, and problems with, the technology. Biannual planning sessions with local NGOs and farmer trainers have also added to an understanding of, and contributed to, improvements in the technology.

In addition, the University of Florida, under a Collaborative Research Support Program funded by USAID, has conducted research in the Eastern Province of Zambia (and Malawi), to map the decision-making criteria farmers use when deciding whether or not to test and adopt improved fallow technologies. Researchers and extension agents have used these criteria, developed into ethnographic decision trees, to design and test more appropriate and successful improved fallow systems and extension programs, and policy-makers have used them to help improve the soil fertility options available to farmers. University researchers have also developed linear programs modelling household resource use and production constraints in the region, to further develop and improve appropriate farming technologies (see References).

The Ministry of Agriculture's Adaptive Research Planning Teams are always seeking ways to develop and improve farmers' soil fertility options. They have conducted informal surveys on degraded lands in eastern Zambia, reviewed the extent and causes of the hungry period in Eastern Province, classified the farming systems in the Province and evaluated the use and role of *dambos* (low-lying, seasonally wet lands) in the area. Although none of their research has evaluated the improved fallow system directly, it has provided critical information that has improved understanding of the farming systems in which the technology is used. Several local projects[2] have also conducted baseline research in the region that has contributed to our understanding of farmer priorities, household constraints and resources.

Despite the significant amount of research already conducted on improved fallows, a number of biophysical and socio-economic research issues need to be addressed in the future. Biophysical research issues include (Kwesiga and Beniest 1998):

- Litter and foliage decomposition rates and nutrient release in the system,
- Changes in soil chemical and physical properties under the system,
- Inoculant interactions in the system,
- Selection of improved fallow species for specific site conditions and
- An evaluation of single species and mixed species fallows.

Socio-economic research questions include:

- For what types of farmers is the technology most suitable?
- What are farmers' reasons for rejecting the technology?
- How does the labour situation on a farm affect adoption, and what are the real economic returns to farmers from different improved fallow systems?
- Is there a way to reduce the amount of labour required to practice improved fallows?
- Which socio-economic or policy factors facilitate or limit large-scale adoption of the technology? How can grazing and fires be controlled?
- What changes in infrastructure will facilitate adoption?
- How can training and extension contribute to the dissemination of the technology?

Other organizations have suggested that further research is needed into the potential viability of using *Tephrosia* as an insecticide to promote small enterprise development in the area and that ways to increase tree survival rates and growth on-farm also need to be explored (Franzel et al.

2002). Scientists and farmers also need to determine the best seed provenances for particular sites and particular problems to evaluate how the technology responds in different environments, assess the interaction of inorganic fertilizers with the technology, determine the optimal use of litter for subsequent crops, devise ways to protect fallows from animals and fire, evaluate alternative ways to space the various species and determine which crops benefit the most from the technology.

All of this research will take time and money and require a coordinated effort on the part of farmers, extension staff, research organizations and local NGOs. While some of these questions are being addressed by current research, others (especially some of the policy questions) have yet to be addressed (Franzel et al. 2002).

In conclusion, the improved fallow system developed in eastern Zambia is an innovative system combining the best of current research with local innovation and adaptation. By involving farmers in the research and dissemination process and working closely with local and international partners, the Zambia-ICRAF team was able to both develop and disseminate the technology effectively and efficiently. Both the improved fallow technology itself and the collaborative mechanisms used in its development serve as examples and models, which can be used by others in the field. It will be interesting to note how the systems change with changing economic, social and political conditions in the country in the future.

NOTES

1. A full adopter could only be expected to plant one quarter of his maize area to improved fallows in a single year, assuming a 4-year cycle of the technology, that is, 2 years in fallow and 2 years in production. The index of adoption is thus the area planted to improved fallow in a given year as a proportion of one quarter of the maize area.
2. The local projects are the Soil Conservation and Agroforestry Extension (SCAFE), World Vision, Co-operative League of the USA (CLUSA), Economic Expansion, Lutheran World Federation (LWF), District Councils with the support of the United Nations Development Programme (UNDP)-Rural Roads, and the Luangwa Integrated Rural Development Project.

REFERENCES

Celis R, Milimo J T, Wanmali S (Eds.) (1991) Adopting improved farm technology: A study of smallholder farmers in Eastern Province, Zambia. Rural Development Studies Bureau-University of Zambia, National Food and Nutrition Commission, Eastern Province Agricultural Development Project and the International Food Policy Research Institute (IFPRI), WA, USA. 232 p.

CSO - Central Statistics Office (1992) Country profile. Lusaka, Zambia.

FEWS - Famine Early Warning System (1999) Zambia monthly report. Associates in Rural Development and United States Agency for International Development (USAID), WA, USA.

Franzel S, Phiri D, Kwesiga F (2002) Assessing the adoption potential of improved tree fallows in eastern Zambia. *In:* Franzel S, Scherr S J (Eds.) Trees on the farm: Assessing the adoption potential of agroforestry practices in Africa. CAB International, Wallingford, UK. p. 37-64.

Katanga R, Phiri D, Böhringer A, Mafongoya P (2002) The Adaptive Research and Dissemination Network for Agroforestry (ARDN): A synthesis of the adaptive workshops. *In:* Kwesiga F, Ayuk E, Agumya A (Eds.) Proceedings of the 14th southern Africa regional review and planning workshop, 3-7 September 2001, Harare, Zimbabwe. Southern Africa Development Community (SADC)- International Centre for Research in Agroforestry (ICRAF) regional programme, Harare, Zimbabwe. p. 93- 99.

Keil A (2001) Improved fallows using leguminous trees in eastern Zambia: Do initial testers adopt the technology? MSc Thesis, Faculty of Agriculture, Institute of Rural Development, Georg-August University, Goettingen, Germany. 94 p.

Kupanda G (1996) Chadiza district profile. Chadiza District Council, Eastern Province, Zambia.

Kwesiga F, Beniest J (1998) *Sesbania* improved fallows for eastern Zambia: An extension guideline. International Centre for Research in Agroforestry (ICRAF), Nairobi, Kenya. 55 p.

Kwesiga F, Franzel S, Place F, Phiri D, Simwanza CP (1999) *Sesbania* improved fallows in eastern Zambia: Their inception, development and farmer enthusiasm. Agrofor. Syst. 47: 49-66.

Ngugi D (Ed.) (1988) Agroforestry research project for the maize-livestock system in the unimodal upland plateau (Chipata and Katete) in Eastern Province of Zambia. Agroforestry Research Networks for Africa (AFRENA) Report No. 10. International Centre for Research in Agroforestry (ICRAF), Nairobi, Kenya.

Peterson J (1999) Kubweletza Nthaka: Ethnographic decision trees and improved fallows in Eastern Province. A report submitted to the Collaborative Research Support Programme on Gender and Soil Fertility in Africa. The University of Florida and the International Centre for Research on Agroforestry (ICRAF), FL, USA. 130 p.

Peterson J, Tembo L, Kawimbe C, Mwang'amba E, Jere I, Tembo M (1999) The Zambia integrated agroforestry project baseline survey: Lessons learned in Chadiza, Chipata, Katete and Mambwe Districts, Eastern Province, Zambia. University of Florida Collaborative Research Support Programme, Gainesville, FL, World Vision International, WA and the United States Agency for International Development (USAID), Lusaka, Zambia. 71 p.

Phiri D, Franzel S, Mafongoya P, Jere I, Katanga R, Phiri S (2003) Who is using the new technology? The association of wealth status and gender with the planting of improved tree fallows in Eastern Province, Zambia. Agric. Syst. In Press.

Raussen T (Ed.) (1997) Integrated soil fertility management on small-scale farms in Eastern Province of Zambia – a review of current knowledge and experiences for agricultural extension workers. Regional Soil Conservation Unit Technical Handbook No. 15, Nairobi, Kenya, and Zambia Ministry of Agriculture, Food and Fisheries-Department of Research and Specialist Services, Chipata, Zambia. Available from the Zambia- International Centre for Research on Agroforestry (ICRAF) project. 174 p.

Zambia-ICRAF - International Centre for Research on Agroforestry (1998) Annual report. Chipata, Zambia.

ZARD-Zambia Association for Research and Development (1996) Zambia today: A gender perspective. Analysis of the public's views. Co-op Printing, Lusaka, Zambia.

Chapter 6

Best-Bet Legumes for Smallholder Maize-Based Cropping Systems of Malawi

R. A. Gilbert

University of Florida, USA

SUMMARY

Malawi's burgeoning population and dependence on maize (*Zea mays* L.) as a staple food had led to continuous maize cultivation and declining soil organic matter levels. Devaluation of the Malawi Kwacha and removal of fertilizer and maize price subsidies have contributed to increased interest in legumes as sources of human nutrition and soil fertility in the N-depleted soils of Malawi. However, the intense land pressure means that any leguminous system must be competitive with continuous maize in terms of calorie production per hectare and economic net benefits. Legumes that can satisfy economic and food security criteria while still improving soil fertility are termed 'best-bet' systems because smallholder maize producers are more likely to adopt them.

Several best-bet leguminous systems have been identified in Malawi based on these criteria: grain legume-maize rotations with groundnut (*Arachis hypogaea* L.) or promiscuous soybean (*Glycine max* [L.] Merr.), legume/maize intercrops with pigeon pea (*Cajanus cajan* [L.] Millsp.) or fishbean (*Tephrosia vogelii* Hook. f.), and velvet bean (*Mucuna pruriens* [L.] DC.)- maize rotations. Recent research in Malawi indicates that grain legume rotations with CG7 groundnut and Magoye soybean (a promiscuous variety able to fix atmospheric N_2 without inoculation) are superior to continuous unfertilized maize in terms of calories produced and economic net benefits (on a cash basis). Constraints to increased adoption of these grain legumes include seed costs for groundnut, fluctuating market prices for soybean and practical difficulties with residue incorporation for both species. Pigeon pea and *Tephrosia* both exhibit a temporal complementarity of resource use with maize because of their use of residual soil moisture. Thus they are well suited to intercropping systems because maize yield is not reduced. Disadvantages

M. Eilittä et al. (eds.), Green Manure/Cover Crop Systems of Smallholder Farmers, 153-174.

for pigeon pea cultivation include a poor seed yield for long-duration pigeon pea in central and northern Malawi and livestock damage in northern Malawi. *Tephrosia* is not browsed but does not have the food value of pigeon pea. *Mucuna* has consistently generated a higher seed yield and total biomass than other annual legumes studied in Malawi. However, if *Mucuna* is managed as a grain legume rather than a green manure, the seeds must be processed extremely carefully for human consumption.

No single leguminous cropping system will be a panacea for a country as agro-ecologically diverse as Malawi. Much recent research and extension effort has gone into comparisons of best-bet legumes that present farmers with as many options as possible so that they can choose cropping systems best suited to their needs. If challenges such as improved human utilization and increased market demand are met, these legumes will play an important role in diversifying Malawian farming systems and ensuring improved household food security.

INTRODUCTION

Malawi is only 118 000 km^2 in area, yet has a very diverse agro-ecology, with 55 natural regions (Benson 1997a). The altitude in agricultural areas varies from 0 to 2000 masl, with average annual precipitation from 600 to 2000 mm (Figure 1). Figure 2 shows the location of the study area. The varied terrain and soil type in hilly areas make uniform soil fertility recommendations impractical and much recent research and extension effort has gone into using geographic information systems (GIS) to generate area-specific recommendations for fertilizer application and organic matter technologies. The precipitation pattern is unimodal, with 4-6 months of rain followed by 6-8 months of drought. High variability of precipitation both within and between growing seasons is typical of southern Africa and makes rain-fed agriculture a risky proposition. The long drought period also makes double or relay cropping of legumes with maize (*Zea mays* L.) problematic, because dry season growth and survival is poor for most species.

Like many African countries, Malawi's burgeoning population (overall population density of 93 people km^{-2}) has led to decreased fallow periods, stagnant food production and declining food production per capita (Figure 3). However, Malawi is unique in its dependence on maize as its staple food crop. Over 90% of total cultivated area is planted to maize, mostly by resource-poor smallholders. Malawians consume more than 150 kg maize $capita^{-1}$ y^{-1}, constituting less than two-thirds of their caloric consumption—the largest per capita consumption in the world (Smale and Heisey 1997). There is evidence that soil organic matter (SOM) is declining as soils are continuously cropped to maize. Mean organic carbon in three regions declined 10-31% over a 20-year period

(Blackie et al. 1998). Devaluation of the Malawi Kwacha (MWK), along with removal of fertilizer and maize price subsidies, has led to a rising fertilizer:maize price ratio and has made fertilizer use on maize largely uneconomic for market sale (Benson 1997b).

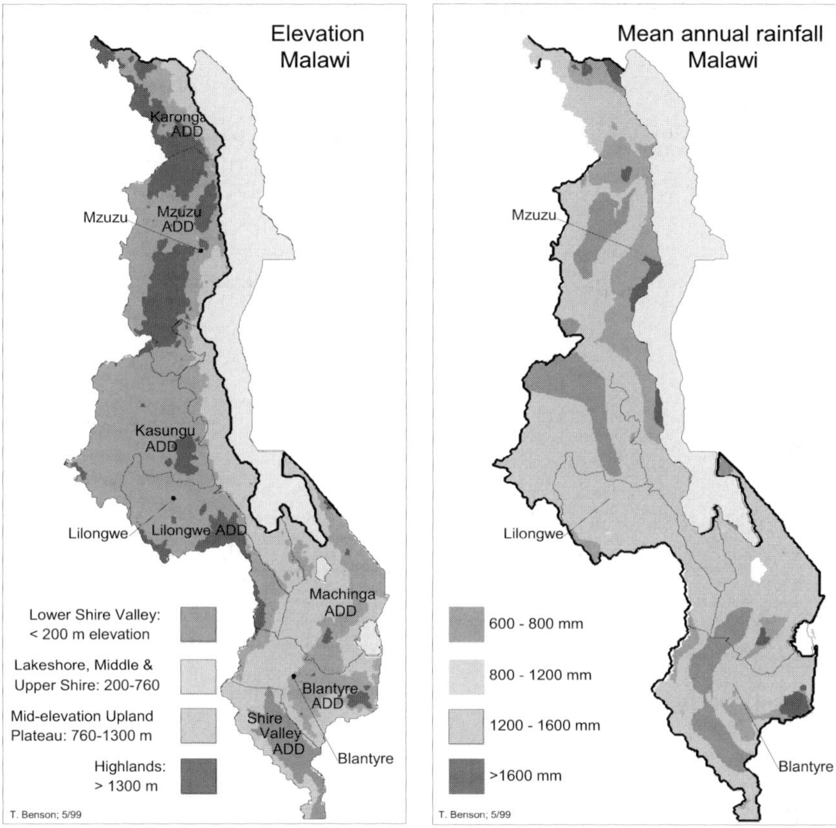

Figure 1. Elevation and mean annual precipitation of Malawi (ADD = Agricultural Development Division).

Maize cultivation is of paramount importance to Malawian smallholders. The practice throughout the country is to plant maize on ridges 90 cm apart. This practice was introduced during the colonial period to reduce soil erosion in the hilly terrain. The ridges are split and remade each year (last year's furrow becomes next year's ridge), keeping the 90-cm spacing. Ridge spacing is generally not changed to accommodate legumes (e.g. to 75 cm for groundnuts, *Arachis hypogaea* L.) because of the extra labour needed to re-make subsequent ridges at 90 cm for maize. Thus many legumes are planted at suboptimal densities or row arrangements.

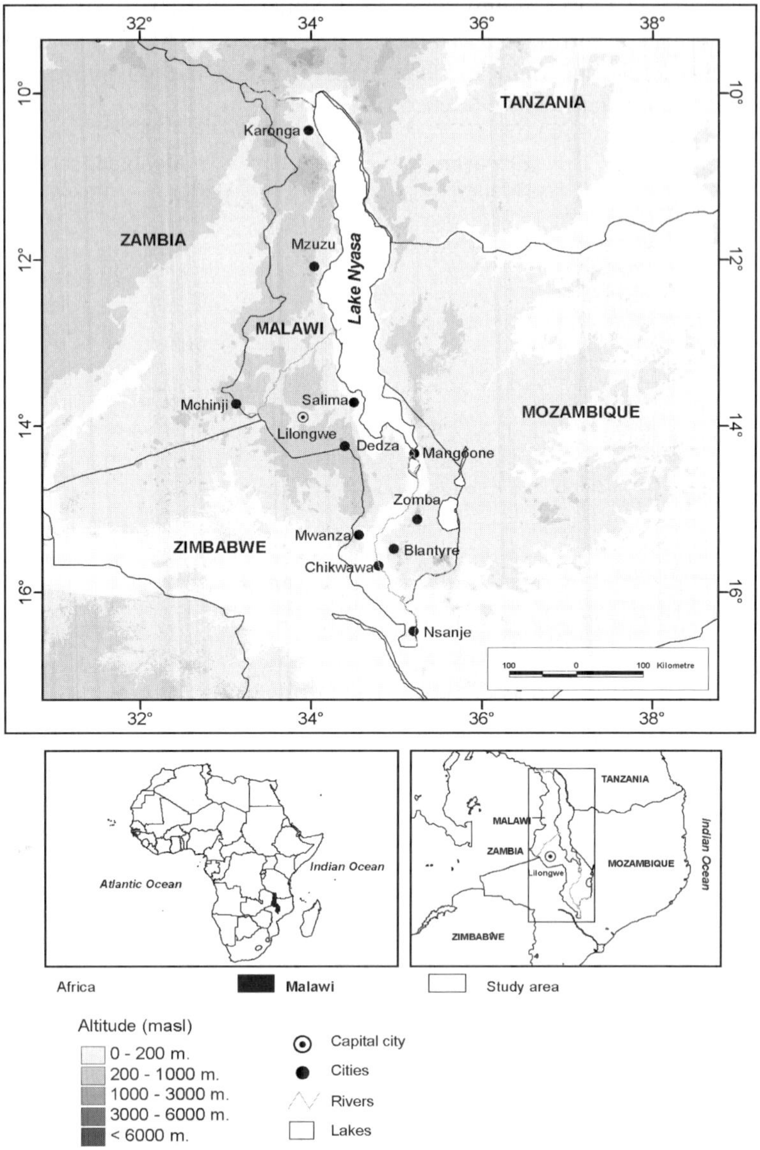

Figure 2. Location of study area, Malawi.

Legumes grown in association or rotation with maize vary with agro-ecology. For example, groundnut is generally planted in hotter, drier and lower elevations near Lake Malawi (also known as Lake Nyasa) while beans (*Phaseolus vulgaris* L.) are usually grown in the cool humid highlands. Livestock density tends to be greater in northern Malawi where human population densities are lower. Livestock are allowed to graze freely in the dry season in northern Malawi (they are tied throughout the year in southern Malawi) and can cause extensive damage to legumes, such as pigeon pea (*Cajanus cajan* [L.] Millsp.), which remain green in the dry season. This serves as a disincentive to adoption of long-duration legumes in the north.

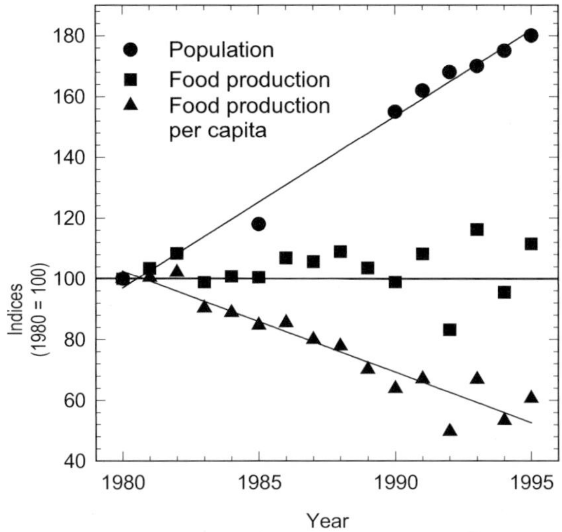

Figure 3. Population and food production trends in Malawi, 1980-95 (*Source:* FAO Production Yearbook, various years).

The socio-economic and biophysical context of Malawi has important implications for leguminous cropping systems. Farmers are searching for ways to ameliorate soil fertility that reduce the need for fertilizers and this provides an important entry point for legumes. However, because of the intense land pressure, any leguminous system must be competitive with continuous maize both in calorie production per hectare and in economic net benefits.

Table 1 lists tentative criteria for success and adoption of leguminous systems in Malawi. The selection criteria for calorie production are based on a standard of continuous maize, which is the most common cropping practice in Malawi. Land Equivalent Ratios are used to compare the maize yield in intercropping systems to sole maize yield. Smallholder farmers in Malawi are chiefly concerned with maize yield, thus leguminous intercropping systems must not reduce maize yield more than 20-30% in the intercrop and must supply greater total maize and legume calories over time. Legume rotational systems also must provide more calories than continuous sole maize. Total calories produced are an extremely relevant criterion in the Malawian context because 65% of the population falls below a poverty line consumption level of 7.7-25.4 MWK d^{-1} (US\$0.30-US\$1.00 d^{-1}) (Benson 2001). The per capita daily calorie requirement is not being met in many of these households and 57-61% of children below the age of 5 are stunted (Benson 2001), a sign of chronic malnutrition.

Economic criteria for leguminous systems are as important as food security criteria. Economic net benefits must exceed those of maize or leguminous systems will not be adopted. Grain legumes such as groundnut, soybean (*Glycine max* [L.] Merr.) and pigeon pea already have established if fluctuating markets for seed. For green manure crops such as fishbean (*Tephrosia vogelii* Hook. f.), and *Mucuna pruriens* (L.) DC., seed market value is low and thus soil fertility benefits and subsequent maize yields must be higher.

Snapp (1998) summarized the soil nutrient status of smallholder farms in Malawi. Soil organic C was > 0.8% in over 75% of the samples analyzed but significantly lower than in non-cultivated soils, buttressing observations of N deficiency documented throughout Malawi. Phosphorus deficiency was locally important but > 60% of all soil samples had P levels above the value of 15 mg kg^{-1}, sufficient for smallholder maize production.

Since N is the most limiting soil nutrient in Malawi, the criterion for sustainability of leguminous cropping systems is based on net N addition to the soil. This criterion is especially stringent for green manure crops because their seed has a lower economic value. The minimum N added to generate a significant maize yield response is 20 kg N ha^{-1}. The maize yield response from 20 kg N ha^{-1} added from legumes is not equivalent to the same amount added as inorganic fertilizer and does not imply fertilizer equivalency values (Carsky et al. 1999). The N additions listed here of 20-80 kg N ha^{-1} are based on realistic estimates of biomass production and N_2 fixation for these species.

This chapter describes the best-bet leguminous systems identified so far, based on these criteria, and discusses future research needs for leguminous cropping systems in southern Africa.

Table 1. Criteria for the success of leguminous cropping systems in Malawi.

System	Malawian example	Caloric criteria[a]	Economic criteria	Soil fertility criteria
Grain legume - maize rotation	1. CG7 groundnut 2. Magoye promiscuous soybean	• Caloric output > continuous maize	• Economic net benefit > continuous maize • Market for legume seed	• N addition > 20 kg ha^{-1} y^{-1}
Maize-grain/legume intercrop	Maize-pigeon pea	• Maize PLER > 0.7 • Total LER > 1.2 • Caloric output > sole maize	• Economic net benefit > sole maize	• Net increase in soil N and organic matter over time
Maize/green manure intercrop	Maize-*Tephrosia vogelii*	• Maize PLER > 0.8 in year 1 • Maize PLER > 1.4 subsequent seasons	• Cost of N produced < equivalent cost of urea • Economic net benefit > sole maize	• N addition > 60 kg N ha^{-1} (>3 t ha^{-1} biomass produced)
Mucuna pruriens - maize systems	*M. pruriens* - maize rotations	• Maize yield following *Mucuna* > 1.5 times continuous maize • *Mucuna* seed yield ≥ unfertilized maize	• Economic net benefit over 2 years > continuous maize • Market for *Mucuna* seed	• N addition > 80 kg ha^{-1} y^{-1} (>4 t ha^{-1} biomass produced)

[a] PLER = Partial Land Equivalent Ratio, the ratio of maize yield in an intercropping system to sole maize yield. A PLER of 0.7 means a 30% maize yield depression in the intercrop. LER = Land Equivalent Ratio, the sum of the PLERs for all crops involved in an intercropping system.

LEGUMINOUS BEST-BET SYSTEMS

The leguminous systems that are presently being promoted and researched on a large scale by the Malawian government fall into three categories: grain legume rotations, maize/legume intercrops and *Mucuna* rotations. Five systems have been found to satisfy the criteria of Table 1.

Two grain legume rotations:
- Groundnut - maize
- Soybean - maize

Two legume/maize intercrops:
- Pigeon pea/maize
- *Tephrosia*/maize

Mucuna - maize rotation

Much of the research on these systems in Malawi is generally conducted under the aegis of the Ministry of Agriculture and Irrigation (MoAI). In 1995, the MoAI set up the Maize Productivity Task Force (MPTF) to increase productivity of maize-based cropping systems (Rukuni et al. 1998). The MPTF was funded by the World Bank, the European Union and The Rockefeller Foundation and divided into four Action Groups:

I. Inorganic fertilizer and integrated nutrient management
II. Seed supply and distribution
III. Marginal farmers-extension
IV. Organic matter technologies

The MPTF's mandate was to coordinate research and diffusion efforts among MoAI, International Agricultural Research Centres (IARCs), non-governmental organizations (NGOs), industry and the donor community. Action Groups (I), (II) and (IV) have carried out research and extension efforts on the leguminous best-bet systems. While there are numerous local efforts to diffuse legumes through various NGOs, the MPTF provided an institutional context of research and diffusion at a national scale. Several of the large-scale MPTF research and extension projects involving legumes are discussed in this chapter. The best-bet leguminous systems are described in more detail below.

Grain Legume-Maize Rotations with Groundnut or Promiscuous Soybean

The rotation of maize with grain legumes such as soybean or groundnut is one of the more promising technological options available for Malawian farmers. The yield benefits of grain/legume rotations are well known and are caused by the following biophysical factors (Karlen et al. 1994):

- Leguminous N_2 fixation leading to residual N for maize;
- Reduced pest and disease pressures (e.g. reduced *Striga asiatica* [L.] Kuntze incidence on maize with the use of trap crops such as groundnut); and
- Incorporation of residues leading to improved soil quality (including increased SOM, water infiltration and aggregate stability), which reduces soil erosion. In addition, increased SOM leads to higher cation exchange capacity and thus more efficient use of applied fertilizer.

Grain legume rotations have been shown to be successful under Malawian conditions. Brown (1958) compared several rotations involving maize and groundnuts at Chitedze, Thuchila and Mbawa research stations. Maize yields following groundnuts were 8-78% higher than after continuous maize, depending on site and year. Groundnut yields following maize were 18-156% higher than continuous groundnut yields. In a follow-up study, Brown (1965) indicated an 1100-1900 kg ha^{-1} yield increase in maize following groundnuts, compared to maize after maize.

MacColl (1989) examined residual N left for maize following legume crops of groundnuts, soybean and *Lablab purpureus* (L.) Sweet near Bunda. Maize yield was increased by 16-37% after groundnut and 13-19% after soybean. *Lablab* rotations provided the greatest maize yield response (53-80%) but *Lablab* was attacked by insects and did not produce seed. The estimated N left for the maize by the legumes ranged from 6 to 41 kg N ha^{-1} (Table 2). A significant maize yield increase was obtained only for values above 25 kg N ha^{-1}. The idea of a threshold or target N value for legumes to improve soil fertility is useful in evaluating different leguminous interventions. For grain legume rotations, 20 kg N ha^{-1} seems to be a reasonable target. Additions of N below this threshold are unlikely to produce a visible yield benefit. For green manure intercrops and rotations, this target would increase because the legume is providing fewer direct cash and caloric benefits (Table 1). MacColl postulated that the amount of N left by a leguminous rotation would depend on:

- Vegetative growth of the legume;
- Legume seed yield - a higher N harvest index leads to a lower soil fertility benefit; and
- Losses of N caused by leaching and gaseous processes.

This study demonstrates that the soil fertility benefit of grain legumes is likely to be much smaller than that of legumes in which grain is not harvested.

Table 2. Nitrogen (kg ha^{-1}) left for maize by previous grain legume crops during 5 years (*Source*: MacColl 1989).

Crop	Average (range)
Lilongwe soil series:	
Groundnut	12.0 (0.0 - 25.8)
Soybean	5.9 (0.0 - 15.3)
Lablab	31.4 (0.0 - 48.9)
Kandiani soil series:	
Groundnut	5.3 (0.0 - 10.3)
Soybean	8.5 (0.0 - 28.7)
Lablab	41.2 (25.2 - 52.2)

The soil fertility benefit for grain legumes is likely to be higher when using a freely nodulating, promiscuous soybean variety, such as Magoye. Magoye's promiscuity allows it to form symbiotic associations with native soil rhizobia; thus it is able to fix atmospheric N_2 in a wide variety of agro-ecosystems without inoculation. In addition, it has a low N harvest index and its leaves stay green longer than traditional soybean varieties, indicating a greater residual soil fertility benefit. Magoye has a seed yield potential of 3000 kg ha^{-1} and has produced yields above 2000 kg ha^{-1} in Zambian soils low in N (Javaheri 1996).

In Malawi, the intense land pressure and need for a continuous supply of calories make it unlikely that legumes will be adopted for soil fertility benefit alone. The legume also must have a nutritional and/or economic value. In the 1996-97 growing season, Action Group I of the MPTF planted a grain legume at 300 sites of the area-specific fertilizer verification trial of 1995-96 (Rukuni et al. 1998). The goal of this trial was to determine the effect of residual phosphorus added in 1995-96 on legume growth and yield in 1996-97. Action Group I also wanted to compare the benefits of a maize-legume rotation to continuous unfertilized maize. The absolute yield of the grain legume was considerably higher than that of continuous unfertilized maize for both Magoye soybean and CG 7 groundnut (Table 3). Soybean yields ranged from 1100 to 1380 kg ha^{-1} in comparison to 940 kg ha^{-1} for maize. Groundnut yields were 1210-1490 kg ha^{-1}, compared with 870 kg ha^{-1} for maize. Legume yield increased with increased residual phosphorus. This residual effect from hybrid maize fertilization therefore provides an important subsidiary benefit to maize-grain legume rotations.

These grain legume yields were from on-farm plots following fertilized maize, planted on time and using improved varieties at optimal plant populations. Farmers often plant legumes late, using local varieties and sparse populations and will obtain much lower yields. An International Crops Research Institute for the Semi-Arid Tropics (ICRISAT) groundnut trial in 1996-97 at 19 locations found that the yield gap between CG7 planted using recommended agronomic practices and Chalimbana (a local variety) planted using farmer practices was 95% (Sieglinde Snapp and Duncan Boughton, personal communication, 1999). The CG7 yields on-station in Malawi averaged 1700 kg ha^{-1} in 1988-92, 71% greater than yields of Chalimbana during the same period. On-farm trial yields of CG7 were 1400 kg ha^{-1}, 117% greater (ICRISAT 1997). Grain legume yields obtained will depend on variety, management, soil and climate. However, legume rotations, particularly with CG7 groundnut and Magoye soybean, clearly are preferable to continuous unfertilized maize.

The large amount of protein and oil in grain legumes provides an important nutritional benefit to the carbohydrate-rich Malawian diet. The grain legume rotation was clearly superior to continuous maize in terms of calories and protein generated. More than four times as much protein was produced in the leguminous cropping systems. Legumes such as soybean and groundnut have an important role to play in establishing household food security in Malawi.

The grain legume rotations also compare favourably in terms of economic net benefit on a cash basis (Table 3). After subtracting for fertilizer and legume seed costs, the groundnut rotation is still 96-157% more profitable using this criterion; the soybean rotation is 44-80% more profitable. The returns to groundnut on a cash basis would be mitigated to some extent if labour costs were included, because groundnut is more labour intensive than maize. If the goal for adoption of a new technology is to have double the net benefits of unfertilized maize, then the soybean rotation does not meet this criterion at present soybean prices. In order to have all soybean treatments meet this goal, either soybean yield or soybean prices must increase by 28-81%, depending on treatment.

Convincing evidence indicates that grain legume rotations provide biological, nutritional and economic benefits in the smallholder farming systems of Malawi. Given the advantages of these rotations, what are the constraints to increased farmer adoption of grain legumes? Groundnut has a low seed multiplication factor and seed storage and planting costs tend to be high. Adoption of promising varieties such as CG7 is often constrained by seed availability at the onset of the rains. The supply of CG7 groundnut would be greater if more effective promotion and seed multiplication efforts were undertaken. CG7 is an improvement over previous varieties in that it is more drought resistant (earlier maturing), higher yielding and less labour intensive in harvesting and shelling. Other promising new varieties combine these qualities with rosette disease resistance. While soybean has a higher seed multiplication factor than groundnut, the market price for soybean has declined considerably in recent seasons. The economic analysis was made using a soybean price of 2.50 MWK kg^{-1}, which is one-fourth the price obtained after the 1995-96 season. Soybean prices were artificially inflated in 1995-96 by donor demand to feed Mozambican refugees. Stable market prices would greatly improve economic benefits and farmer adoption rates of soybean. In addition, a lack of familiarity with processing procedures constrains soybean value and utilization.

Table 3. Economic net benefits (on a cash basis) of a maize-soybean rotation (132 sites), or a maize-groundnut rotation (39 sites) (*Source*: Maize Productivity Task Force unpublished data).

Crop	Treatment ($N:P_2O_5:S$ added in 1995-96) ($kg\ ha^{-1}$)	Yield[a] ($kg\ ha^{-1}$)		Output ($MWK\ ha^{-1}$)		Costs ($MWK\ ha^{-1}$)		Economic net benefit ($MWK\ ha^{-1}$)
		Maize 1995-96	Legume 1996-97	1995-96	1996-97	Seed 1995-97	Fertilizer 1995-96	
Magoye soybean	1. (96:40:0)	3190	1380	7180	3440	465	2070	8080
	2. (35:0:0)	2250	1100	5070	2760	465	580	6790
	3. (35:10:2)	2410	1190	5430	2990	465	780	7170
	4. (69:21:4)	3050	1280	6850	3210	465	1560	8040
	5. (92:21:4)	3330	1350	7500	3370	465	1950	8460
	6. (0-Maize)	1420	940	3200	2110	600	0	4710
CG7 groundnut	1. (96:40:0)	3040	1490	6830	7470	1290	2070	10940
	2. (35:0:0)	2060	1210	4640	6060	1290	580	8830
	3. (35:10:2)	2440	1330	5500	6670	1290	780	10100
	4. (69:21:4)	2810	1380	6310	6910	1290	1560	10370
	5. (92:21:4)	3230	1510	7260	7560	1290	1950	11580
	6. (0-Maize)	1400	870[b]	3160	1960	600	0	4520

[a] Yields adjusted downward by 20% for small plot size and extension agent management.
[b] Maize yield for continuous maize treatment in 1996-97.

Residue incorporation is another constraint to soil fertility benefit in grain legume rotation systems. The residues, which are much richer in N than are maize residues, need to be put back into the soil to provide N for the subsequent maize crop. Often, farmers take grain legumes such as soybean and groundnut home for threshing and the residues never are returned to the field, which results in a minimal soil fertility benefit in these systems. Pigeon pea is advantageous in this regard, because it sheds its leaves in the field before harvest. The current management of legume residue incorporation will constrain soil fertility benefits from grain legume-maize rotations.

Intercrops of Pigeon Pea or *Tephrosia* with Maize

In unimodal rainfall systems, the length of the potential growing season is constrained by moisture availability. Maize-legume intercropping systems must minimize competition for water, light and soil nutrients. Successful leguminous intercrops, such as pigeon pea and *Tephrosia*, exhibit a temporal complementarity of resource use with maize. These crops grow slowly early in the season and use residual moisture to remain green in the dry season. Thus they are only slightly competitive with maize but still produce enough biomass to have a significant beneficial impact on soil fertility.

Pigeon pea has long been used as an intercrop with maize in southern Malawi. MacColl (1989) found that sole-cropped pigeon pea grown for 2 years left 24-107 kg N ha^{-1} in the soil. Cropping system trials have found that the planting of pigeon pea in the same ridge and at the same time as maize did not reduce maize yields (Jones 1990; Sakala 1994). Delaying pigeon pea planting dates 2 to 4 weeks after maize did not improve maize yields; and pigeon pea seed yield declined over 70% with the later planting dates. Ratooning pigeon pea in year two of the intercrop, rather than reseeding, made pigeon pea more competitive. Maize yields were reduced by 500 to 1000 kg ha^{-1} in the ratooned system but pigeon pea seed yields were significantly greater when ratooned (Sakala 1993).

Sakala et al. (2000) conducted a detailed study on the N dynamics of maize/pigeon pea intercropping systems in Malawi. He found that pigeon pea was not competitive with maize at two high fertility sites but that on sandy soils at Lisasadzi, maize yields were significantly reduced in the maize/pigeon pea intercropping system. ^{15}N natural abundance methods showed that over 60% of pigeon pea N came from N_2 fixation in these intercrops. In leaching tube decomposition studies, Sakala found that N from green pigeon pea leaves was immediately mineralized, while senesced pigeon pea leaves, of lower N content, had an N immobilization phase of 14-28 days. In comparison, the immobilization phase of maize residue was over 500 days. When maize residue was combined with green or senesced pigeon pea leaves in a 1:1 ratio, the immobilization phase was still more than 250 days. This has important implications for the management of crop residues for soil fertility. The wide C:N ratio of maize means that the residues have a tremendous immobilization capacity and should be removed from the field for any crop yield benefit to be seen in the second season of maize/pigeon pea intercrops.

Advantages of maize/pigeon pea associations include their suitability for intercropping and a ready market for seed for dhal processing and export to India. Disadvantages include a poor seed yield for long-duration pigeon pea in central and northern Malawi (most likely because of the absence of winter rains in these regions) and the devastating effect of free livestock grazing in the dry season in northern Malawi. The development of medium-duration pigeon pea varieties, which maintain the temporal complementarity with maize yet are able to set seed sooner, should improve the suitability of maize-pigeon pea systems throughout Malawi.

The intercropping of green manures has only been recently studied. However, the promotion of green manures in Malawi is certainly not new. Davy (1925) names *Tephrosia*, *Crotalaria juncea* L. and *Mucuna* as promising green manure species for Nyasaland. *Crotalaria* and *Mucuna* were found to have a greater positive effect on maize yields than other green manures used extensively in crop rotations in Zimbabwe through the 1950s (Rattray and Ellis 1952). *Crotalaria* and *Mucuna* have been shown to be excellent N_2-fixers in a wide range of environments (Yost et al. 1985; Bowen et al. 1988; MacColl 1990; Buckles et al. 1998). *Tephrosia* has been less well researched. It is renowned as a fish poison and has the advantage (for soil fertility improvement) of remaining green in the dry season but being unpalatable to livestock.

Green manures can potentially accumulate up to 250 kg N ha^{-1} y^{-1} (Peoples and Herridge 1990; Giller and Wilson 1991), resulting in subsequent cereal yield increases of 600-4100 kg ha^{-1} (Peoples and Herridge 1990). However, these figures are for sole-cropped green manure species. The N benefits and best management options for green manure species undersown to maize in Malawi have not been elucidated as yet. The key question for undersown green manure technology is whether it can increase maize yields using agronomic management practices that fit into existing farming systems.

Gilbert (1997) tested four green manure *species (Tephrosia, Mucuna, Lablab* and *Crotalaria*) intercropped to maize at 2 or 6 weeks after planting and at two seeding rates. He found that the different growth habits of the green manures have practical implications for their management. *Mucuna* produced copious biomass but tended to smother the maize (see below). While *Tephrosia* was not competitive with maize because of its slow growth rate, neither did it out-compete weeds and thus was difficult to keep weed-free. The growth habit of *Crotalaria* appeared ideal for maize intercropping because it is not overly competitive with the maize, yet it also shades out weeds. However, its short growth duration (to maximum green biomass) makes incorporation problematic because it has to occur while the maize is still growing. The early incorporation of

Crotalaria also means that potential for loss of N is higher because of mineralization and leaching under late-season rainfall.

When the biomass data for *Tephrosia* were compared across sites, it became clear that biomass production was linearly correlated with precipitation received (Figure 4). Average biomass when undersown early at Bvumbwe (1760 mm rainfall) was 2.9 times that at Chitedze (787 mm). When averaged across all sites, *Tephrosia* undersown early by broadcasting at 20-40 kg ha^{-1} produced the greatest biomass of the treatments tested. This was achieved with a reduction of maize yield of less than 20%, because of the temporal complementarity in the growth habits of *Tephrosia* and maize. In contrast, *Mucuna*, especially when sown early, can be overly competitive with maize because of its aggressive climbing growth habit, reducing maize yield by up to 60%. *Mucuna* thus may be better suited to rotational systems.

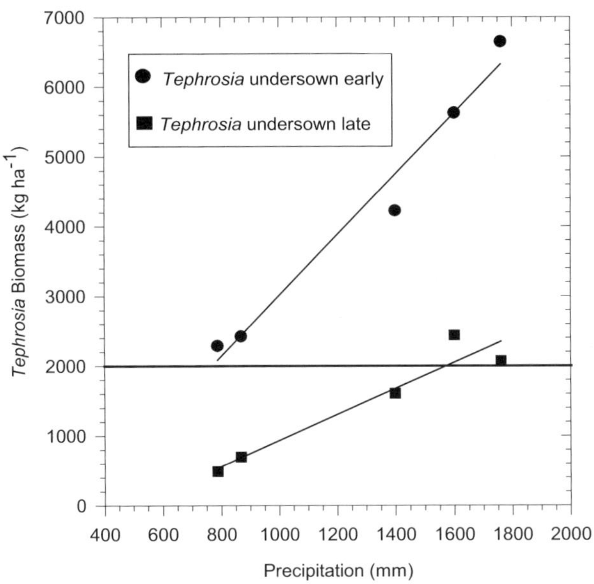

Figure 4. *Tephrosia* biomass production as a function of rainfall received at a site (*Source*: Gilbert 1997).

In 1997, the Malawi Agroforestry Extension project (MAFE) initiated a demonstration at 200 sites aimed at promoting the intercropping of *Tephrosia* and maize. *Tephrosia* seed was not broadcast but was direct-sown at the same time as maize in two planting stations in between maize hills. Preliminary data from the first season from 195 sites indicated that *Tephrosia* was not competitive with maize: average maize yield in the maize/*Tephrosia* intercrop was not significantly less than the control (MAFE–PROSCARP 1997). MAFE earlier reported an average *Tephrosia* biomass of 3000-5000 kg ha^{-1} leaves and 7000-10 000 kg ha^{-1} wood at the end of the second season. The direct-seeded system has the advantage of earlier planting, lower seeding rate and easier weeding than the broadcast seed method. However, the fallow year is a disadvantage in areas of land scarcity. It now appears that a fallow year may not be necessary in areas of high precipitation if *Tephrosia* is planted early (Trent Bunderson, personal communication, 2000).

Tephrosia, like pigeon pea, is well suited to intercropping systems because of its temporal complementarity of resource use with maize. Unlike pigeon pea, however, it does not have a food value and thus, for it to be attractive to small-scale farmers, other benefits such as fuel wood and soil fertility will have to be much greater. It may be best suited for land that is so degraded that farmers are about to abandon it.

Mucuna-Maize Rotations

Mucuna has been widely promoted as a superior green manure crop in Mexico, Benin and elsewhere (see other chapters in this volume). In Malawi, Davy (1925) regarded it as 'the finest green manuring plant for Nyasaland.' Green manure crops such as *Mucuna* and *Crotalaria* were planted on 600 000 acres annually in Zimbabwe in the 1950s (Rattray and Ellis 1952). Traditionally, *Mucuna* has been used in Malawi as an intercrop with maize. A survey of farmers in southern Malawi by Kabambe et al. (1998) found that 57% planted *Mucuna* at 5-7 weeks after maize to avoid deleterious effects on maize yield. Unfortunately, Gilbert (1997) found that late planting of *Mucuna* did not produce enough biomass to significantly enhance soil fertility. All the surveyed farmers used *Mucuna* for food, often boiling the beans for 8 hours or more. Seventy-one percent of farmers said that *Mucuna* improved soil fertility.

While *Mucuna* intercropping systems have been practiced for generations in Malawi, the rotation of *Mucuna* with maize is relatively new. Kumwenda and Gilbert (1997) found *Mucuna* (5700 kg ha^{-1} of biomass, 1770 kg ha^{-1} of seed) rotations superior to those of *Crotalaria* (4300 kg ha^{-1}, 800 kg ha^{-1}) and *Tephrosia* (2970 kg ha^{-1}, 0 kg ha^{-1}) in terms of biomass production and seed yield when rotated with maize. Early incorporation of the green manures at flowering led to a significantly greater maize yield response than did late incorporation after seed harvest. However, an economic analysis of the same data showed that late-incorporated *Mucuna* was the most profitable treatment if *Mucuna* seed was priced for sale.

In further experiments, Kumwenda et al. (1998) reported that the average biomass of *Mucuna* (6670 kg ha^{-1}) was greater than that of *Crotalaria* (4870 kg ha^{-1}) or *Lablab* (4200 kg ha^{-1}) at eight on-farm sites. Gilbert (1998a) found that *Mucuna* seed yield (2130 kg ha^{-1}) was greater than unfertilized maize (1680 kg ha^{-1}) grown at the same site. In addition, *Mucuna* has a copious amount of N-rich litter, which falls during the season. *Mucuna* leaf litter fall was double that of soybean in the same experiment. Even when the seed is removed, *Mucuna* rotations were found to add 100 kg N ha^{-1} into the soil for subsequent maize crops. This is higher than any other annual legume tested in Malawi (R. Gilbert, unpublished data).

The experiences of researchers with *Mucuna* in West Africa provide some insights into the opportunities and constraints of using *Mucuna* in southern Africa. Vissoh et al. (1998) report that the reduction of the weed *Imperata cylindrica* (L.) Palisot was the most important determinant of adoption of *Mucuna* in Benin. This illustrates an important point: it is unlikely that leguminous crops will be adopted purely for a soil fertility benefit. Farmers must perceive additional benefits. In Malawi, preliminary evidence shows that *Mucuna* rotations reduce the incidence of the parasitic weed, *Striga*, which would be an important corollary benefit. Vissoh et al. (1998) also report that *Mucuna* rotations were most profitable when *Mucuna* seed was priced for sale—findings similar to those in Malawi. With extremely high biomass production and seed yields, and a large soil fertility benefit for subsequent cereals, *Mucuna* managed as a grain legume rather than a green manure is a promising, sustainable cropping system for southern Africa.

However, if *Mucuna* is managed as a grain legume, then extreme care must be taken in processing the seeds for human consumption. *Mucuna* seeds contain L-dopa, which can be toxic to humans if the seeds are not prepared properly. Lorenzetti et al. (1998) report a range of L-dopa levels of 2.2-6.2% in 36 accessions of seed and a maximum tolerable limit for L-dopa of 1.5 g $person^{-1}$ d^{-1}. Thus 100 g of unprocessed *Mucuna* seed would contain intolerable amounts of L-dopa. In Malawi, the traditional cooking method for *Mucuna* involves repeatedly boiling the seed and discarding the water, for up to 8 hours. Preliminary calculations indicate that the cost of fuel wood and water in this method is up to 10 times the economic value of the seed—clearly providing an impediment to scaling-up this technology for resource-poor smallholders. Fortunately, Versteeg et al. (1998) report a *Mucuna*-maize recipe based on traditional cooking methods in Benin that can reduce L-dopa levels to 0.08-0.10% with total cooking time of 1 hour. Improved *Mucuna* recipes that render seed both safe to eat and economic to prepare will assist in promoting large-scale adoption of *Mucuna* in southern Africa.

COMPARISON OF BEST-BET LEGUMES

Several research and extension efforts are now underway to compare these best-bet leguminous systems in the same field and the same year to allow for direct comparison and farmer evaluation. Action Group I of the MPTF initiated a verification trial with all extension agents in Malawi (2000 sites) comparing fertilized maize, soybean or groundnut rotations, maize/pigeon pea intercrops and *Mucuna* rotations in 1998. Results from this trial confirm that *Mucuna* has a higher seed yield than soybean, groundnut or pigeon pea in Malawi (Gilbert et al. 2002). Pigeon pea intercropped with maize did not reduce maize yield compared to the sole maize control, indicating that pigeon pea was a bonus crop. In 1998-99, 106 000 farmers attended field days associated with this trial.

Other best-bet comparisons include a farmer-managed single replicate 'baby' trial design (three leguminous treatments plus a farmer control) in conjunction with researcher-managed replicated 'mother' trials to increase farmer feedback, testing and adoption of legumes using participatory research methods (Kanyama-Phiri et al. 2000). The MAFE project has initiated 200 long-term minimum tillage trials, which include maize/pigeon pea intercrops, soybean rotations and *Tephrosia* undersown to maize. These comparisons are an effective way to present farmers with as many options as possible to allow them to choose cropping systems that best suit their needs.

FUTURE RESEARCH NEEDS FOR BEST-BET LEGUMES

Although several promising best-bet leguminous systems have been identified for Malawi, several hurdles still remain to be overcome to ensure greater adoption and impact of these systems. Some of these obstacles are discussed below.

Obstacles to Greater Adoption and Impact

Legume Screening and Germplasm Exchange

Changing pest and disease pressures can render dependence on just a few leguminous accessions extremely risky. In addition, changing economic scenarios can render promoted legumes less attractive to farmers. Active legume-screening efforts are valuable to identify and feed future best-bet legumes into the extension network. Several green manure species such as *Canavalia ensiformis* (L.) DC., *Crotalaria grahamiana* Wight & Arn and *Crotalaria ochroleuca* G. Don have been identified as good performers in the agro-ecology of the Lilongwe plain (Gilbert 1998b). Similar efforts in Kenya have identified suitable legumes for diverse agro-ecologies throughout Kenya. There is a need for a technical clearinghouse of information and germplasm exchange for promising legume species.

Multiple Uses for Legumes

As mentioned above, legumes in land-constrained areas are unlikely to be adopted solely for soil fertility benefits. Clearly, further study is needed on the human consumption of species such as *Mucuna* and *Canavalia*. These species have demonstrated superior growth and yield in Malawi, to some extent because of toxins that give them partial resistance to insect attack but also render the seed difficult to prepare safely. *Tephrosia* also contains the toxins tephrosin and rotenone, which has led to interest in its use as insecticide. Quantifying the effect of these legumes on the incidence of the noxious weed, *Striga,* is also important, because this is a major biotic constraint to maize production in southern Africa.

Pest and Disease Complexes on Legumes

Leguminous technology is not without risk for the smallholder farmer. Just as a cash investment in mineral fertilizer can be lost through drought, an investment in land, labour and seed to legumes can be wiped out to pests and diseases. Some diseases, such as *Fusarium* wilt on pigeon pea, *Curvularia* leafspot on *Mucuna* and rosette on groundnut, have manifested themselves in Malawi in recent years and can drastically reduce plant biomass, yield and soil fertility benefit. A diverse selection of leguminous germplasm is needed in order to investigate resistance to these diseases and present a wide array of legumes for use by small-scale farmers.

Economic and Policy Issues

As noted in this chapter, several leguminous technologies have performed well based on biophysical and socio-economic criteria in Malawi. However, some key constraints to their adoption remain. Seed availability and market for seed are crucial factors. The lack of availability of improved legume varieties at the village level, and the long lead time for seed multiplication of legumes such as groundnut, frustrate increased demand for seed. Also, farmers are uncertain of the price they will receive for legumes, which reduces the amount of land they are willing to commit to legume production. Adding value to legumes such as pigeon pea and soybean through processing (e.g. pigeon pea dhal) needs to be addressed.

Integrated Nutrient Management Recommendations

For a country such as Malawi, where farmers have neither enough land to devote to leguminous fallows nor enough cash to use high rates of inorganic fertilizer, the combination of lesser amounts of both may be the way forward. Palm (1997) discusses the state of knowledge on the combination of organic and inorganic sources of fertility and concludes that a systematic framework for guidelines on integrated nutrient management (INM) is needed. An interaction between the two nutrient sources is not always seen. The yield responses generated are often due to simple additive effects of nutrients. However, even an additive effect might be helpful if the combination more efficiently uses scarce farm resources (e.g. by increasing the impact of small amounts of cash and land devoted to soil fertility improvement). Another topic requiring further research is the fate of nutrients added in leguminous biomass. Ideally, farmers should know the fertilizer equivalency for subsequent cereals of planting different legumes on different soils in different climates. This is difficult to do, because there are numerous pathways for nutrient release during decomposition of leguminous biomass. Nitrogen added in biomass, for example, can be lost due to processes such as leaching, volatilization of ammonia, denitrification and termite offtake. Finding practical management practices that minimize N losses would make adoption of leguminous technologies more attractive to resource-poor smallholders.

CONCLUSIONS

Malawi's biophysical and socio-economic environment has led to greater farmer interest in legumes in recent years. Several best-bet leguminous cropping systems have been identified that are superior to continuous unfertilized maize in terms of returns to cash, calories per hectare and soil fertility improvement. However, challenges remain for research and extension personnel to improve human utilization, reduce pests and diseases and identify promising leguminous accessions. If these challenges are met, legumes will play an important role in diversifying Malawian farming systems and ensuring improved household food security.

REFERENCES

Benson T D (1997a) Developing flexible fertilizer recommendations for smallholder maize production in Malawi. *In*: Waddington S D, Murwira H K, Kumwenda J D T, Hikwa D, Taguira F (Eds.) Soil fertility research for maize-based farming systems in Malawi and Zimbabwe. Soil Fertility Net (SFNET) - Centro Internacional de Mejoramiento de Maíz y Trigo (CIMMYT), Harare, Zimbabwe. p. 275-85.

Benson T D (1997b) The 1995/96 fertilizer verification trial in Malawi: Economic analyses of results for policy discussion. Ministry of Agriculture and Livestock Development, Lilongwe, Malawi. 27 p.

Benson T D (2001) Profile of poverty in Malawi, 1998. National Economic Council - International Food Policy Research Institute (IFPRI), Washington, USA. 110 p.

Blackie M J, Benson T D, Conroy A, Gilbert R A, Kanyama-Phiri G, Kumwenda J D T, Mann C, Mughogho S, Phiri A (1998) Malawi: Soil fertility issues and options. Rockefeller Foundation, Lilongwe, Malawi. 55 p.

Bowen W T, Quintana J O, Pereira J C, Bouldin D R, Reid W S, Lathwell D J (1988) Screening legume green manures as nitrogen sources to succeeding non-legume crops. Plant Soil 111:75-80.

Brown P (1958) The results of some short-term rotation experiments in Nyasaland. Rhod. Agric. J. 55:626-633.

Brown P (1965) A review of groundnut experiments in Malawi. Rhod. J. Agric. Res. 3:39-50.

Buckles D, Triomphe B, Sain G (1998) Cover crops in hillside agriculture: Farmer innovation with *Mucuna*. International Development Research Centre - Centro Internacional de Mejoramiento de Maíz y Trigo (IDRC-CIMMYT), Ottawa and Mexico City. 218 p.

Carsky R J, Oyewole B, Tian G (1999) Integrated soil management for the savanna zone of W. Africa: Legume rotation and fertilizer N. Nutr. Cycl. Agroecosys. 55:95-105.

Davy E W (1925) Crops for green manuring and soil protection. Nyasaland Protectorate Dept. Bull. 3:1-4.

FAO – Food and Agriculture Organisation (Various years) FAO Production Yearbook. FAO, Rome, Italy.

Gilbert R A (1997) Undersowing green manures for soil fertility enhancement in the maize-based cropping systems of Malawi. *In*: Waddington S D, Murwira H K, Kumwenda J D T, Hikwa D, Taguira F (Eds.) Soil fertility research for maize-based farming systems in Malawi and Zimbabwe. Soil Fertility Net (SFNET) and the Centro Internacional de Mejoramiento de Maíz y Trigo (CIMMYT), Harare, Zimbabwe. p. 73-80.

Gilbert R A (1998a) Comparison of best-bet soil fertility interventions: Preliminary results. *In*: Annual report of the cereals commodity group for 1997-98. Ministry of Agriculture and Irrigation, Lilongwe, Malawi. p. 225-227.

Gilbert R A (1998b) The SFNET legume screening trial: Large-seeded legumes perform well in Malawi. TARGET 17:5-6.

Gilbert R A, Sakala W D, Benson T D (2002) Gender analysis of a nationwide cropping system trial survey in Malawi. African Studies Quarterly 6, No. 1: [online] URL: http://web.africa.ufl.edu/asq/v6/v6i1a9.htm.

Giller K E, Wilson K J (1991) Nitrogen fixation in tropical cropping systems. CAB International, Wallingford, UK. 313 p.

ICRISAT – International Crops Research Institute for the Semi-Arid Tropics (1997) ICRISAT – southern and eastern Africa region highlights 1996. ICRISAT, Bulawayo, Zimbabwe.

Javaheri F (1996) Naturally nodulating (promiscuous) soybean varieties for smallholder maize farmers: The Zambian experience. *In*: Mpepereki S, Giller K E, Makonese, F. (Eds.) Soybean in smallholder cropping systems of Zimbabwe: Potential contributions from biological nitrogen fixation. Proceedings of a preparatory workshop, 8-9 February 1996. University of Zimbabwe with Soil Fertility Net (SFNET), Harare, Zimbabwe. p. 63-69.

Jones R B (1990) Maize/pigeon pea intercropping trial. Maize commodity team annual report: 1989/1990. Ministry of Agriculture and Livestock Development, Lilongwe, Malawi. p. 40-51.

Kabambe V H, Ganunga R P, Kumwenda J D T (1998) *Mucuna* management in an intercrop with maize in southern Malawi: Preliminary survey results. *In*: Annual report of the cereals commodity group for 1997/98. Ministry of Agriculture and Irrigation, Lilongwe, Malawi. p. 187-189.

Kanyama-Phiri G, Snapp S, Kamanga B, Wellard K (2000) Towards integrated soil fertility management in Malawi: Incorporating participatory approaches in agricultural research. Managing African Soils No. 11, Russell Press, Nottingham, UK. 27 p.

Karlen D L, Varvel G E, Bullock D G, Cruse R M (1994) Crop rotations for the 21st century. Adv. Agron. 53:1-45.

Kumwenda J D T, Gilbert R A (1997) Biomass production by legume green manures on exhausted soils in Malawi: A soil fertility network trial. *In*: Waddington S D, Murwira H K, Kumwenda J D T, Hikwa D, Taguira F (Eds) Soil fertility research for maize-based farming systems in Malawi and Zimbabwe. Soil Fertility Net (SFNET) and the Centro Internacional de Mejoramiento de Maíz y Trigo (CIMMYT), Harare, Zimbabwe. p. 85-86.

Kumwenda J D T, Saka A R, Kabambe V H, Ganunga R P (1998) Legume manure crop rotation trial for the 1997/98 cropping season. *In*: Annual report of the cereals commodity group for 1997/98. Ministry of Agriculture and Irrigation, Lilongwe, Malawi. p. 161-169.

Lorenzetti F, MacIsaac S, Arnoson J T, Awong D V C, Buckles D (1998) The phytochemistry, toxicology and food potential of velvetbean (*Mucuna* spp. Adans, *Fabaceae*). *In*: Buckles D, Etèka A, Osiname O, Galiba M, Galiono G (Eds.) Cover crops in West Africa: Contributing to sustainable agriculture. International Development Research Centre (IDRC) - International Institute of Tropical Agriculture (IITA), Ottawa, Canada. p. 67-84.

MacColl D (1989) Studies on maize at Bunda, Malawi. II. Yield in short rotations with legumes. Expl. Agric. 25:367-374.

MacColl D (1990) Studies on maize at Bunda, Malawi. III. Yield in rotations with pasture legumes. Expl. Agric. 26:263-271.

MAFE-PROSCARP – Malawi Agroforestry Extension - Promotion of Soil Conservation and Rural Production project (1997) Reduced tillage/agroforestry/legume demonstration trials: Summary of 1st year 1996/97 season preliminary results. Lilongwe, Malawi. 19 p.

Palm C A, Myers R J K, Nandwa S M (1997) Combined use of organic and inorganic nutrient sources for soil fertility maintenance and replenishment. *In*: Buresh R J, Sanchez P A, Calhoun F (Eds.) Replenishing soil fertility in Africa. Special publication, Soil Sci. Soc. Am. J. 51:193-218.

Peoples M B, Herridge D F (1990) Nitrogen fixation by legumes in tropical and subtropical agriculture. Adv. Agron. 44:155-223.

Rattray A G H, Ellis B S (1952) Maize and green manuring in southern Rhodesia. Rhod. Agric. J. 49:188-197.

Rukuni M, Blackie M J, Eicher C K (1998) Crafting smallholder-driven agricultural research systems in southern Africa. World Dev. 56:1073-1087.

Sakala W D (1993) Maize/pigeon pea intercropping and ratooning trial. *In*: Maize commodity team annual report: 1992/93. Ministry of Agriculture and Livestock Development, Lilongwe, Malawi. p. 135-144.

Sakala W D (1994) Maize/pigeon pea intercropping at two fertility levels. *In*: Maize commodity team annual report: 1993/94. Ministry of Agriculture and Livestock Development, Lilongwe, Malawi. p. 156-169.

Sakala W D, Cadisch G, Giller K E (2000) Interactions between residues of maize and pigeon pea and mineral N fertilizers during decomposition and N mineralization. Soil Biol. Biochem. 32:679-688.

Smale M, Heisey P W (1997) Maize technology and productivity in Malawi. *In*: Byerlee D, Eicher C K (Eds.) Africa's emerging maize revolution. Lynne Reiner, Boulder, Colorado, USA. p. 63-79.

Snapp S (1998) Soil nutrient status of smallholder farms in Malawi. Comm. Soil Sci. Plant Anal. 17/18:2572-2588.

Versteeg M N, Amadji F, Etèka A, Houndékon V A, Manyong V M (1998a) Collaboration to increase the use of *Mucuna* in production systems in Benin. *In:* Buckles D, Etèka A, Osiname O, Galiba M, Galiano G (Eds.) Cover crops in West Africa: Contributing to sustainable agriculture. International Development Research Centre (IDRC)-International Institute of Tropical Agriculture (IITA)- Sasakawa-Global 2000 (SG2000), Canada. p. 33-43.

Vissoh P, Manyong V M, Carsky R J, Osei-Bonsu P, Galiba M (1998) Experiences with *Mucuna* in West Africa. *In*: Buckles D, Etèka A, Osiname O, Galiba M, Galiono G (Eds.) Cover crops in West Africa: Contributing to sustainable agriculture. International Development Research Centre (IDRC) – International Institute of Tropical Agriculture (IITA)- Sasakawa-Global 2000 (SG2000), Ottawa. p. 1-32.

Yost R S, Evans D O, Saidy N A (1985) Tropical legumes for N production: Growth and N content in relation to soil pH. Trop. Agric. 62:20-24.

Chapter 7

Promotion of *Marejea* Cultivation in the Ruvuma Region of Tanzania: Experiences of the Catholic Missionaries at Peramiho Mission Centre

C. K. K. Gachene

University of Nairobi, Kenya

SUMMARY

The existing information on *marejea* (*Crotalaria ochroleuca* G. Don) work in the Ruvuma region of Tanzania indicates that the crop has the potential to improve soil fertility and control weeds. It can also be used for feeding animals. Although the main purpose of promoting *marejea* in the region was for improving soil fertility, because of the incentives offered, it rather became an exercise of seed bulking. Farmers were enthusiastic to cultivate *marejea* as long as incentives were provided. Once they were withdrawn, interest waned. Two workshops, held in 1986 and 1987, were important in creating a forum to discuss a number of issues related to the use and adoption of *marejea* in the region. However, there was no follow-up after the workshops. The number of farmers currently growing *marejea* and their reasons for continuing the practice are not known. A study documenting this could be a starting point for efforts to reintroduce the crop in the region.

INTRODUCTION

Increased food production without adequate soil conservation measures has led to a deterioration of land quality in Tanzania. Most smallholders cultivate cereals and legumes continuously on sloping land without carrying out adequate soil conservation measures.

The Southern Highlands is the major cereal-producing region in Tanzania. In the area, the use of modern technologies, such as fertilizers, herbicides, improved seed varieties and tillage methods, has not been adequately adopted because of adverse socio-economic factors. Farmers acknowledge that crop yields are decreasing and attribute this to declining

M. Eilittä et al. (eds.), Green Manure/Cover Crop Systems of Smallholder Farmers, 175-193.

soil fertility. There is therefore a need to explore alternative ways of increasing and sustaining food production using resources that are within the reach of most smallholder farmers.

Opportunities for maintaining and improving soil fertility through the use of green manure/cover crops (GMCCs) have been reported in the Ruvuma region of the Southern Highlands (Temu and Aune 1995) and elsewhere in Tanzania. In the early 1940s, the Wakala of Ukara Island at Lake Victoria were deliberately growing and incorporating *Crotalaria striata* DC. to restore soil fertility (Baijukya 1999). At the same time, in the Mwanza region, *marejea* (*Crotalaria ochroleuca* G. Don) was incorporated into the soil prior to planting cassava. A long-term trial involving *Crotalaria zanzibarica* Benth. (now known as *Crotalaria trichotoma* Bojer.) in rotation with maize (*Zea mays* L.) was initiated in 1983-84 at Mbozi Maize Farms Ltd., Mbeya region. More recently (1996), green manure trials based on experiences from the International Center for Tropical Agriculture (CIAT, its Spanish acronym)-Uganda with *Mucuna pruriens* (L.) D.C. and *Canavalia ensiformis* (L.) D.C. were initiated in the Bukoba district.

In Peramiho village, Ruvuma region, *marejea* cultivation (see Box 1) was started in 1942. From the 1970s onwards, its cultivation increased, because it was actively promoted to the region's farmers. This experience, led by the Catholic Missionaries based at the Peramiho Mission Centre, merits special attention in that the incentives provided seemed to be sufficient motive to generate adoption but, when they were withdrawn in the early 1990s, almost all farmers stopped growing *marejea* in the region.

Box 1. Some background information on _marejea_

Crotalaria ochroleuca G. Don belongs to the family *Leguminosae* (*Fabaceae*). More than 500 species are recognized in Africa (66% of which are in Madagascar). While 20 species of *Crotalaria* have toxic compounds, *marejea* is known to contain only a small amount of them. *Marejea* has the capacity to fix up to 170 N kg ha^{-1}. Growth is limited by a high light requirement, sensitivity to frost and extreme acidity. The species is widespread in tropical Africa, excluding the north-eastern and the most southern parts. It is well adapted to well-drained soils and altitudes of 1300-1800 masl. Seeds are reddish-white in colour. The plant is erect, generally annual, up to 1.5 to 3 m high, with lax ascending ribbed branches. The flowers are large, yellow, reddish-veined and the petioles are shorter than the leaflets. Each pod has 30 to 60 seeds (Rupper 1997). The flowers last for 4 or 5 days, then fade. The crop takes 3 to 5 months to mature.

It should be noted that little written information is available on this experience because only limited research was conducted during the time the system was promoted to the region's farmers and no other retrospective studies on it have been done. This study attempts to piecemeal and analyse the information (some of which is contradictory) collected by the author from the few written sources available. These have been complemented by in-depth interviews with various people familiar with the system, particularly with the late Father Gerold Rupper of Peramiho Mission Centre.

RUVUMA REGION OF TANZANIA

Biophysical Environment

Ruvuma region is located in the south-western part of Tanzania at lat 6-9.2°S, long 34.5-38.2°E (Figure 1). The region borders Mozambique in the south, Lake Nyasa (also known as Lake Malawi) in the west, Mtwara region in the east and Iringa and Morogoro regions in the north and is divided into three districts, Tunduru, Songea and Mbinga, which are further divided into several divisions. Songea is the largest town and serves as the regional/district headquarters. *Marejea* cultivation was largely carried out in villages within Songea district, particularly in Peramiho division where the Peramiho Catholic Mission Centre is located in the town of Peramiho, 25 km west of Songea town.

The land is rolling to hilly in the north and gently undulating to undulating in the southern part. The altitude ranges from 500 to more than 2000 masl; Peramiho is at an altitude of 1000 masl (ARI 1999). The mean annual rainfall for the area around the Peramiho Mission is 1180 mm. In most of the Ruvuma region, the rainfall pattern is unimodal, allowing for one growing season in the year from November to May (Table 1). The rains usually start between late October and early December. January, February and March are the wettest months. The rainfall starts tapering off in May and little rain falls between June and mid-September. Flooding is experienced in low-lying areas during the months of heavy rain; these areas are mostly used for growing paddy rice (*Oryza sativa* L.).

Detailed information on the soils of the region is lacking. According to the early work of Scott (1972), soils are mainly lateritic in nature. As expected, soil type varies greatly because of differences in relief and parent materials. Soils range from shallow (Leptosols) to very deep, highly weathered Ferralsols of varying texture and largely developed on volcanic and Basement Systems rocks. Vertisols and Gleysols characterize the flat to very gently undulating areas. In the Peramiho area, the soils are Lixisols and Ferralsols with sandy clay to sandy clay loam texture. They are acidic and have low organic matter content, P and N (Method Kilasara, personal communication, 1998). They have a hard consistency when dry and are therefore difficult to till.

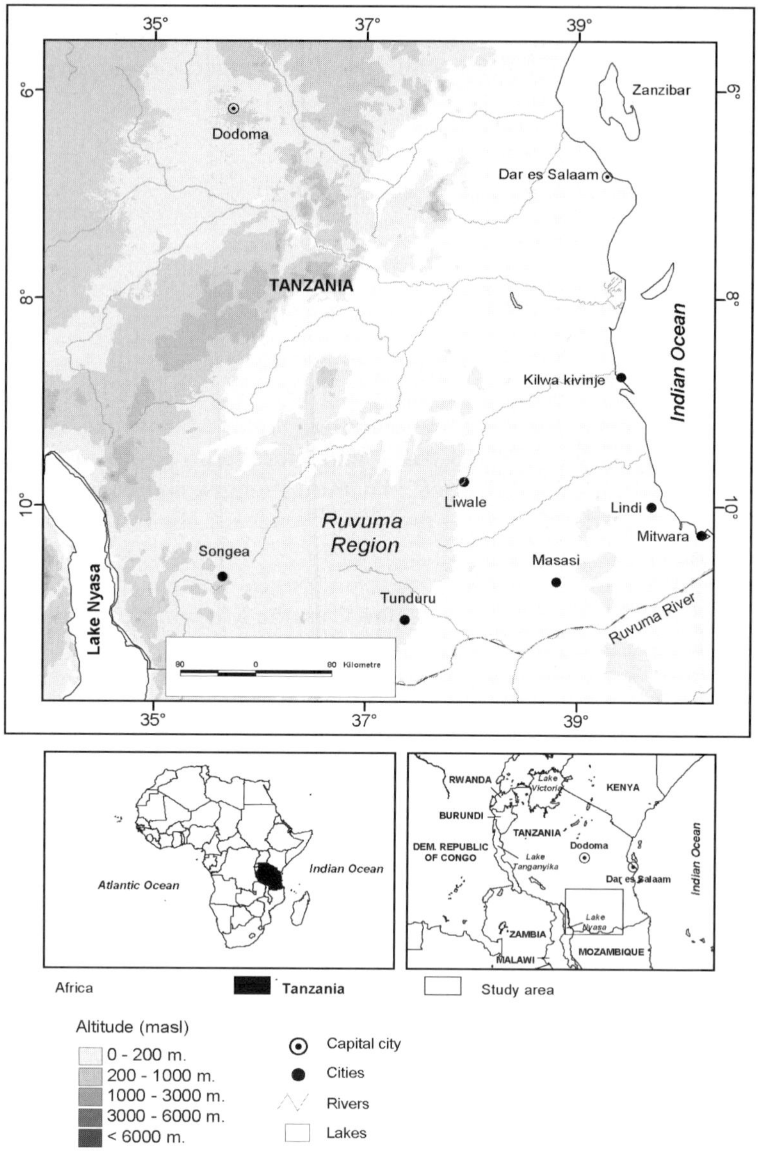

Figure 1. The study area, Ruvuma region of Tanzania.

Farmers in the Ruvuma region, like farmers generally in the Southern Highlands, recognize that crop yields are declining because of soil nutrient depletion. The traditional fallow periods are no longer feasible because land is cultivated throughout the year. Since most farmers do not keep animals, the maize stover is burned (by about 60% of farmers), used as firewood or, less commonly, left on the soil surface so that it can be incorporated into the soil during the following season. Soil conservation measures have been inadequate even in the northern part of Ruvuma region where cultivation is being carried out on very steep slopes.

Table 1. Average monthly rainfall in Peramiho (*Source*: Peramiho Catholic Mission, unpublished data for the period 1929-30 to 1996-97).

Month	Rainfall (mm)	Month	Rainfall (mm)
January	255.0	July	0.0
February	221.8	August	0.0
March	270.8	September	4.6
April	134.4	October	9.8
May	21.8	November	65.7
June	0.0	December	195.3
	Total: 1180 mm y^{-1}		

The Ruvuma region has few natural resources. In the western part lies Lake Nyasa, which is used for fishing. The few rivers in the area are occasionally used for irrigating vegetables. Because of intensive cultivation, the miombo vegetation, which once covered most of the region, has been almost fully replaced by maize fields (Gerold Rupper, personal communication, 1998). Despite the government's effort to curb tree felling, the communities still seemingly find access to the forest, as witnessed by numerous bags of charcoal and firewood bundles displayed for sale along the main road.

Human Environment

The Wangoni, who are farmers by tradition, largely dominate the region. In the Peramiho area the majority of the inhabitants are Catholics, most certainly as a result of the influence of the Peramiho Catholic Mission Centre. Subsistence farming is the main activity in the region, while fishing predominates around Lake Nyasa. Land is still state owned and farmers typically cultivate an average of 1 to 2 ha. Land availability is quite good, and the right to use land is obtained from a committee of village elders, which is chaired by a political appointee. Women are responsible for most of the household work and farming. Men carry out certain farming activities, such as wood felling and charcoal burning, in addition to assisting with land preparation (both slashing and hoeing) and harvesting.

The region has a network of non-tarmac roads; a single tarmac road connects the area with Iringa region. Off-farm income opportunities in the area are very limited and the Peramiho Mission, with its several enterprises, is a major provider of work. The Mission Centre offers a modern boarding school, a big in- and outpatient hospital, bookshop/printing press, trade school for tailoring, carpentry and shoemaking, a farm for organic farming methods and livestock keeping and a rehabilitation centre for lepers at Morogoro village. Extension services are not effective in the region, especially because of transport constraints. Farmers have little access to credit.

In the late 1980s and early 1990s, most smallholders utilized fertilizers because of the vigorous promotion and credit programmes supported by the International Fund for Agricultural Development (IFAD)-Food and Agriculture Organisation (FAO). This coincided with the period in which *marejea* also was being promoted as an alternative means of improving soil fertility. However, the credit system collapsed in 1994 because many of the fertilizer stockists (i.e. traders) were unable to repay their loans; in addition, the mechanisms for collecting debts were not effective. Many farmers agree, however, that without inputs (either organic or inorganic) to the soil a good maize crop cannot be realized.

Cropping Systems

Maize is the staple food crop in the Ruvuma region and almost all the smallholders cultivate it for subsistence. Additionally, the region is among the leaders in maize production in the country. In the smallholder farms, the crop is grown on ridges, typically intercropped with bean (*Phaseolus vulgaris* L.). The most commonly used maize varieties are Songea Selected and Katumani. The maize recommend spacing is 30 by 75 cm, although spacing and density vary from farm to farm. The yields are low, averaging around 1 t ha^{-1} and usually ranging from below 1 to 2 t ha^{-1}, mainly because of lack of nutrient inputs (Soko 1991a; b). Other subsistence crops are produced in varied associations: cowpeas (*Vigna unguiculata* [L.] Walp.), sesame (*Sesamum indicum* L.), sunflower (*Helianthus annuus* L.), sweet potatoes (*Ipomoea batatas* [L.] Lam.), sorghum (*Sorghum bicolor* [L.] Moench), cassava (*Manihot esculenta* Crantz), pumpkin (*Cucurbita* sp. L.), finger millet (*Eleusine coracana* [L.] Gaertn.), bambara nuts (*Vigna subterranean* [L.] Verdc.), groundnuts (*Arachis hypogaea* L.), potatoes (*Solanum tuberosum* L.) and garden peas (*Pisum sativum* L.). With the exception of the maize/bean intercrop, it is difficult to predict the types of crops a farmer would intercrop during any one season (Gabriel Mhagama, personal communication, 1998) but almost all fields include other crops besides maize. Cash crops typically include coffee (*Coffea* L.) and tobacco (*Tabacum nicotianum* Berchtold & Opiz. – most commonly plantation crops but also grown by some smallholders), sunflower (often intercropped with subsistence crops) and paddy rice.

By tradition, the local communities do not keep domestic animals and this tradition has been largely maintained despite the successful keeping of livestock at the Peramiho Mission Abbey Farm. Those few farmers with animals use farmyard manure and some farmers have been exposed to composting at the Peramiho Mission Centre. Fertilizers (particularly sulphate of ammonium, SA) became accessible to farmers cultivating *marejea* in the 1980s as well as to those participating in the fertilizer credit schemes in the 1990s. More recently, high fertilizer prices and the poor profitability of maize farming have led to a decline in their use. The current cost of fertilizer is US$13.5 (US$1 = Tsh. 740), while that of a bag of maize is US$8.0.

Most farmers incorporate crop residues (i.e. maize stover, weed and any other plant residue) into the soil before planting the subsequent crop. Land preparation is done using a *jembe*, or hoe. In almost all fields, farmers make ridges or furrows during land preparation; they are typically made along the contours and have two main benefits. First, the crops are planted along the ridges where the soil has been loosened, thus permitting rooting in an otherwise hard soil. Second, during land preparation, the residues are buried along the furrow and a new ridge is made on top of it, allowing the new crop to make use of the nutrients released by the buried material. Farmers cultivating *marejea* planted it along the furrow between the maize rows and later incorporated it into the soil as normal.

Weeding is usually done twice in a season but the frequency depends on weed incidence. The first weeding is carried out 3 to 4 weeks after the maize planting; the second is done when maize is flowering. The crop is normally harvested after 8 months (Figure 2).

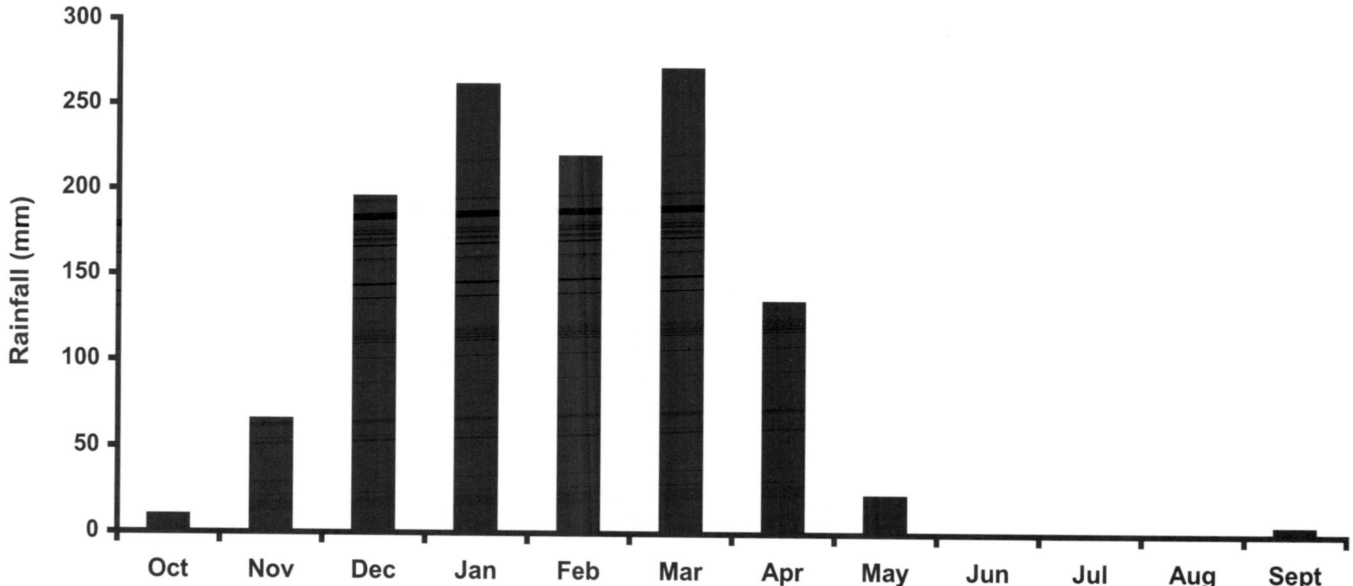

	Oct	Nov	Dec	Jan	Feb	Mar	Apr	May	Jun	Jul	Aug	Sept
Simultaneous intercropping	Land preparation	Maize & *marejea* planting	Weeding	*Marejea* incorpo-ration	Weeding					Maize harvesting		Land preparation
Intercropping at first weeding	Land preparation	Maize planting	Maize weeding; *marejea* planting		Weeding					Harvest of maize grain and *marejea* seed		Land preparation

Figure 2. Typical cropping calendar for the Ruwuma region of Tanzania.

PROMOTION OF *MAREJEA* IN THE RUVUMA REGION

Already, in the early 1940s, Othmar Morger, the Principal of the Teacher Training College in Peramiho, cultivated *marejea* at the mission farm for improving soil fertility and for feeding cattle (Rupper 1997). It is said that he received the first seeds from South Rhodesian agriculturalists working on a tobacco farm at Songea. Later, the interest in *marejea* decreased because of Trypanosomiasis, which affected the cattle. However, one of the missionaries, Crispin Schulz, planted *marejea* each year for feeding the pigs at the Mission Farm (Hermann Mayr, personal communication, 1998).

There is relatively little knowledge of *marejea* cultivation in the region during the 1950s but it is known that, in 1953, C. Schulz from Peramiho gave some seeds to the sisters at the Chipole farm and to Likonde Secondary School for use as green manure (Rupper 1987). In the early 1960s, some of this seed (6 kg) was given back to Hermann Mayr, the manager of Peramiho Abbey Farm, who started planting *marejea* in rotation with maize, a practice that continues until today (Hermann Mayr, personal communication, 1998). On average, 8 to 12 ha are still planted each year with *marejea* and about 20 bags of seeds are collected for replanting.

In the 1970s, *marejea* cultivation assumed increased importance in the area. With funds from the Dietikon Parish, Zurich, Gerold Rupper provided fertilizer to several families with the condition that they should plant about 3 kg of *marejea* and give back the seeds. The objective of the scheme was to distribute the received seed to other farmers, ensuring a continuous seed supply (Gerold Rupper, personal communication, 1998). The *marejea* cultivation was introduced to the region's farmers mainly to promote practices favouring soil fertility improvement in the region. It was also done to assist the lepers living in the Mission's rehabilitation centre to gain meaningful employment in the Seed Bank.

Marejea was quickly turned into a cash crop. Based on the amounts of seeds the farmers were harvesting, the farmers doubtless were sole cropping it instead of intercropping it with maize. It is said that women soon found out that *marejea* was a way to be free from hunger. Some would grow it and sell the seeds to their neighbours (so that they could plant it) or exchange the seeds for fertilizer at the seed bank, where the farmer had to first provide evidence that s/he had planted *marejea* seed, before being allowed to buy fertilizers. Additional fertilizer meant more food to the farmers. Some farmers were able to get 10 or more bags of SA fertilizer in exchange for *marejea*. With an exchange rate of 100 kg of *marejea* for six bags of SA fertilizer, large amounts of seed were received. Others were able to sell the seed to the seed bank. It should be noted that there was little follow up at the farmer level and that the local extension system was not involved.

In the 1980s, a number of organizations became involved in the efforts. The Peramiho Mission newspaper, *Mwenge* (The Torch), started to report on *marejea*. Farmers in some five villages around Peramiho began planting it on their farms. The Freedom from Hunger Council of Ireland (Gorta) assisted in the spreading of *marejea* cultivation between 1983 and 1985. A pamphlet called *Marejea, Rafiki ya Wakulima* (*Marejea*, the Farmer's Friend) was published during this time. Gorta also started a *marejea* seed bank at Peramiho in 1983. In 1985, 11.2 tons of seeds were out on loan to farmers and 18.9 tons were in stock.

At the village level, a promoter with a small storeroom, having obtained the *marejea* seeds from the Diocesan headquarters, would contact at least 20 farmers, providing each of them with literature and 10 kg of seed. Others joined the group, having bought the seeds from the initial farmers. Where farmers were unable to harvest the seeds, they would allow neighbouring farmers to collect them without charge. In Namanguli village alone, 40 ha of *marejea* were cultivated and after a bumper harvest in 1984 the village was able to buy a lorry and a tractor. However, this later stopped because of financial mismanagement by the village committee.

The promotion was further enhanced when high-ranking politicians, such as the regional commissioner of the Ruvuma region and the President of Tanzania, Mwalimu Julius K. Nyerere, endorsed *marejea* cultivation. Even further enhancement came when, in April 1984, it was announced on the radio that President Nyerere had presented Gerold Rupper with the medal of the United Republic for propagating *marejea*. For several years, Radio Tanzania included *marejea* in its agricultural program, *Ana kwa Ana*.

In October 1986 and April 1987, the German Foundation for International Development (DSE) financed two workshops on *marejea* (Rupper 1987). The workshops were initiated by Gerold Rupper and involved staff from Peramiho Mission and the Sokoine University of Agriculture. Topics included the use of *marejea* for weed control and soil fertility improvement, its taxonomy, its value as a livestock feed and factors affecting its adoption in Peramiho. After the workshop, it was hoped that the researchers would address some of the research gaps identified during the workshop. A few efforts were completed (Sarwatt 1992; Temu and Aune 1995; Kullaya et al. 1998) but unfortunately no thorough follow-up was ever made (Gerold Rupper and Gabriel Mhagama, personal communication, 1998).

In 1987 and 1988, 200 tons of *marejea* seeds were distributed in Tanzania using a grant from MISEREOR (Rupper 1997), a German Catholic Church development agency. In addition, seed samples were sent to several tropical countries including Papua New Guinea, The Philippines and Nicaragua. More recently, the Uluguru Mountain Agricultural Development Project (UMADEP) and Kilimo Hai Tanzania (KIHATA, a national organization on organic farming), having obtained seed from Peramiho, are promoting it for organic farming systems.

In the early 1990s, interest in *marejea* cultivation in the Ruvuma region waned. Because of this, in 1993-94, 60 tons of seeds from Ruvuma were distributed in other regions of Tanzania. Unfortunately, no follow-up of this effort was made.

Two factors particularly caused the decreased interest in *marejea* cultivation. First, the incentives for its cultivation were withdrawn (Gerold Rupper, personal communication, 1998). Second, after Rupper's retirement in 1991, the *marejea* project was incorporated into the Organic Farming Project and, in this context, it became only one of many inputs used in organic farming.

STRUCTURE OF THE *MAREJEA* SYSTEMS

The maize-*marejea* systems in the Ruvuma region involved the following (Rupper 1997).

Simultaneous Intercrop of *Marejea* with Maize

This system was carried out at the demonstration farm at Peramiho. *Marejea* was sown at the same time with maize, which at the start would grow more quickly. When *marejea* reached the size of maize, it was incorporated into the soil. The advantages of this method included better maize nutrition and improved control of erosion and weeds. Depending on the fertility of the field, inorganic fertilizers were added at times.

Intercropping *Marejea* during the First Weeding of Maize

Marejea was sown in the maize field during the first weeding. For ease of harvesting maize, farmers were advised to leave every second furrow free from *marejea*. After maize harvest, *marejea* was left in the field to mature and its seed was harvested. In the following season, it was ploughed into the soil. Depending on soil fertility, SA fertilizer may have been added but typically after only 1 year little or no fertilizer was needed.

Sole-cropped *Marejea*

This system was mainly practiced in the fields at the Peramiho farm that were heavily infested with weeds or where soil fertility had been exhausted. The best time to plant *marejea* was after the first rains. The seeding rate was high, about 50 kg ha^{-1} (ASPRO 1994); for seed production the recommended seed rate was 25 kg ha^{-1}). At the end of the season, *marejea* was incorporated into the soil, after which maize was planted. Depending on the fertility of the field, maize was grown either without or with only little SA fertilizer or calcium ammonium nitrate. The method could be repeated the following year. Some farmers practiced a 3-

year rotation that involved *marejea* as a single stand during the first year, maize during the second year and a crop with lower nutrient demand (e.g. cassava, sweet potato or sorghum) during the third year. In the fourth year, the cycle would be started over again.

Planting of *Marejea* in a (Paddy) Rice Field

Following rice harvesting, the field was ploughed and planted with *marejea*. The planting took place usually during the month of May, when the rains were gradually diminishing. After 3 months, the crop was incorporated into the soil and beans were planted. Once the beans were harvested, the field reverted to the growing of paddy rice. This method was effective only in fields with sufficient moisture at the time of *marejea* planting. Only a few farmers practised this system.

Marejea in the Pit System (*Ingoro*)

The Matengo of Mbinga district used this cultivation system. Several pits were dug in the fields. Ridges were built around them that controlled soil erosion in this hilly region. The *marejea* seeds were planted inside the pit and covered with a thin layer of soil. Once harvested for seeds, the stems were returned to the pit and buried together with other plant residues. New pits were dug allowing the stems to decay in the previous pits. Crop performance was always better in the areas where the stems were buried (Rupper 1997). Very few farmers practised this system.

PRODUCTIVITY OF THE MAIZE-*MAREJEA* SYSTEMS

Unfortunately, no on-farm studies have been made on the productivity of the maize-*marejea* systems. However, at the demonstration farm in Peramiho, some studies have been carried out on the effect of relay-cropping *marejea* on soil fertility and crop growth. In addition, only scanty economic data are available on the system.

Maize and *Marejea* Productivity

Perhaps the best indication of the sustainability and productivity of the *marejea* system is that the system has been practiced at the Peramiho demonstration farm since the 1960s. On the farm, annual rotation has resulted in sustained maize yields of 4.5-6.0 t ha^{-1}. Both farmyard manure and fertilizers are sometimes added. The system has been noted to be effective in eradicating *Elytrigia repens* (L.) Desv. Ex W.D. Jacks (couch grass), controlling weeds and improving soil fertility; as previously mentioned, *marejea* has also been fed to cattle. As mentioned, 8 to 12 ha are still planted with *marejea* each year and about 20 bags of seeds are collected for replanting. Because cattle manure and inorganic fertilizers have been used in the maize-*marejea* fields, the effect of *marejea* alone on soil properties and maize yield cannot be quantified.

Currently, at the Mission farm, *marejea* is planted as a pure stand at the beginning of the rainy season. After 2 or 3 months, it is cut and ferried to the cowshed. In the morning, the cows are fed maize stover while in the afternoon they are fed *marejea* forage. According to Hermann Mayr, the animals have been in good health and their milk production levels have been normal.

Studies carried out beyond the Ruvuma region have assessed the effect of *marejea* on maize growth and its potential as a livestock feed (Sarwatt 1992; Temu and Aune 1995; Kullaya et al. 1998; Sarwatt and Mkiwa 1988).

In on-station conditions, *marejea* biomass production has usually been around 8 t DM ha^{-1} when grown as a sole crop and 5 t DM ha^{-1} when intercropped with maize. According to Gerold Rupper (personal communication, 1998), *marejea* biomass in farmers' fields was always highly variable because of differences in fertility and in management practices, such as planting density. It was recommended that, to obtain enough biomass and seeds, a farmer should plant 20 to 30 kg of *marejea* seeds per 0.5 ha.

According to farmers, maize grain yields in *marejea* systems were always higher than in those without *marejea.* This has hardly been quantified under on-farm conditions. A farmer in the region (Wolfram Ngairo, personal communication, 1998) indicated that there was always a good harvest after rotating *marejea* with maize. Even in the early 1980s, some farmers indicated high maize yields with *marejea* (Rupper 1987). Results obtained elsewhere in eastern African have indicated increased maize grain yield after the incorporation of *Crotalaria* into the soil (Temu and Aune 1995; Kullaya et al. 1998; Gachene et al. 2002; Mureithi et al. 2002).

All in all, the maize-*marejea* system seemingly performs well in the region both in on-station conditions and on farmer fields. A number of factors appear to contribute to this:

- The rainfall pattern favours maize and *marejea* production. The region receives an average annual rainfall of about 1200 mm with 3 dry months in a year.
- The ridge-and-furrow method, which is widely used in the region, favours residue decomposition, moisture conservation and rooting of crops.
- In most areas, the soils are low in N and P.

Costs and Benefits

No detailed studies have been conducted on the costs and benefits associated with the *marejea* -maize systems. Typically, with GMCCs the benefits include soil fertility improvement, higher main crop yield, erosion, pest and weed control and fodder production. The negative aspects include reduced land for crop production (Gachene et al. 2000). A detailed study on costs and benefits has been conducted in Uganda with *Mucuna*, *marejea* and *Lablab purpureus* (L.) Sweet (Fischler 1997), indicating that production systems with *marejea* can be of low profitability, especially if labour for incorporating biomass in the soil is valued. Silenge and Bertram (1996) have made a brief study on the costs and benefits of using *marejea* in Ruvuma (Table 2). It should be noted, however, that the study only considered the seed cost, fertilizer cost and yield benefit. Other costs, such as labour and benefits beyond yield impact, were not taken into account. The assessment of costs and benefits indicated that, while in the first year the *marejea* system would bring a net loss, in the second year it would bring a large net benefit. The average benefit during the 2-year period would be higher than in a system utilizing inorganic fertilizer.

Table 2. Costs and benefits (Tsh. ha^{-1})[a] of cultivating *marejea* in comparison to utilizing inorganic fertilizer (*Source*: Silenge and Bertram 1996).

Inputs and outputs	Costs	Benefits
Inorganic fertilizer:		
1994-95		
7.4 bags at Tsh. 7000	51850	
Yield of 24.7 bags of maize at Tsh. 3500		86420
Profit		34570
1995-96		
7.4 bags at Tsh. 12 000	88890	
Yield of 24.7 bags of maize at Tsh. 5000		123460
Profit		34570
Total profit for 2 years		69140
Marejea:		
1994-95		
49.4 kg of seed at Tsh. 150	7410	
No yield		0
Profit	Loss of 7410	
1995-96		
No fertilizer input	0	
Yield 34.6 bags at Tsh 5000		172840
Profit		172840
Total profit for 2 years		165430

[a] US$1 = Tsh. 740; calculations from original acre values.

ADOPTION OF THE *MAREJEA* SYSTEMS

Systems Adopted

Lupanga et al. (1987) attempted to find out the social and economic factors influencing the adoption of *marejea* in the surrounding areas of Peramiho. The study was conducted in eight selected villages in the Ruvuma region. Seven were around the Peramiho Mission and the remaining one was in the vicinity of Mpitimbi Mission. Both structured and unstructured interviews were used in collecting data from *marejea* growers selected by Gerold Rupper and community leaders.

The method of choice for cultivating *marejea* was a single stand. Only 10% of the 30 farmers interviewed grew it intercropped with maize. The farmers were reported to have favoured single-stand cultivation to enrich the soil (Lupanga et al. 1987). However, this statement seems dubious, as it appears that at least a part, if not all, of sole-cropping of *marejea* was done to harvest as much seed as possible. Farmers had not developed any fixed cropping patterns with regard to frequency of *marejea* cultivation but rather the frequency was determined by the harvest of the food crop. If the harvest were below expectation, the farmer would grow *marejea* in the next season. Note, however, that although not reported in this study, expected markets for *marejea* seed must have played a role in determining whether a farmer was to plant it or not in any given year. About half of the farmers grew *marejea* every third year in the same field.

Factors Affecting Adoption

A number of factors caused the adoption of *marejea* cultivation in the 1980s and early 1990s and changes in them resulted in its abandonment in the 1990s. They included those outlined below.

Seed Purchases

Seed purchasing by the Peramiho Mission Centre seems to have most influenced farmers to adopt *marejea* and, indeed, was probably alone sufficient to cause its adoption. Farmers who had cultivated *marejea* pointed out that although a lot of work was involved in its cultivation, they adopted the crop because of the seed purchases and other incentives. Already, in 1987, Lupanga et al. (1987) argued that if there was no market for the seed, it was likely that some farmers would discontinue the practice. This did happen 7 years later. In 1994, the practice of buying seeds or exchanging them with fertilizer was discontinued, which seemed to have closed the chapter of *marejea* cultivation in the Ruvuma region (Gerold Rupper, personal communication, 1998). The importance of the seed purchases was also evident in the farmers' practices that favoured seed production. Moreover, once the practice was terminated in 1994, no farmers went back to buy seeds until before 1998, when a few farmers started requesting *marejea* seeds, apparently as a response to high increases in inorganic fertilizer prices (Gerold Rupper and Gabriel Mhagama, personal communication, 1998).

Access to Fertilizer and Other Incentives

Because of the fertilizer 'subsidy', many farmers felt that they stood to gain by planting *marejea*. The Seed Bank went a step further by storing fertilizers, thus bringing them closer to the farmers. To be allowed to purchase fertilizers, a farmer had to first prove that s/he was growing *marejea* (Gabriel Mhagama, personal communication, 1998). However, by the early 1990s, most smallholders were able to acquire fertilizers through a credit programme supported by IFAD-FAO and most farmers preferred the short-term effects and lower labour demand of inorganic fertilizer to *marejea* as a means of improving soil fertility. Unfortunately, the credit system collapsed in 1994 because of non-payment of loans.

The Mission's Strong Focus on *Marejea*

The Mission's sole focus on *marejea* undoubtedly contributed to its adoption in many ways. It seems that most adopters were Catholics (Gerold Rupper, personal communication, 1998). Personal contact with Gerold Rupper significantly influenced the adoption of *marejea* and many farmers felt proud when he visited them. The study by Lupanga et al. (1987) also documented the impact of the Mission. The villages near Peramiho, Morogoro, Peramiho A, Peramiho B and Magima were aware of *marejea* cultivation already in the 1970s. Other villages farther from Peramiho, such as Litapwasi, Mpitimbi, Mdunduwalo and Mwungano-Zomba, became aware of the benefits of *marejea* only in the early 1980s. Lupanga et al. (1987) hypothesized that once Gerold Rupper stopped visiting the farmers, the adopters might also stop growing *marejea*. After the retirement of Gerold Rupper, his successor integrated *marejea* into the Organic Farming Project, giving it less emphasis.

Efforts through Mass Media and Involvement of Prominent Personalities

In the study by Lupanga et al. (1987), one-third of the farmers became aware of the crop through the monthly news magazine *Mwenge* published in Peramiho. This magazine is still in existence but promotion of *marejea* cultivation is no longer featured strongly.

The promotion of *marejea* cultivation was at its highest when the country's president and party leaders made it a public issue.

Other Factors

Farmers with small plots of land were unwilling to adopt *marejea*. This observation agrees with similar work carried out elsewhere in the tropics (Raquet 1990; Drechsels et al. 1996; Fischler 1997; Fischler and Wortmann 1999). Gerold Rupper suggested early and late intercropping of *marejea* with food crops where land was a limitation but these practices were little adopted.

Surprisingly, in the survey of 30 farmers' motives for *marejea* adoption (Lupanga et al. 1987), the most dominant driving force to use *marejea* was the felt need to improve soil fertility. Eighteen out of 30 farmers ranked soil fertility in first position of importance for improved yields. Ninety-three percent of the farmers claimed that they would keep *marejea* within their farming system even if the incentives were discontinued. Whether these farmers have continued its cultivation is not known as there was no follow-up but it seems unlikely. After the incentives were discontinued in 1994, there have been almost no farmer visits to the Seed Bank.

A number of farmers indicated that they adopted *marejea* as a result of being pressurized by their peers, particularly in women's groups.

THE WAY FORWARD

Issues that would be worthwhile considering in efforts to reintroduce *marejea* in the region include those outlined below.

Seed Availability

What alternative arrangements could be made to ensure continuity in seed multiplication and seed distribution in other parts of the country? As already mentioned, 60 tons of *marejea* seeds were distributed to other regions of Tanzania in 1993-94 but it is not known what use was made of these seeds. It is well known that lack of seed is one of the constraints affecting the utilization of GMCCs in the tropics.

Labour-saving Systems

How could *marejea* be incorporated in a way that would put the least demand on family labour? Although a few studies in Tanzania have assessed the effect of intercropping *marejea* with food crops (Temu and Aune 1995; Kullaya et al. 1998), the effect on labour demand is yet to be assessed.

Involvement of Diverse Institutions

What other institutions could be used to popularize the growing of *marejea* in Tanzania? Apparently, extension officers were not involved in its promotion in Ruvuma and they also the lacked information needed to popularize it. In Kenya, diverse organizations including research institutions and non-governmental organizations (NGOs) have been involved with the popularization of GMCCs. Future GMCCs in Tanzania should incorporate different organizations in efforts to popularize *marejea*.

On-farm Research

Research of a participatory nature has been shown to have worked well elsewhere (Fischler 1997; Fischler and Wortmann 1999) and is needed if *marejea* cultivation is to be reintroduced in the region.

ACKNOWLEDGEMENTS

I wish to thank all those whom I contacted during the execution of this case study. In particular, the following merit especial thanks: The late Father Gerold Rupper and Mr. Gabriel Mhagama of Peramiho Mission Centre, Tanzania; Dr. Method Kilasara and Mr. Malipuro, Sokoine University of Agriculture, Tanzania; Ms. Edith Shayo and Mr. Abassi Rehani, Regional Agricultural Office, Morogoro and Dr. Joseph G. Mureithi, Coordinator, Legume Research Network, Kenya Agricultural Research Institute (KARI), Kenya, for giving useful suggestions and comments during the writing of this report. Thanks are also due to Caroline Maina and Beth Wangeci for typing the manuscript. The Rockefeller Foundation Office in Mexico provided financial support.

REFERENCES

ARI - Agricultural Research Institute (1999) Southern highlands farmings systems zonation. Map 2: agro-ecological zones (draft). ARI, Mlingano, Tanzania. 1 p.

ASPRO - Agro Silvo Pastoral Project (1994) Sunnhemp-*Marejea*: a farmer's friend. ASPRO Peramiho 6/94, Peramiho, Tanzania. 2 p.

Baijukya F P (1999) Herbaceous legume cover crops in the lake zone of Tanzania: A review on the practices, research development and dissemination. Paper presented during a mini-workshop on legume cover crops for soil fertility improvement in the Eastern African Region, 18-19 February 1999. Tropical Soils Biology and Fertility Program (TSBF) HQ, Nairobi, Kenya.

Dreschsel P, Steiner K G, Hagedorn F (1996) A review on the potential of improved fallows and green manures in Rwanda. Agrofor. Syst. 33:109-136.

Fischler M (1997) Legume green manures in the management of maize-bean cropping systems in Eastern Africa with special reference to *Crotalaria* (*C. ochroleuca* G. Don). Ph.D thesis No. 12099, Eidgenössische Technische Hochschule-Zentrum (ETH) Zurich, Switzerland. 211 p.

Fischler M, Wortmann C S (1999) Green manure for maize-bean systems in eastern Uganda. Agronomic performance and farmers' perception. Agrofor. Syst. 47:123-138.

Gachene C K K, Palm C A, Mureithi J G (2000) Legume cover crops for soil fertility improvement in Eastern Africa Region. *In*: Proceedings of a workshop held at the Tropical Soils Biology and Fertility Program (TSBF) HQ, Nairobi, 18-19th February 1999. 21 p.

Gachene C K K, Mureithi J G, Anyika F, Makau M (2002) Incorporation of green manure cover crops in maize-based cropping system in semi-arid and sub-humid environments of Kenya. *In*: (Eds.) Mureithi J G, Gachene C K K, Muyekho F N, Onyango M, Mose L, Magenya O. Participatory technology development for soil management by smallholders in Kenya. Kenya Agricultural Research Institute (KARI) Nairobi, Kenya. p. 145–151.

Kullaya I K, Kilasara M, Aune J B (1998) The potential of *marejea* (*Crotalaria ochroleuca*) as green manure in maize production in the Kilimanjaro region of Tanzania. Soil Use Manage 14:117-118.

Lupanga J J, Matee A Z, Mvena Z S K, Weber J J (1987) Social economic factors influencing the adoption of *Crotalaria ochroleuca* in farming communities around Peramiho - A preliminary study. *In*: Proceedings of a writers' workshop held at Peramiho, Songea, Tanzania. Benedictine Publications, Ndanda Peramiho, Tanzania. p. 48-65.

Mureithi J G, Mwaura P M, Nekesa C O (2002) Introduction of legume cover crops smallholder farms in Gatanga, Central Kenya. *In*: (Eds.) Mureithi J G, Gachene C K K, Muyekho F N, Onyango M, Mose L, Magenya O. Participatory technology development for soil management by smallholders in Kenya. Kenya Agricultural Research Institute (KARI), Nairobi, Kenya. p. 153–164.

Raquet K (1990) Green manuring with fast-growing shrub fallow in the tropical highlands of Rwanda. *In:* Kotschi J (Ed.) Ecofarming practices for tropical smallholdings. Trop. Agroecol. 5:55-80.

Rupper G (1987) Cultivation of *marejea* (*Crotalaria ochroleuca*). Experience of Peramiho. Proceedings of a writers' workshop held at Peramiho, Songea, Tanzania. Benedictine Publications, Ndanda Peramiho, Tanzania. p. 9-12.

Rupper G (1997) Sunnhemp *Marejea*. Benedictine Publications, Ndanda Peramiho, Tanzania.

Sarwatt S V (1992) Effects of replacing sunflower seed cake with *Crotalaria ochroleuca* hay on feed intake digestibility and growth rate of grazing sheep. Small Ruminant Res. 7:21-28.

Sarwatt S V, Mkiwa F E J (1988) The current status of knowledge on the feed value of *Crotalaria* species. *In:* Dzowela B H (Ed.) African forage plant genetic resources, evaluation of forage germplasm and extensive livestock production systems. International Livestock Centre for Africa, Addis Ababa, Ethiopia. p. 196-203.

Scott R M (1972) Soils. *In*: Morgan W T W (Ed.) East Africa: Its peoples and resources. Oxford Univ. Press, Nairobi. p. 95-105.

Silenge F H, Bertram T M (1996) Kilimo Hai. Peramiho Printing Press, Peramiho, Tanzania.

Soko E (1991a) Alley cropping with *marejea*. Mruma Centre Kilimo cha Mseto. Artisanal Seed Production (ASP) 1990-91, Peramiho, Tanzania.

Soko E (1991b) One season of alley cropping with *marejea*. Mruma Centre Kilimo cha Mseto Artisanal Seed Production (ASP) 1990-91, Peramiho, Tanzania.

Temu A E M, Aune J B (1995). Effect of green manuring and rotation on maize yield in the Southern Highlands of Tanzania. Norw. J. Agric. Sci. Suppl. 21:93-98.

Chapter 8

Forage Production Systems for Dairy Production in the Coastal Lowlands of Kenya

D. M. G. Njarui and J. G. Mureithi

Kenya Agricultural Research Institute (KARI), Kenya

SUMMARY

The coastal lowlands of Kenya lie in the south-eastern part of the country and cover an area of about 80 000 km^2. The poor nutrition of dairy cows is a major constraint facing the dairy industry in the region. Available feed resources are inadequate in both quantity and quality and rarely meet the nutrient requirements of lactating cows. Because improved grasses and legumes can potentially contribute significantly to the nutrition of dairy cows, efforts have been undertaken since the early 1970s to identify and introduce suitable species to farmers. *Pennisetum purpureum* Schumach. was identified as a suitable fodder grass for the region, *Clitoria ternatea* L. as a suitable herbaceous legume and *Leucaena leucocephala* (Lam.) De Wit and *Gliricidia sepium* (Jacq.) Walp. as suitable fodder shrubs. Forage productions systems were developed using these species and were promoted in smallholder dairy farms in the region. Adoption of the forage production systems was not high. Farmers tended to modify the research recommendations for production of the forages to suit their farming systems and practices. Availability of planting materials, extension methodology and institutional support appeared critical to dissemination and adoption of the forages.

INTRODUCTION

As in most parts of the tropics, dairy development in the coastal lowlands of Kenya is mainly constrained by inadequate feeding. Small-scale farmers have limited resources for feeding their ruminant livestock (Muinga 1992; Abdulrazak 1995). The available feeds are inadequate in quantity and quality to meet the year-round nutrient requirements of animals (Reynolds et

M. Eilittä et al. (eds.), Green Manure/Cover Crop Systems of Smallholder Farmers, 195-218.

al. 1993). The main livestock feeds available at farm level are natural pastures, which are mainly grasses, and crop residues such as maize (*Zea mays* L.) stover. Generally, these feeds are of poor quality. Commercial concentrates are available but only a few farmers can afford to buy sufficient quantities for feeding their dairy cows (Thorpe et al. 1993). However, the region has great potential for dairy development because of the high demand for milk and other dairy products (Staal and Mullins 1996). The urban centres in the region, such as Mombasa, offer good markets for the dairy products and currently the price of milk is at US$0.40 per litre (US$1 = Ksh75). The region is milk deficient and more than 50% of the dairy products consumed in the region are supplied by other parts of Kenya (Staal and Mullins 1996).

A considerable amount of research has been carried out in the region to identify productive forages and develop forage production systems to alleviate the feed shortage. In the mid-1970s, the Scientific Research Division of the Ministry of Agriculture, now the Kenya Agricultural Research Institute, (KARI), in collaboration with the Food and Agriculture Organisation (FAO), began a project to evaluate suitable forage species for the coastal region (Njunie et al. 1995). The first stage involved collection, introduction and screening of forage germplasm to identify adapted and productive species. This work was carried out at KARI's Regional Research Centre (RRC) at Mtwapa, located just outside Mombasa. In 1988, the KARI and the International Livestock Centre for Africa (ILCA), now the International Livestock Research Institute (ILRI), initiated a collaborative project to develop dairying in the coastal region. The collaborative project used the promising species identified by KARI-FAO to develop forage production systems for the dairy industry in the region. On-farm research on the most promising treatments began in 1990 and researchers, extension staff and farmers were active partners. These efforts led to the development of forage production systems to support the dairy enterprise in the coastal region.

Maintenance of soil fertility was an important consideration during the development of the forage production systems. The region is characterized by free-draining, sandy soils that are very low in inherent fertility. The use of inorganic fertilizers in the region is limited mainly by their high prices and irregular availability. During the development of the forage production systems, inexpensive alternatives such as animal manure, biological nitrogen fixation and recycling of nutrients received particular emphasis. In this chapter we present and discuss the main species selected from the numerous screening trials and the major forage production systems developed.

REGIONAL CONTEXT

The Biophysical Environment

Location and Landmarks

The Coast Province accounts for about 15% (80 000 km^2) of the country's surface area and includes seven districts: Taita Taveta, Tana River, Lamu, Malindi, Kilifi, Mombasa and Kwale (Figure 1). The total landmass of the Coast Province covers a wide range of physical features including the gently rolling Shimba hills to the south and the relatively flat Yatta plateau to the north. The coastal lowlands mainly lie within three districts, Malindi, Kilifi and Kwale, which cover 1 275 800 ha. The coastal lowlands region is situated at lat 0-4°S and long 38-44°E and extends inland from the coast bordering the Indian Ocean to the east and rising to 300 masl toward the west.

Climate

Rainfall is varied along the lowland coastal region and it is highest (1200 mm) in the south-west in ecological zone Coastal Lowland 3 (CL3, coconut [*Cocos nucifera* L.]– cassava [*Manihot esculenta* Crantz] zone). Further north, in the Kilifi-Malindi area (CL4, cashew nut [*Anacardium occidentale* L.]– cassava region) annual rainfall decreases to 1000 m and in Lamu (CL5, livestock-millet [*Panicum miliaceum* L.] zone) it further drops to 900 mm (van Einnatten 1979; Jaetzold and Schmidt 1983). Similarly, rainfall decreases from the coast to the hinterland. Consequently, the number of wet months (≥ 100 mm m^{-1}) decreases from 5 months on the southern coast to 3 months in Lamu. The El Niño rains of 1997-98 resulted in flooding of the low-lying areas and destruction of infrastructure.

Rainfall is bimodal with a long rainy season from April to June and the short rains from October to December. There are two distinct dry seasons, which normally occur from January to March and from June to September. This bimodal rainfall allows for two crop growing seasons: the more reliable long rains and the short rains, which are seldom enough to support a good crop. Access to irrigation in the lowland region is limited. However, in the bordering district of Tana River, cotton (*Gossypium hirsutum* L.) is grown under irrigation. Potential evapotranspiration is high, 1900-2300 mm, and exceeds annual precipitation in most months, resulting in water deficits. Mean annual temperatures are also high, ranging from 22 to 35 °C. The warmest period is from January to April when the daily temperatures average more than 30 °C. Relative humidity is high and ranges from 70 to 90% for most of the year.

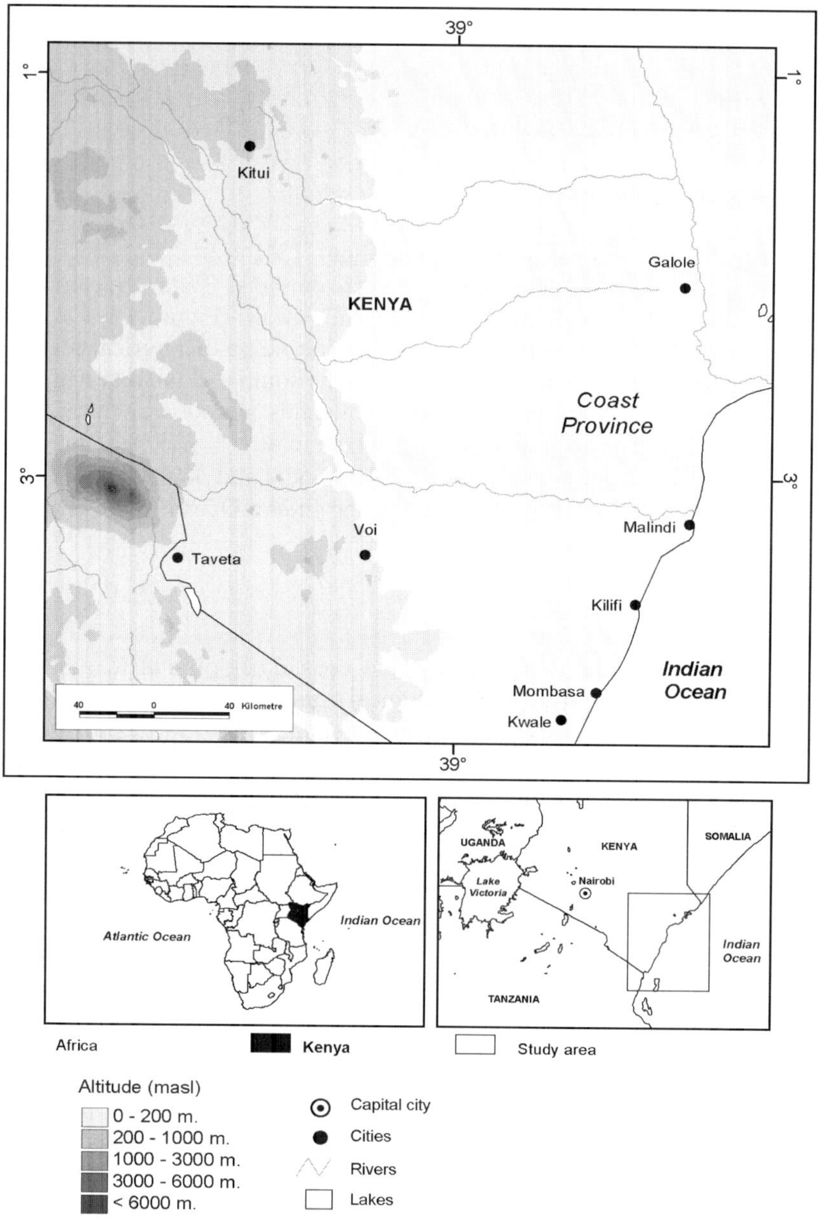

Figure 1. Location of study area, coastal region of Kenya.

Soils and Soil Fertility

The dominant soil types in the coastal lowlands are derived from sandy depositions, such as the Magarini sands, and unconsolidated lagoonal deposits, such as the Kilindini sands. These soils include

Ferralsols, which are sandy clay soils, strongly weathered and red brown in colour, and Cambisols, which are shallow to moderately deep, little-weathered clay soils. Other soils include Arenosols, which are very deep sandy soils, and Acrisols, which have sandy clay subsoil and very sandy topsoil. Because of their sandy topsoil, most soils have high infiltration rates and are prone to nutrient leaching; additionally, their low organic matter content makes them of poor to very poor fertility (Table 1). Soils are deficient in nitrogen (N), phosphorus (P) and potassium (K). Generally, soil P is below the critical level of 20 ppm.

Table 1. Texture class and chemical composition of some common soils of the coastal lowlands of Kenya (*Source*: Michieka et al. 1978).

Soil type	Texture class[a]	pH	N	C	K	Ca	Na	Mg
		(H_2O)	(%)		(meq $100g^{-1}$)			
Ferralsol	SCl	5.9	0.04	0.49	0.11	1.40	Trace	0.50
Cambisol	Cl	5.9	0.14	1.23	0.85	12.20	1.60	6.80
Arenosol	LS	5.4	0.06	0.38	0.05	0.50	Trace	0.45
Acrisol	SCl	5.8	0.06	0.54	0.30	2.60	0.13	0.90

[a] SCl = sandy clay, Cl = clay, LS = loamy sand.

In most smallholder farms, soil fertility is declining rapidly because of soil erosion, burning of plant residues, overgrazing, lack of rotation and continuous cultivation without the replenishment of soil nutrients. Use of inorganic fertilizers is limited by shortage of cash and, sometimes, by timely availability of fertilizers. Farmers with livestock return cattle manure to their fields and intercropping with leguminous crops is a common method used to add nitrogen to the soil.

The Human Environment

Population

The 1989 census gave total population for the region as 2.4 million. Most of the inhabitants belong to a group collectively known as Mijikenda, which is composed of nine subgroups. The region has many migrants, mainly from western, central and eastern Kenya, who were attracted by employment opportunities in the tourism and shipping industries (Maarse et al. 1998). The port city of Mombasa is the largest town in the region, with an estimated population of 650 000 people (Mullins et al. 1996). Because of its linkage with the Arabian Peninsula and India, Mombasa is a home of Muslim, Hindu and Christian cultures.

Farm Income

Farm income is limited mainly by low crop yields coupled with poor market infrastructure. Off-farm employment is an important source of income, contributing about 60% of total smallholder household income in Kilifi district. However, off-farm income has been declining because of the low number of tourists visiting the region in recent years (Maarse et al. 1998). Also, employment opportunities from manufacturing and tertiary industries have been few because of the poor growth and performance of these sectors.

Farming Systems

There are few studies with detailed household characterization including wealth ranking. However, Thorpe et al. (1993) conducted a survey to characterize households in Kaloleni Division in the region. About 75% of the households had less than 6 ha of land and the norm was 2 ha. Over 90% of households had houses with temporary roofs and walls but 75% had access to piped water. About 65% of households kept livestock and less than 20% had some cattle. The households with livestock had 13.2 members while those without had 9.4 members. Mean total annual farm income was US$980 and livestock contributed 15-65% of total farm income. On average, farm income was only 30% of the total household income. About 33% of all households, especially those without cattle, were seeking credit at the time of the survey. However, credit from commercial banks is expensive and most farmers are not willing to use land title deeds as collateral for the credit.

Of late, the traditional land ownership has changed to a freehold system. Land has been demarcated and individuals issued with title deeds. Landholdings in the two major settlement schemes, Mtwapa and Tezo of Kilifi District, were initially 12 acres but they have been subdivided into smaller units because of increasing population pressure. A few large-scale farms are found in the region, mainly growing sisal (*Agave sisalana* Perrine ex Engelm.), coconut or keeping dairy animals. Mixed farming forms the base of subsistence for most of the rural people. Cultivation is carried out with hand tools although the ox plough is also used in some areas. Both men and women carry out farming activities. Young people of school age usually help during school holidays.

Maize, cassava and cowpea (*Vigna unguiculata* [L.] Walp.) are the main staple food crops. They are almost always cultivated as intercrops. Other minor but important food crops include upland rice (*Oryza sativa* L.) and beans (*Phaseolus vulgaris* L.). Tomatoes (*Lycopersicon esculentum* Mill.), pepper (*Capsicum* sp. L.) and *Amaranthus* sp. L. are popular vegetables grown for cash. Major cash crops are coconut, cashew nut, bixa (*Bixa orellana* L.) and horticultural tree crops, such as mangoes (*Mangifera indica* L.), citrus (*Citrus* sp. L.) and bananas (*Musa* spp. L.). The tree crops, coconut and cashew nuts are generally planted without any pattern and the age of the trees ranges widely, indicating continuous planting of trees by farmers. Pineapples (*Ananas comosus* [L.] Merr.), tobacco (*Nicotiana tabacum* L.) and paw paw (*Carica papaya* L.) are also grown in limited areas. Cotton is cultivated in the north.

Crop yields in farmers' fields are generally low (Table 2), mainly because of poor soil fertility and use of inappropriate agronomic practices and unsuitable crop varieties. Maize, the staple crop, hardly produces more than 1.5 t ha^{-1} y^{-1}. The region is food deficient and imports staple foods from other parts of the country.

Table 2. Typical farmer yields of crops in the coastal lowlands of Kenya.

Crop	Yield (t ha^{-1})
Sesame (*Sesamum indicum* L.)	0.5 (dry grain)
Cotton (*Gossypium hirsutum* L.)	1.0 (fibre)
Tomato (*Lycopersicon esculentum* Mill.)	11.0 (fresh fruits)
Water melon (*Citrullus lanatus* [Thunb.] Matsum. & Nakai)	8.0 (fresh fruits)
Cashew nut (*Anacardium occidentale* L.)	0.6 (nuts)
Bixa (*Bixa orellana* L.)	0.5 (dry grain)
Citrus (*Citrus* spp. L.)	4.0 (fresh fruits)
Mango (*Mangifera indica* L.)	9.0 (fresh fruits)
Paw paw (*Carica papaya* L.)	7.0 (fresh fruits)
Pineapple (*Ananas comosus* [L.] Merr.)	12.0 (fresh fruits)
Coconut (*Cocos nucifera* L.)	1.2 (nuts)
Subsistence:	
Maize (*Zea mays* L.):	
Local	1.0 (dry grain)
Improved	1.5 (dry grain)
Cowpea (*Vigna unguiculata* [L.] Walp.)	0.6 (dry grain)
Sweet potato (*Ipomoea batatas* [L.] Lam.)	6.0 (fresh tubers)
Cassava (*Manihot esculenta* Crantz)	4.0 (fresh tubers)
Green gram, mung bean (*Vigna radiata* [L.] R. Wilczek)	0.4 (dry grain)

The cropping calendar follows the rainfall pattern, which is bimodal (Figure 2). January to March and July to September are relatively dry months when food crops are harvested and land is prepared for the next food crop. Planting usually coincides with the onset of the rains in April-May and October-November. The crop is weeded during the growing period. Coconuts are harvested in April to July and October to December.

Livestock includes sheep, goats, cattle and poultry. Sheep and goats are common and are kept under a free-grazing system. They are sold to provide cash for school fees, food and other household needs. Most of the cattle are local Zebus (*Bos indicus*), which are mainly kept as a source of cash and security for the household.

Dairy herds are few and are mainly kept in CL3 and CL4 ecological zones. Two-thirds of the dairy herds in CL3 and CL4 are under a free-grazing system (Thorpe et al. 1993). The other one-third is kept under a zero-grazing system, which was introduced by the Dutch-supported National Dairy Development Program (NDDP) in the early 1980s to boost milk production. In this system, animals are confined to a shed, with separate feeding and sleeping areas. Farmers who practice the zero-grazing system have planted Napier grass (*Pennisetum purpureum* Schumach.), which they cut and carry to their cows. Some farmers have opted to practice a semi-zero-grazing system where animals graze in the field during the day and are stall-fed additional forages in the evenings. Farmers supplement their dairy animals with small quantities of maize bran alone or in a mixture with copra cake (from *C. nucifera*). Crop residues, mainly maize stover and grain legume haulms, are important livestock feeds.

The major livestock diseases in the region are East Coast Fever (ECF), a tick-borne disease, and trypanosomiasis, a tsetse-fly-borne disease. The cost of controlling these diseases is high and sometimes unaffordable. Free grazing exposes animals to these diseases, while the zero-grazing system minimizes the exposure. At present, dairy cows are available from upcountry but at US$700 per cow they are expensive. The government supported a program of artificial insemination to upgrade the local cows but it is currently undergoing restructuring and the services are not readily available.

Market and Infrastructure

Large urban centres such as Mombasa provide the main market for the agricultural produce from this region. However, marketing of farm produce is difficult and expensive because of the poor road network. Only a single tarmac road passes through the region; the rest are all-weather roads. Prices of the major farm produce are typically low. At farm-level, the price of 90 kg of maize is about US$10 (KSh. 750) and this is only slightly higher than the break-even price. The prices of coconut and cashew nut are also low. However, the prices of milk and dairy products have remained high because of the strong demand from the increasing urban population.

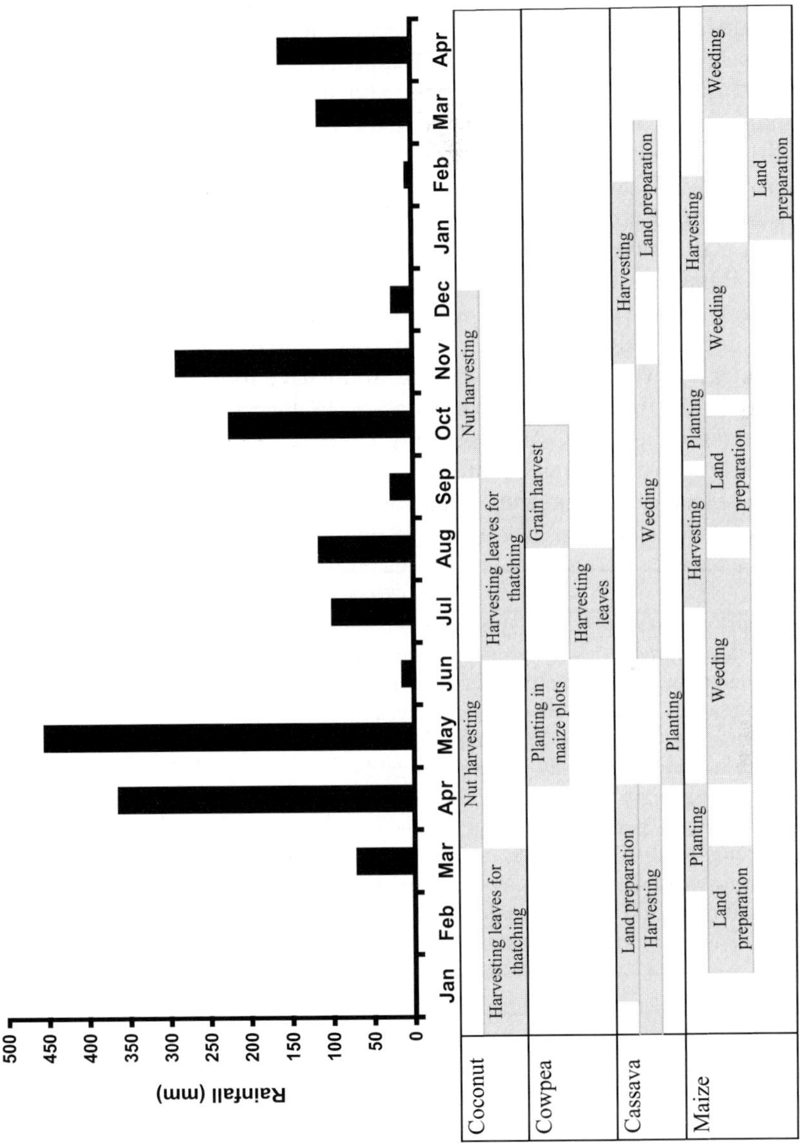

Figure 2. Typical cropping calendar for subhumid coastal Kenya.

SCREENING OF BEST-BET FORAGE SPECIES

Major efforts to screen and identify best-bet forage species began in 1974 when the FAO-Ministry of Agriculture Forage Project established a large forage nursery at RRC-Mtwapa. This nursery included legume and grass species that both had been collected in the region and introduced from outside the region. Over 200 species were planted (CARS Annual Reports 1974-80). Among the legume species collected in the region were *Clitoria ternatea* L., *Capsicum* spp. and *Crotalaria emarginata* Benth. Among those introduced from outside were *Centrosema pubescens* Benth., *Macroptilium atropurpureum* (DC.) Urban and *Leucaena leucocephala* (Lam.) De Wit.

The project also evaluated a range of forage grasses and, in one of its studies, 32 accessions of Napier grass. From this study, Napier grass variety Bana was found to be the most productive. In 1989, 11 *Leucaena* accessions were evaluated and *Leucaena* cultivar K28 was recommended for the region (Mureithi et al. 1994). The NDDP introduced *Gliricidia sepium* (Jacq.) Walp. in 1987-88 as an alternative fodder tree to *Leucaena* (Gelder 1988). In 1991, 62 accessions of 18 herbaceous legume species were assessed at RRC-Mtwapa and *Clitoria* and *Macroptilium* were among the most productive among them (Njunie et al. 1996). Forage species selected and recommended for the region included Napier grass var. Bana, *Clitoria, Leucaena* var. K28 and *Gliricidia* (Boxes 1-4).

Box 1. Clitoria

Known as butterfly pea in Australia and blue pea on the East Africa coast, *Clitoria ternatea* L. is native to tropical America and commonly grown as an ornamental in the world's warmer regions. It is a perennial climbing legume or trailing herb, shrubby at the base; it needs a minimum annual rainfall of 400 mm but performs best with 1500 mm. It is drought tolerant and grows in soils ranging widely from sandy to deep alluvial loams to heavy black cracking clays. It tolerates salinity and can grow from sea level to 1800 masl. In coastal Kenya, it has shown persistence under frequent harvesting (Njunie et al. 1996), producing up to 3.3 t ha^{-1} DM at frequent cuttings as a sole crop. CP content in DM ranges from 10 to 26%. It has a good feeding value and daily live-weight gains of 0.68 kg d^{-1} were recorded on cattle grazing *Brachiaria mutica* (Forssk.) Stapf (Para grass)-*Clitoria* pastures in Australia (Skerman et al. 1988).

Box 2. Gliricidia

Gliricidia sepium (Jacq.) Walp is native to Mexico and the West Indies but has spread in the tropics. It is a leguminous tree growing to 15 m in height in a wide range of soils and tolerates acidity better than *Leucaena* (Simons and Stewart 1994). It is very productive and has a high leaf quality with up to 29.6% CP. It is believed to be the most widely cultivated multi-purpose tree after *Leucaena* (Simons and Stewart 1994). It has great potential as a fodder tree in cut-and-carry systems and is a good browse for goats. However, the pungent smell of the fresh leaves reduces its palatability.

Box 3. Leucaena

Leucaena leucocephala (Lam.) de Wit is native to Central and South America and the Pacific Islands but has been naturalized in much of the tropics. This small leguminous tree grows up to 20 m high. Deep rooted and drought tolerant, it remains green during the dry season. It is adapted to tropical areas with an annual average rainfall over 650 mm and thrives in average temperatures from 20 to 30 °C. It grows from sea level to 1800 masl but, at higher elevation, growth is slower and it tends to remain shrubby. It grows best in deep, well-drained soils, which have a neutral to slightly alkaline pH. It can thrive in heavy clay soils with adequate internal drainage. It is a vigorous plant and produces high DM (up to 20 t ha^{-1} recorded in Hawaii). In coastal Kenya, yields range from 8 to 13 t ha^{-1} of DM and average CP content is 27% (Mureithi et al. 1994). It is highly palatable and live-weight gains of up to 0.7 kg d^{-1} were reported with steers grazing almost pure stands of *Leucaena* in Hawaii. However, it has an alkaloid, mimosine, causing hair loss and reduced appetite in ruminants.

***Box 4.* Napier grass**

Napier grass (*Pennisetum purpureum* K. Schum), commonly known as elephant grass, is native to eastern and central Africa. Its natural habitat is in riverbed areas where it grows up to 10 m high (dwarf types <4 m). It is a perennial producing up to 15 tillers at maturity. Stems are thick and the leaf sheath has numerous stiff hairs, which fall off early. Its optimum temperature for growth is 25 to 40 °C and it stops growing at 10 °C. It needs over 1000 mm of well-distributed rainfall and deep fertile soils for best growth. In subhumid areas of Kenya, it remains green in the dry season but has poor survival in areas with drought period > 4 months. It is the most important fodder grown on smallholder farms (Goldson 1977) and has contributed greatly to dairy cattle feeding in Kenya, where 10-40 t ha^{-1} DM yields have been recorded and the mean CP level is 7.6% (Wouters 1987). In the coastal area, <20 t ha^{-1} yields over a period of 12 months were recorded (Mureithi et al. 1996b) and CP content in DM ranges from 6.4 to 7.2%.

THE DEVELOPMENT OF FORAGE PRODUCTION SYSTEMS

In response to a need recognized by the Ministry of Livestock Development and Marketing (MoLD; now the Ministry of Agriculture and Rural Development), KARI and ILCA established a collaborative dairy research project in 1988. The aim was to identify and address biological, social and economic constraints to the development, adoption and productivity of smallholder dairy systems in the coastal lowlands. The project was based at KARI's RRC-Mtwapa and was planned and carried out in close collaboration with MoLD's extension services, particularly with the NDDP. It covered on-farm and on-station research and involved farming system description, constraint identification and technology development and testing. The major research areas included the development of forage production and feeding systems, disease risk to dairy cattle and dairy product consumption and marketing. Here, we report on the development and transfer of forage production technologies. These forage production systems were targeting farmers practicing zero- and semi-zero-grazing systems.

Structure of the Production Systems

The development of forage systems was based on the alley cropping system on which a significant amount of research was conducted in West Africa in the early 1980s. According to Kang et al. (1990), the alley cropping system is an agroforestry practice in which woody plants, preferably leguminous trees or shrubs, are grown simultaneously with a crop (Kang et al. 1990). The tree prunings can be used as a green manure to fertilize the companion crop or are fed to livestock. The trees are grown in wide-spaced rows with the crops planted in alleys between the hedgerows. Because of their extensive root system, the trees extract nutrients and moisture from deep soil layers and provide quality forage even during the dry season. When planted on steep slopes, they effectively control soil erosion.

Four production systems were developed that were based on Napier grass intercropped with the tree legumes or the herbaceous legume *Clitoria*:

(1) Napier grass-*Leucaena*: Here we present the most important aspects of the system (see Mureithi [1992] and Mureithi et al. [1995] for greater detail of the research that led to development of this system in coastal Kenya). *Leucaena* hedgerows are spaced 5 m apart and have an intra-row spacing of 25 cm. This gives a plant population of 8000 trees ha^{-1}. Napier grass cv. Bana is planted between *Leucaena* hedgerows at a spacing of 1 m between rows and 0.5 m within rows. The distance between Napier grass rows and *Leucaena* hedgerows is 1 m; thus a 5-m wide *Leucaena* alley can accommodate four rows of Napier grass. Because it does not produce viable seed, Napier grass is propagated vegetatively. Planting is normally carried out at the beginning of the rains, using rooted splits or cane cuttings. The latter are planted at an angle of 45° to the ground. *Leucaena* seedlings are raised in a nursery for about 6 weeks before transplanting, when they are about 30 cm high. *Leucaena* hedgerows can also be established by direct seeding but they are slow to develop.

(2) Napier grass-*Gliricidia*: In this system, the arrangement for Napier grass and *Gliricidia* is similar to that of the Napier-*Leucaena* production system. *Gliricidia* is propagated vegetatively using cuttings because it is a poor seed producer.

(3) Napier grass-*Clitoria*: Because of its climbing habit, *Clitoria* mixes well with grasses. The method of planting Napier grass is similar to that described above. After planting Napier grass, *Clitoria* seeds are

drilled between rows of the grass at a rate of 3-5 kg ha^{-1}. No seed scarification or inoculation is needed.

(4) Napier grass-*Leucaena* or *Gliricidia-Clitoria*: In this system, the arrangement for Napier grass and *Leucaena* (or *Gliricidia*) is similar to that used in the Napier-*Leucaena* production system. After planting both Napier grass and *Leucaena* (or *Gliricidia*), *Clitoria* seeds are drilled between Napier grass rows at the same rate as in the above system (3-5 kg ha^{-1}).

Management and Productivity of the Systems

Generally, the tree legumes take at least 1 year to establish. *Leucaena* hedgerows establish much faster but raising seedlings in a nursery can be expensive. Most farmers use direct seeding, which is easier, but the trees take longer to establish. *Gliricidia* hedgerows are easily established from stem cuttings. Napier grass is planted together with the trees but is cut frequently to avoid shading of the tree seedlings.

After establishment, the tree hedgerows are cut to a stump height of 50 cm and the subsequent harvests are made at the same time as the harvesting of Napier grass. The grass is harvested when it attains a height of 1 m and should be cut at 10 cm above the ground. *Clitoria* is also harvested at the time of harvesting Napier grass. Farmers are advised to fertilize Napier grass using inorganic N at the rate of 75 kg ha^{-1}. The recommended rate of slurry application is 55 t ha^{-1} and it should be buried 10 cm between Napier grass rows to reduce loss of nitrogen by volatilization (Mureithi et al. 1996a).

Studies at the RRC have indicated that forage production systems combining Napier grass and legumes yield more total DM than Napier grass grown alone (Table 3). Napier grass is tolerant of frequent cutting, while *Leucaena* is a prolific leaf producer in fertile soil. However, the performance of the intercropped forages suffers, especially during stress conditions. The arrival of the psyllid pest (*Heteropsylla cubana* Crawford) in the region in 1992 (Reynolds and Bimbuzi 1993) severely affected the productivity of *Leucaena* but it has now stabilized, probably because of the establishment of natural enemies. The productivity of the various forage production systems in farmers' fields has not been well studied. However, careful management can maintain high biomass yield of both Napier grass and legumes over several years.

Table 3. Dry matter (DM) yield of various forage production systems for eight harvests taken between June 1990 and February 1992 (*Source*: Mureithi et al. 1995).

Treatment[a]		Forage DM yield (t ha^{-1})[b]			
	Source of N	Napier grass	*Leucaena*	*Clitoria*	Total
Without *L*	-	16.4	-	-	16.4
With *L*	-	14.3	5.7	-	20.0
With *L*	*Clitoria*	13.7	4.9	4.8	23.4
With *L*	Slurry	22.3	6.1	-	28.4
Without *L*	Fertilizer (NPK)	26.2	-	-	26.2
SED		2.36	0.47	-	2.59
F-test probability		**	NS		**

[a] *L* = *Leucaena.* Rate of fertilizer application (kg ha^{-1}): N 75, P 20 and K 25.
[b] NS = Not significant; ** = P<0.01.

Labour Demand of the Systems

Despite being high yielding, the forage systems require additional labour for the establishment and management of the hedgerows. The dairy unit requires labour for feeding, cleaning and collection of slurry. Good cooperation is required between family members for the dairy enterprise to succeed. Mullins et al. (1996) studied the role of gender in labour contribution to the dairy enterprise. The study sampled smallholder farmers in Kilifi district. Considering all the activities in a dairy enterprise, they found that women performed on average 50% of the work (Table 4). However, according to Nicholson et al. (1998), labour supply in dairy farms averaged 429 days per year with twice as much female as male labour.

Table 4. Labour contribution by role for selected activities in the dairy enterprise, Kilifi district of Kenya (*Source*: Adapted from Mullins et al. 1996).

Activity	Share of work (%) performed			
	Wife	Hired labour	Children	Husband
Planting	69	16	0	15
Weeding	67	33	0	0
Cleaning shed	53	26	16	5
Cutting grass	48	28	17	7
Milking	33	30	33	4
Herding	25	50	25	0

Benefits of the Forage Systems

Benefits to Soil

Studies conducted by Mureithi (1992) showed that intercropping *Clitoria* with Napier grass increased the N level in the soil (Table 5), possibly increasing the availability of N to Napier grass. *Clitoria* sheds its leaves throughout its growth but more so during the dry season (Njunie et al.

1995). Decomposition of *Clitoria* is likely to increase the soil N supply. On the other hand, the presence of *Leucaena* hedgerows significantly depressed P and K levels in the soil. However, this depression did not seem to affect the performance of Napier grass. A soil moisture study conducted to assess moisture extraction in the alley system by trees and Napier grass revealed that the soil below the *Leucaena* hedgerows had higher moisture content at all depths up to 2 m than the soil away from the hedgerows (Mureithi 1992). This was attributed to the harvest of rainfall by the trees and the ability of their root systems to extract soil moisture from deeper soil layers. This moisture could have boosted the growth of Napier grass rows adjacent to the hedgerows. Field observations also suggest that, when planted across slopes, the trees are very efficient in soil erosion control.

The Ministry of Agriculture has recommended Napier grass for establishing grass strips and stabilizing terrace banks. Also, most farmers planted these forages in parts of their farms requiring soil erosion control.

Table 5. Effect of *Leucaena* hedgerows, *Clitoria* and slurry on soil organic matter (SOM), total nitrogen, phosphorous and potassium after 3 years of cropping (*Source*: Adapted from Mureithi 1992).

Treatment	SOM (%)	N (%)	P (ppm)	K (cmol (+) $100g^{-1}$ of soil)
Without *Leucaena*	0.89	0.04	7.87	0.14
With *Leucaena*	0.82	0.03	6.63	0.12
SED	0.036	0.002	1.083	0.010
F-test probability[a]	NS	NS	**	*
Nitrogen source:				
None	0.78	0.03	6.09	0.12
Clitoria	0.84	0.04	5.06	0.11
Slurry	0.94	0.04	10.59	0.16
SED	0.043	0.002	1.326	0.012
F-test probability[a]	**	**	***	***

[a] NS = Not significant; * = $P<0.05$; ** = $P<0.01$; *** = $P<0.001$.

Benefits to Livestock

The fodder production systems produced high amounts of good-quality biomass (Mureithi 1992). Muinga (1992) reported increases in DM intake and milk production when Napier grass was supplemented with *Leucaena*. A study conducted at the RRC-Mtwapa found that supplementing dairy cows fed Napier grass basal diet with 2 kg DM *Leucaena* per day increased daily milk yield from 6.1 to 7.8 kg d^{-1} cow^{-1} (Muinga et al. 1992). Njunie et al. (1995) reported that 88% of farmers who fed *Clitoria* forage to dairy cows recorded increased milk production. Abdulrazak et al. (1996) reported that crossbred steers fed on a maize stover basal diet increased DM intake from 2.3 to 2.5 kg d^{-1} when offered *Gliricidia* forage and from 2.3 to 2.7 kg d^{-1} when offered *Leucaena* forage at 15 g DM kg per $W^{0.75}$ of the diet. Live-weight gains increased linearly to about 700 g d^{-1} with increasing proportion of *Leucaena* in the diet. Benefits accruing from feeding these forages have not been widely studied in on-farm conditions but farmers have reported better performance of animals with them.

Benefits to the Household

Thorpe et al. (1993) reported that livestock contributed 15-65% of the total farm income and that, on average, dairying contributed 72% of the income derived from livestock. An attempt to estimate economic returns of a system combining maize-*Leucaena* and Napier-*Leucaena* systems is reported by Mureithi et al. (1995). Maize in the alley was intercropped with cowpea and fertilized with 50% of the harvested tree prunings. Napier grass was fertilized with cattle slurry from a zero-grazing unit. Figure 3 shows the products from the combined system and economic calculations. Milk sales contributed more than 74% of the total gross income, which emphasizes the importance of dairy cows in smallholder agriculture in the region.

Nicholson et al. (1999) conducted a study to assess the impact of dairy adoption on household income and nutritional status. The results showed that average income per month from dairying was remarkably higher for households that had adopted dairy farming than for non-adopting households (Table 6). Also, the nutritional status of household members was likely to improve from consuming milk and food bought with cash from milk sales.

Table 6. Monthly cash income (US$) of households that adopted dairying compared to non-adopters[a] (*Source*: Unpublished data, Kenya Agricultural Research Institute-International Livestock Research Institute adoption survey of 202 households, June-July 1997).

Income source	Adopters	Non-adopters
Dairying	110	2
Poultry or eggs	32	8
Crops	26	15
Wages, salaries or non-farm activities	150	46
Other income	13	2
Total income	331	73

[a] Adopters are households currently owning at least one dairy cow. Non-adopters own no grade or crossbred cattle.

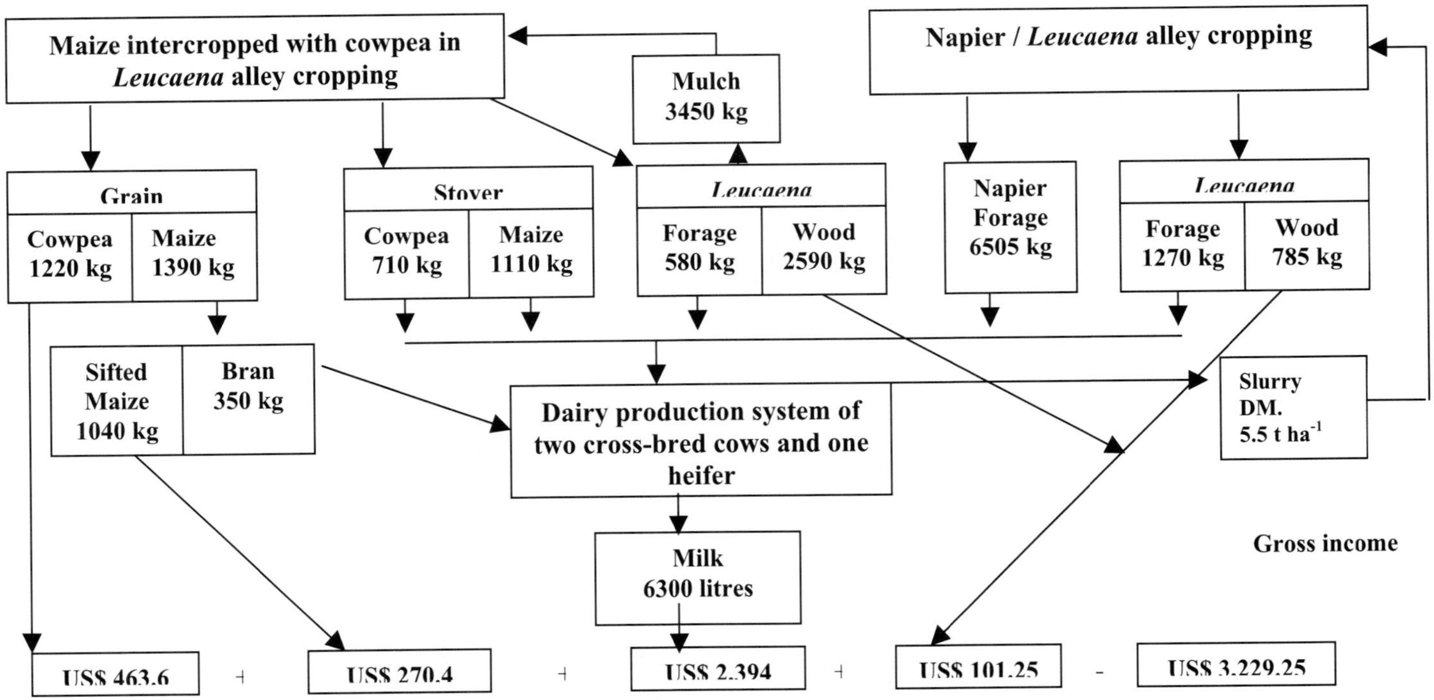

Figure 3. Estimated gross income from a hectare of *Leucaena* alley cropping system, half of which is planted with maize intercropped with cowpea and the other half planted with Napier grass fertilized with slurry. *Source*: Mureithi et al. 1995. (Prices as at July 1997 in US$: sifted maize = 0.26 kg^{-1}, cowpea = 0.38 kg^{-1}, wood = 0.03 kg^{-1} and milk = 0.38 L^{-1}.)

DIFFUSION STRATEGIES

Several institutions have been involved in forage introduction and dissemination in the region. The NDDP introduced Napier grass for dairy cattle feeding as a high-quality basal feed in 1981. The project provided planting material and assisted farmers to secure loans to buy dairy cows. Later, in 1987, the project encouraged planting of *Leucaena* and *Gliricidia* as fodder trees to bridge the protein gap in feeds.

The NDDP was an active partner in the KARI-ILCA collaborative project, which fostered strong research-extension-farmer linkages through on-farm trials and regular meetings. Because of continuous interaction, good relationships developed among researchers, extension staff and farmers. Monthly seminars and regular workshops were held, mainly to review results from field studies and on-station experiments. From 1991 onwards, meetings of researchers and senior extension staff were held to review program activities and new proposals. In these meetings, research-extension working groups were formed to supervise on-farm trials. One such trial was the evaluation of forage production systems on farm. Prior to the implementation of this trial, farmers were exposed to different forage production systems in meetings and through demonstrations. Those interested were further trained on the implementation of the trials and participated in their planning, monitoring and evaluation. Field days and demonstrations were held in the fields of participating farmers to expose more farmers to available forage technologies.

Use and Adoption by Farmers

The forage production systems described above target the dairy producers who keep animals in zero-grazing or semi-zero-grazing systems. More farmers adopted forages, particularly Napier grass, after the initiation of the NDDP in 1980 because the project was aggressive in its extension approach (Mureithi et al. 1998). Since 1992, *Clitoria* seed has been distributed to about 80 smallholder farms (Thorpe et al. 1993). Njunie et al. (1995) reported that 60% of the farmers followed the recommended agronomic practices 2 years after being exposed to the technologies and over 95% of the farmers recommended the legumes to their neighbours. In contrast, a survey by NDDP reported that only 37% of the farmers had plots well weeded and 36% of the farmers applied slurry to Napier grass (Mureithi et al. 1998). About 40% of the farmers used cattle manure to make compost. An adoption survey by Nicholson et al. (1999) interviewed 75 adopters and 125 non-adopters of dairy cattle using a stratified random sampling method. They reported that the amount of Napier grass planted per farm was highest in the Kilifi district. They also reported that the number of farmers planting Napier grass has declined since 1993. Several reasons may be attributed to this decline, including extra labour required to manage the dairy enterprise, loss of cows because of diseases and the inability to purchase cows.

Some farmers tended to modify the research recommendations for the production of the forages. For example, a study by Maarse et al. (1998) reports that only about 60% of farmers harvested Napier grass when it attained a height of 60-120 cm (recommended height is 1 m). *Leucaena* and/or *Gliricidia* hedgerows were established along farm boundaries instead of being established in alleys and some farmers planted Napier grass at a wider spacing than the recommended one. In some farms, the trees were established in coconut and/or cashew nut fields, which resulted in poor performance of trees caused by shading.

Factors Impacting Adoption and Performance of the Forage Systems

In general, a great deal of research has been conducted to develop forage technologies for the coastal lowlands region and both researchers and extension officers have undertaken efforts to transfer the technologies to farmers. The researchers have demonstrated the benefits of practicing dairying—most importantly, improved health and food security and increased income. However, challenges also face the performance and adoption potential of forage production systems in the coastal lowland. Several authors have discussed these factors, which are summarized here.

One such limiting factor is that dairying is an expensive enterprise and requires capital for the purchase of, for example, dairy cows, for materials for the construction of sheds and for the planting of forages. Most of the households are poor and credit facilities are not readily available, including from the Agricultural Finance Corporation (AFC). Commercial banks operating in the region can provide loans but few farmers are willing to use them because of high interest rates. Clearly the availability of credit at lower interest rate would enhance the adoption of dairying in the region. Additionally, ECF and trypanosomiasis are prevalent in the region, causing high cattle losses, and many blame the diseases for their decision to abandon their dairy enterprise. Finally, poor infrastructure and long distances to market centres limit the potential of dairying in the region. There is no organized system of milk marketing and only producers near urban centres have markets for their milk.

Other factors specifically affect the performance and adoption potential of forages, including two that related to issues of land and labour. Also, a number of issues related to dissemination affect the performance and adoption potential of the forage system. Consequently, they can be quite readily impacted through improved efforts in extension.

Allocation of Labour

Compared to food crops, farmers give low priority to the management of forage production (e.g. fertilizing and weeding of forages). For example, during the project, most farmers seemingly had difficulties in appreciating that improved forages are like any other crop that requires good agronomic practices, including fertilization, in order to be productive. Additionally, because of labour shortage, farmers are not able to manage the forages as recommended, which results in their low productivity. More work needs to be done on encouraging farmers to grow high-quality feed for their livestock and to educate farmers on other benefits of the legumes, for example, the potential use of *Leucaena* as a multipurpose tree for firewood and the benefits of legumes to soil.

Allocation of Land among Different Cropping Systems

Maize and horticultural crops are planted on the fertile part of the farm, while the rest is left for other activities such as forage production. Coconut and cashew nut trees are planted all over the farm without any pattern and in some farms the forages are planted under them, reducing forage productivity because of shading and competition for nutrients.

Availability of Planting Materials

Each species differs in its requirements for planting materials. Napier grass is not a major problem because cane cuttings and splits are abundant in the wet season. The Ministry of Agriculture and RRC-Mtwapa used to multiply Napier grass for distribution to farmers for free but now the Ministry sells it for a nominal fee. The Agroforestry Centre at Mtwapa and the Ministry also used to give out *Leucaena* and *Gliricidia* seedlings for free but now farmers are charged for them. Farmers willing to multiply *Gliricidia* seeds have expressed disappointment because *Gliricidia* flowers poorly in the region. The only major source of the *Clitoria* seeds is RRC-Mtwapa and only small quantities are available. If methods of accelerating seed production of *Clitoria* are developed, these systems will most likely become more popular. Clearly, the above-mentioned projects improved the availability of planting material, which fostered the adoption of the technologies. In addition, some farmers have multiplied their own seed, which they have sold and sometimes given free to their neighbours. At present, the Legume Research Network Project, supported by the Rockefeller Foundation, is bulking legume seeds mainly for green manuring to alleviate the shortage in the region.

Extension Methodology

The method of delivery of technologies greatly influences adoption. A methodology that involves regular training and the distribution of simple leaflets describing the various technologies has potential to enhance technology adoption. The current awareness that farmers have on the forages and the production systems can be attributed to training and sustained extension advice through workshops, seminars, farm visits, field days, agricultural shows and on-station demonstrations by the NDDP and the KARI-ILRI collaborative project. Booklets and technical manuals are also important for dissemination. One manual produced in the region, Growing Fodder Crops in Coastal Kenya (Mureithi et al. 1996b), was released to farmers in 1996. The manual was well illustrated but was written in English, a language most farmers do not understand. In contrast, the NDDP manuals were in English and in Kiswahili, the local language in the area, which is expected to foster technology transfer.

Institutional Support

Farmers receive institutional support from the government and a few non-governmental organizations (NGOs) operating in the region. Extension services of the Ministry of Agriculture are active but the region is vast and therefore the coverage is limited. The RCC-Mtwapa is the only governmental research institute that carries out agricultural research in the region. These two institutions have maintained close ties with each other and with the farmers who have improved prospects for technology adoption. However, in the past, transfer of these technologies on farm depended more on researchers than the extension officers. The NGOs operating in the region, Heifer Project International (HPI), World Vision and Care International, provide loans to dairy farmers for purchasing animals. One of the conditions to qualify for loans is that farmers must establish forages. The Coast Development Authority is also involved in technology transfer. It has set up several Farmer Field Schools in strategic areas to educate farmers on better methods of farming. Farmers attend these schools from time of planting to harvesting.

Finally, the biophysical environment of the coastal lowlands has a negative impact on the performance and adoption potential of forages in the region. The region suffers from erratic and prolonged drought periods, which often decreases productivity of forage systems. Napier grass in particular dries up during prolonged droughts. Most soils are sandy and low in organic matter and plant nutrients. Without fertilization and the addition of organic matter to conserve moisture, the productivity of the forages can be affected.

FUTURE RESEARCH

A number of research issues need to be addressed to increase the positive impact and adoption of improved forage systems in the lowlands of coastal Kenya. Improving soil fertility through means that are available to the farmers will be a key factor in increasing the yields of the various forage production systems. In particular, better on-farm manure management practices need to be developed because inorganic fertilizers are too expensive. Another important factor is to establish reasons for the decline in the adoption of planted forages. As mentioned earlier, Nicholson et al. (1999) reported a decline in the adoption of Napier grass in the region. Further studies are needed to affirm the causes behind this disadoption.

ACKNOWLEDGEMENTS

During the preparation of this case study, several people were consulted and the following are gratefully acknowledged for providing valuable assistance: Dr. R.W. Muinga, Centre Director KARI, RRC-Mtwapa; M.N. Njunie and R. Ali, forage agronomists at the Centre; Elizabeth Wekesa, socio-economist also at the centre; E.M. Kangunu, District Livestock Production Officer, Kwale District and F. Jefwa, District Livestock Extension Coordinator, Kilifi District. Special thanks go to the Centre Director for organizing the visits and to R. Ali for arranging field visits. We are thankful to the farmers in both Kwale and Kilifi districts who provided useful information. The Green Manure-Cover Crop Exploration Project of the Rockefeller Foundation provided funds for the preparation of this case study and we are indeed grateful for this support.

REFERENCES

Abdulrazak S A (1995). The effects of supplementary roughage diets with leguminous tree forages on intake, digestion and performance of cross-bred cattle in coastal Kenya. Ph. D. thesis, University of Aberdeen, UK. 217 p.

Abdulrazak S A, Muinga R W, Thorpe W, Ørskov E R (1996) *Gliricidia sepium* as alternative to *Leucaena leucocephala* in livestock feeding. *In*: Focus on agricultural research for sustainable development in a changing economic environment. Proceedings of the 5th Kenya Agricultural Research Institute (KARI) Scientific Conference, 14-16th October 1996, KARI headquarters, Nairobi, Kenya. p. 732-39.

CARS – Coast Agricultural Research Station (1974-80) Annual Reports 1974-80. Kenya Agricultural Research Institute (KARI), Regional Research Centre, Mtwapa.

Gelder G R (1998) A guide for the integration of fodder trees in NDDP. National Dairy Development Project (NDDP) Headquarter Report, Ministry of Agriculture, Livestock Development and Marketing (MoALDM), Nairobi, Kenya.

Goldson J R (1977) Napier grass (*Pennisetum purpureum* K. Schum) in East Africa: A review. Technical Report 24. Ministry of Agriculture, Pasture Research Project, National Agricultural Research Station, Kitale, Kenya. 30 p.

Jaetzold R, Schmidt H (1983) Farm management handbook of Kenya. Volume IIc. East Kenya (Eastern and Coast Provinces). Farm Management Branch, Ministry of Agriculture, Nairobi, Kenya. 411 p.

Kang B T, Reynolds L, Atta-krah A (1990) Alley farming. Adv Agron 43: 315-358.

Maarse L, Wentholt W, Chibudu A N (1998) Making change strategies work: Gender-sensitive client-oriented livestock extension in coastal Province, Kenya. Royal Tropical Institute, Amsterdam, the Netherlands.114 p.

Michieka D O, van der Pouw B J A, Vleeshouwer J J (1978) Soils of the Kwale-Mombasa-Lungalunga area. Reconnaissance Soil Survey Report No. R3, Kenya Soil Survey, Republic of Kenya. 300 p.

Muinga R W (1992) Nutrition and performance of dairy cows in coastal lowland Kenya in relation to utilization of local feed resources. Ph.D. thesis, University of Aberdeen, UK. 255 p.

Muinga R W, Thorpe W, Topps J H (1992) Voluntary feed intake, live-weight change and lactation performance of crossbred dairy cows given *ad libitum Pennisetum purpureum* (Napier grass var. Bana) supplemented with *Leucaena leucocephala* forage in the lowland semi-humid tropics. Anim. Prod. 55: 331-337.

Mullins G, Wahome P, Tsangari P, Maarse L (1996) Impacts of intensive dairy production on smallholder farm women in coastal Kenya. Hum. Ecol. 24: 231-253.

Mureithi J G (1992) Alley cropping with *Leucaena* for food and forage production in smallholder farms in lowland coastal Kenya. Ph.D. thesis, Department of Agriculture, University of Reading, UK. 173 p.

Mureithi J G, Thorpe W (1996) The effects of herbaceous legumes intercropping and mulching on the productivity of Napier grass (*Pennisetum purpureum*) and total forage yield in coastal lowland Kenya. *In*: Ndikumana J, de Leew P (Eds.) Sustainable feed production and utilization for smallholder livestock enterprises in sub-Saharan Africa. African Feed Resources Network (AFRNET) Workshop, 6-10 December 1993, Zimbabwe. AFRNET, Nairobi, Kenya. p. 45-50.

Mureithi J G, Thorpe W, Tayler R S, Reynolds L (1994) Evaluation of *Leucaena* accessions for semi-humid lowland tropics of East Africa. Trop. Agric. (Trinidad) 71: 83-87.

Mureithi J G, Njunie M N, Tayler R S, Thorpe W (1995) The effect of *Leucaena* alley cropping and different management practices on production of maize and Napier grass in lowland coastal Kenya. *In*: Proceedings of the Fourth Kenya Agricultural Research Institute (KARI) Scientific Conference, 25-28th October 1994, Nairobi, Kenya. p. 46-62.

Mureithi J G, Tayler R S, Thorpe W (1996a) Effect of the application of dairy cattle slurry and intercropping with cowpea (*Vigna unguiculata*) on the performance of maize (*Zea mays*) in coastal lowland Kenya. Afr. Crop Sci. J. 43:315-324.

Mureithi J G, Njunie M N, Muinga R W, Ramadhan A, Thorpe W, Blokland A (1996b) Growing fodder crops in coastal Kenya. Kenya Agricultural Research Institute (KARI), Regional Research Centre Mtwapa, Information Bulletin No. 1. KARI, Kenya. 19 p.

Mureithi J G, Njunie M N, Muinga R W, Ali R, Thorpe W, Mwatate C D (1998). Adoption of planted forages for smallholder dairy farmers in coastal Kenya. Trop. Grassland 32: 221-229.

Nicholson C F, Thornton P K, Mohamed L, Muinga R W, Mwamachi D M, Elbasha E H, Staal S J, Thorpe W (1998) Smallholder dairy technology in coastal Kenya: An adoption and impact study. A collaborative study by Kenya Agricultural Research Institute (KARI), Ministry of Agriculture (MOA) and International Livestock Research Institute (ILRI). Summary paper for the Impact Assessment and Evaluation Group (IAEG), August 1998. 18 p.

Nicholson C F, Thornton P K, Mohamed L, Muinga R W, Mwanamachi D M, Elbasha E H, Staal S J, Thorpe W (1999) Smallholder dairy technology in coastal Kenya. *In*: Sechrest L, Stewart M, Stickle T (Eds.) A synthesis of findings concerning CGIAR case studies on the adoption of technological innovation. Consultative Group on International Agricultural Research (CGIAR), Washington, USA. p. 65-9.

Njunie M N, Mureithi J G, Ali R, Muinga R W, Thorpe W, Chibudu A N, Maarse L (1995) Development and transfer of forage production technologies for small-holder dairying: Experiences of on-farm trials with legumes in coastal lowland Kenya. *In*: Proceedings of the Fourth Kenya Agricultural Research Institute (KARI) Scientific Conference, 25-28th October 1994, Nairobi, Kenya. p. 32-45.

Njunie M N, Reynolds L, Mureithi J G, Thorpe W (1996) Evaluation of herbaceous legume germplasm for coastal lowland East Africa. *In*: Ndikumana J, de Leew P (Eds.) Sustainable feed production and utilization for smallholder livestock enterprises in sub-Saharan Africa. African Feed Resources Network (AFRNET) Workshop, 6-10 December 1993, Zimbabwe. AFRNET, Nairobi, Kenya. p. 11-17.

Reynolds L, Bimbuzi B (1993) The *Leucaena psyllid* in Coast Province, Kenya. Nitrogen Fixing Tree Research Report Vol. 11. 103 p.

Reynolds L, Mureithi J G, Mullins G, Thorpe W (1993) Feed resource availability and milk production on smallholder dairy farms in the sub-humid coastal region of Kenya. *In*: Gill M, Owen E, Pollott G E, Lawrence T L J (Eds.) Animal production in developing countries. Symposium on animal production in developing countries, September 1991, Wye, UK. Occasional Publication of the British Society of Animal Production, No 16. p. 158-159.

Simons A J, Stewarts J L (1994) *Gliricidia sepium* - a multipurpose forage tree legume. *In*: Gutteridge R C, Shelton H M (Eds.) Forage tree legumes in tropical agriculture. CAB International, Wallingford, UK. p. 30-48.

Skerman P J, Cameron D G, Riveros F (1988) Tropical forage legumes. Second edition. Food and Agriculture Organization of the United Nations (FAO) Plant Production and Protection Series No. 2. FAO, Rome. Italy. 691 p.

Staal S J, Mullins G (1996) Dairy consumption and its determinants in coastal Kenya. Kenya Agricultural Research Institute (KARI) / International Livestock Research Institute (ILRI) Collaborative Project Report, Kenya. 47 p.

Thorpe W, Chabari F, Maloo S M, Muinga R W, Mukhebi G, Mullins G, Mureithi J G, Mussukuya E, Nyambaka R, ole Maki, Otieno L, Perry B, Rugema E, Wekesa E (1993) Smallholder dairy cattle production in coastal lowland Kenya. Resource base assessment and constraints identification. *In*: Gill M, Owen E, Pollott G E, Lawrence T L J (Eds.) Animal production in developing countries. Symposium on animal production in developing countries, September 1991, Wye, UK. Occasional Publication of the British Society of Animal Production, No 16. p. 167-168.

van Einnatten C L M (1979) Summary of information of coconut in Coast Province and proposal for future work. Coastal Agricultural Research Station Communication No. 4. 3 p.

Wouters A P (1987) Dry matter yield and quality of Napier grass on farm level 1983-1986. Research Report, Ministry of Livestock Development, Kenya. 39 p.

Chapter 9

Green Manure/Cover Crop Technology in Eastern and Central Uganda: Development and Dissemination

C. K. K. Gachene[1] and C. S. Wortmann[2]

[1] *University of Nairobi, Kenya*
[2] *University of Nebraska, USA*

SUMMARY

In 1992, Ugandan National Agricultural Research Organization along with International Center for Tropical Agriculture researchers from Kawanda Agricultural Research Institute initiated collaborative research with farmers in Uganda. Five villages in the vicinity of the Ikulwe District Farmers Institute in Iganga district of eastern Uganda were included. This area represents the traditional banana (*Musa* sp. L.)-coffee (*Coffea* L.)-based systems of the Lake Victoria Crescent agro-ecological zone. It has a mean annual rainfall of 1255 mm in a bimodal distribution and soils that are variable but typically have low total soil N as well as low P availability in Ikulwe. Banana, bean (*Phaseolus vulgaris* L.), sweet potato (*Ipomoea batatas* [L.] Lam.), cassava (*Manihot esculenta* Crantz), groundnuts (*Arachis hypogaea* L.) and maize (*Zea mays* L.) are important food crops. The approach was participatory, systems oriented and inter-disciplinary. Research activities on green manure/cover crops (GMCCs) consisted of some designed by farmers and researchers, and some farmers' own experimentation. Farmer-researcher trials indicated that yields of the GMCC species were reduced 40-70% when intercropped with a food crop as compared to sole crop production and that yields of food crops were reduced 61-87% when intercropped with *Crotalaria ochroleuca* G. Don. In contrast, maize grain yield response in the first season following sole-crop GMCC production ranged from 0 to 240%. These trials also indicated that *Mucuna pruriens* (L.) DC. and *Lablab purpureus* (L) Sweet were best for weed suppression and for control of soil erosion, that tillage and weeding requirements can be much less following GMCCs and that maize can often be planted directly in the holes left from uprooting *Mucuna* and *Lablab,* reducing labour

M. Eilittä et al. (eds.), Green Manure/Cover Crop Systems of Smallholder Farmers, 219-236.

requirements the following season. Land productivity was not improved with alley cropping. Farmers initiated trials on their own, such as on *Tephrosia vogelii* Hook. f. to control mole rats, which indicated effective control and resulted in significant adoption by neighbouring farmers. Other farmer experimentation focused on intercropping GMCCs with coffee, banana with *Canavalia ensiformis* (L.) DC., cassava and sweet potato, and *Mucuna* with maize.

Researchers' interviews with 19 of the farmer-researchers indicated that improvement of soil fertility followed by suppression of weeds was the most frequently mentioned positive feature of GMCCs. Uprooting of the mature GMCC plants was easiest for *Canavalia. Lablab* was most preferred for forage production. The legumes were similarly mentioned as effective in reducing soil erosion. Farmers frequently expressed concern about the climbing tendency of *Mucuna* as well as *Lablab. Lablab* produced little, but edible, seed, while *Canavalia* and *Mucuna* produced much inedible seed. *Crotalaria* was frequently observed to be laborious to cultivate. *Canavalia*, *Crotalaria*, *Lablab* and *Mucuna* differed little from one another in intercrop compatibility with banana, maize and cassava. *Canavalia* was more compatible with sweet potato than were other legumes. Information from on-station trials, formal on-farm trials and farmers' own experimentation was integrated to develop a decision guide on the use of GMCC legumes. Researchers facilitated the dissemination of GMCC technology through informal seed exchanges, printed materials, farmer group visits to Ikulwe, agricultural shows, government and non-government extension, farmer experimentation mini-kits and provision of materials to stockists of agricultural inputs. A banana-*Mucuna*-dairy system and the *Tephrosia* system for mole rat control have been the most adopted.

INTRODUCTION

Levels of crop productivity are low for smallholder agriculture in sub-Saharan Africa and nutrient balances are typically negative. Nutrient depletion is attributed primarily to erosion and to crop harvest with little return of nutrients to the annual crop fields (Smaling and Braun 1996; Gachene et al. 1997; Wortmann and Kaizzi 1998). Little fertilizer is used because small-scale farmers generally have poor or no access to credit facilities. Their investment in fertilizer use is often of insufficient profitability to prevail over other opportunities to use scarce monetary resources. Integrated soil fertility management may be improved by using green manure/cover crops (GMCCs).

Soil-improving legumes were not effective in improving soil productivity in earlier research in Uganda (Martin and Biggs 1937), but promising GMCC options have been identified recently (Wortmann et al. 1994; Fischler 1997; Fischler and Wortmann 1999). Adaptive research has continued in various parts of Uganda. The International Centre for Research in Agroforestry/Agroforestry Research Network for Africa (ICRAF/AFRENA) investigated *Mucuna pruriens* (L.) DC. and *Canavalia ensiformis* (L.) DC. in a highland area of Kabale district. CARE International is evaluating vetches (*Vicia* spp. L.) and lupins (*Lupinus* spp. L.) in Kabale district. In 1994, Action Aid Uganda (AAU) and the Natural Resources Institute (NRI) experimented with *Mucuna*, *Crotalaria ochroleuca* G. Don and cowpea (*Vigna unguiculata* [L.] Walp.) in Mubende district. Both International Center for Tropical Agriculture (CIAT, the Spanish acronym) and National Agricultural Research Organization (NARO) researchers collaborated with farmers in the Ikulwe area of Iganga district to investigate the integration of GMCCs in the traditional banana (*Musa* sp. L.)-coffee (*Coffea* L.)-based system. In general, however, the adoption of GMCCs by smallholder farmers has been slow (Lupanga et al. 1987; Rupper 1987; Raquet 1990; Dreschsel et al. 1996; Fischler 1997; Wortmann and Kirungu 1999). Factors constraining adoption include inadequate flow of information to farmers; demands for labour, land and capital; scarcity of seed; lack of short-term economic benefits; and inadequate targeting of technical options.

In 1992, NARO, along with CIAT researchers from Kawanda Agricultural Research Institute (ARI), initiated collaborative research with farmers at several locations in Uganda to improve their agricultural systems. Five villages in the vicinity of the Ikulwe District Farmers Institute in Iganga District of eastern Uganda were included to represent the traditional banana-coffee based systems of the Lake Victoria Crescent agro-ecological zone (AEZ) (Wortmann and Eledu 1999). The approach was systems oriented and inter-disciplinary, with farmers, extension agents and researchers jointly working in characterization and diagnosis, prioritization of problems, planning and implementing the research and evaluating the results. Research activities addressed genetic and management aspects of several crops, soil fertility management, integrated pest and weed management and food storage.

This chapter elaborates on the development and adoption of GMCC options with the Ikulwe farmers.

IKULWE AND THE LAKE VICTORIA CRESCENT AGRO-ECOLOGICAL ZONE

The Ikulwe community, lat 0°26'N, long 33°28'E, is representative of the Lake Victoria Crescent AEZ, which is large, with an area of 14 797 km^2 (Figure 1). Eighty-one percent of the land is classed as cropland and the remainder as grassland (8%), woodland (8%) and wetland (3%). Much of the following information is from Wortmann and Kaizzi (1998), Wortmann et al. (1998) and Wortmann and Eledu (1999).

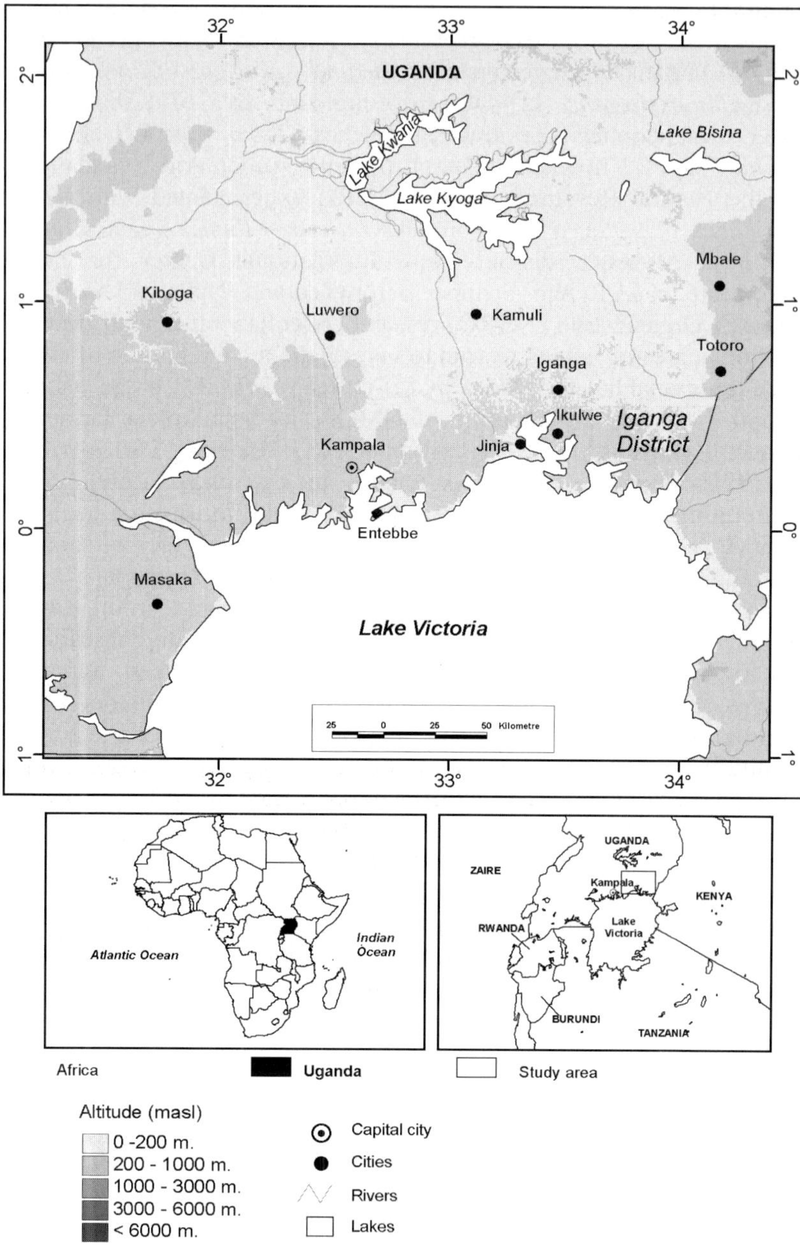

Figure 1. Location of the study area, Uganda.

Biophysical Characteristics

Nearly all of the land area of this AEZ lies within 50 km of the equator, with a mean altitude of 1174 masl. Mean monthly minimum and maximum temperatures vary little throughout the year from the annual means of 16 and 28 °C respectively. Mean annual rainfall for the AEZ is 1255 mm (1345 mm for Ikulwe) in a bimodal distribution. There are two similar cropping seasons, February to June and August to January. The mean rainfall for the driest month is more than 50 mm. Water deficits are common but drought is uncommon. Annual rainfall at Kawanda ARI, which lies in this AEZ, was below 1000 mm y^{-1} in only 8 of the last 60 years. Flooding can be a problem in the valleys but reduced radiation rather than excess water may be the greater constraint to crop growth when rainfall is excessive. Little irrigation is practiced except on some large farms oriented to the export of high-value crops. Generally, the amount and distribution of rainfall is compatible with GMCC requirements in that annuals such as *Canavalia*, *Mucuna* and *Crotalaria* can easily produce seed in one season, while many tree and shrub species tolerate the relatively short dry periods.

The land is undulating to rolling, with wide valleys that are often swampy. The soils are variable but generally reddish-brown sandy loams or sandy clay loams, often with clay loam subsoil (common for Ikulwe); sandy loams are more common in the eastern part of the AEZ. The soils are primarily plinthosols, and underlying petroplinthite often inhibits root growth (Harrop 1970). In the western part of the AEZ, soil on the ridge tops is often too shallow for the land to be arable. The moderately acidic soils can be productive when well managed. Soils in the Ikulwe area have lower fertility levels than those in much of the AEZ. Fischler (1997) found 80% of sampled fields in Ikulwe were below the critical values of soil organic matter (SOM) content, total N and available P (Table 1). Median soil fertility levels are higher in banana than in annual crop fields and have significantly higher levels of available P, Ca and Mg as well as more total N and K and a higher pH. Low total soil N and low P availability in Ikulwe are the greatest nutritional constraints in the area. Empirical evidence is lacking of any change in soil quality but farmers perceive that soil productivity has declined.

Table 1. Soil properties for 25 samples collected on 19 farms in Ikulwe village (*Source*: Fischler 1997).

Property	Minimum	Maximum
pH (H_20)	5.1	6.5
Organic matter	1.1	3.1
% total N	0.015	0.293
Available P (ppm)	<1	19
K (mg 100 g^{-1})	3	44
Ca (mg 100 g^{-1})	24	109
Clay (%)	22	38
Silt (%)	9	16
Sand (%)	51	65

Agronomic and Socio-economic Characteristics

The farming systems in Ikulwe and throughout Lake Victoria Crescent AEZ are typically at near subsistence levels with much biological and agronomic diversity. Banana, bean (*Phaseolus vulgaris* L.), sweet potato (*Ipomoea batatas* [L.] Lam.), cassava (*Manihot esculenta* Crantz), groundnuts (*Arachis hypogaea* L.) and maize (*Zea mays* L.) are important food crops (Table 2). *Robusta* coffee is the major cash crop. In addition, there are several lesser crops and livestock. The crops are produced in various intercrop associations and rotation sequences. Bean is usually produced by intercropping with maize, banana, cassava, sweet potato or coffee, while groundnut is intercropped with maize and banana. Land is tilled using hand hoes. Fertilizer use is negligible. Annual crops are weeded with hoes 3 to 4 weeks after planting and again a few weeks later. Yields are low; maize yields are typically less than 1500 kg ha^{-1} and bean yields, 800 kg ha^{-1}.

Table 2. Percentage of agricultural land[a] occupied by major food crops in the Lake Victoria Crescent (LVC) agro-ecological zone and in the Ikulwe community (*Source*: Wortmann and Eledu 1999).

Region	Crop					
	Maize	Cassava	Sweet potato	Banana	Bean	Groundnut
LVC	6	4	4	13	5	2
Ikulwe	23	21	3	12	9	4

[a] Excludes government protected parks, reserves, wetlands, etc.

Banana is a traditional staple crop but nowadays the major proportion of the crop is of brewing cultivars used to produce *Tonto*, which is often marketed. Banana is commonly intercropped (54%) in Ikulwe, especially in newly planted banana fields. Coffee and bean are the most frequently associated crops in mature stands of banana fields and scattered trees are common. Intercropping with bananas in Uganda is less prevalent in areas where the crop is more commercialized (Ssekabembe 1994).

A few farmers, especially near Kampala, have dairy cows of mixed or exotic breeds that may be stall-fed with cut grass. More farmers have Zebu cattle, sheep, goats, chickens and pigs. In 1995, the average numbers per household for Ikulwe were 1.5 cows, 1.7 sheep and goats, 12 chickens and 0.9 pigs (Wortmann and Kaizzi 1998). The Zebu cattle, sheep and goats graze fallow and grassland as well as annual cropland following harvest. Manure is typically applied to banana and coffee. Most maize and sweet potato residues are left in the field where livestock might graze them. A significant proportion of the residues of bean, groundnut and other crops are eventually transferred to banana and coffee fields (Bekunda and Woomer 1996).

Labour demand is especially high during planting and weeding times, i.e. late February through April and late August through October. Some roles are gender specific but men and women share responsibility for most agricultural operations. Health greatly affects labour availability and HIV-AIDS (Human Immunodeficiency Virus – Acquired Immunodeficiency Syndrome) is prevalent but evidence is lacking of how adoption of alternative practices is related to the health of a household.

The population density for the rural areas, excluding Kampala and the district headquarter towns, is 280 persons km^{-2}. The average farm size for the AEZ is about 2.1 ha (1.8 ha for Ikulwe) and less than 5% of the farms are larger than 10 ha. Few farmers have official titles to their land but their tenure is generally secure and they do buy and sell land.

Market access is good because the AEZ is bisected by Uganda's main tarmac road, which runs from Rwanda to Kenya, and the density of roads is higher for this AEZ than for any other in the country. Kampala is centrally located and about 90 km from Ikulwe. The AEZ is bordered by Lake Victoria and bisected by the Nile River; the Lake is also important for transport.

Commodity sales are probably the main source of income for most households followed by off-farm labour but average cash income may not be more than US$1 per day. About 30% of the food produce and all of the coffee may be marketed. Small-scale traders play a major role, buying and consolidating small amounts from farmers; they may sell to other traders or directly deliver to retailers in the towns and Kampala, to relief organizations, or to others. Women may play major roles in marketing at the village level by selling directly to consumers, while men more often handle larger consignments.

Fertilizer is often available through suppliers of agricultural inputs but the sales volume is typically low and the prices are high. Good quality advice generally does not accompany the sale of fertilizers but efforts are underway to improve this. Formal credit services are not available unless a person is well connected to credit providers.

Scarcity of resources constrains government extension services, which yet remain the main source of information to farmers on agricultural alternatives. Non-governmental organizations (NGOs) also provide significant extension services. The national extension system was revised in 1999 to enable sub-counties to contract extension services as needed from the government extension system or from NGOs; implementation of this system is only just beginning.

Adoption of GMCC technology is related to the quality of the extension services. Improved markets may increase demand for GMCC technology and lower fertilizer prices may decrease it. Generally, land tenure is sufficiently secure that it does not hinder adoption of GMCC technology on at least some of a farmer's land. Good extension services can greatly facilitate adoption of GMCC technology but promotion of GMCCs is often of low priority. Several, but not all, NGOs have been effective in promoting GMCCs.

Significant migration of diverse ethnic groups to the Ikulwe area occurred three to four decades ago but the population is still primarily Basoga, a people who have inhabited the area for centuries. The main migratory force is the movement of young people to towns. Ethnicity and migration do not appear to affect GMCC adoption.

Land resources, including water for domestic use and wood for fuel and construction, are relatively abundant in this AEZ compared to many parts of eastern Africa. Over 20% of the arable land is typically fallow. Wildlife (e.g. birds, feral pigs, monkeys, antelopes and mole rats) is present and most often seen as a pest. Generally, these factors do not affect GMCC adoption, except when *Tephrosia vogelii* Hook. f. *vogelii* is planted to rid the fields of mole rats (or root rats, *Tachyoryctes splendens*).

DEVELOPMENT OF GREEN MANURE/COVER CROP TECHNOLOGY WITH IKULWE FARMERS

Research to develop GMCC technology for central and eastern Uganda had several complementary aspects. On-station research was important for the identification of likely options and to better understand the effects of GMCCs on nutrient dynamics and crop productivity (Wortmann et al. 1994; Fischler et al. 1999; Wortmann and Kaizzi 2000; Wortmann et al. 2000). Collaboration with farmers in farmer participatory research (FPR) to test GMCC alternatives under farmer management enabled valuable feedback from farmers. Farmers eventually experimented a lot on their own to investigate options for the integration of GMCCs into their cropping systems; such FPR might be considered to be of a collegial nature (Biggs 1989). Eventually, factors affecting adoption were studied, although more work in this area is needed.

The Farmer Participatory Research Process

Farmer participatory research was initiated in Ikulwe in 1992 to better integrate farmer knowledge with researcher knowledge, to make on-farm testing more effective, improve feedback from farmers on research priorities and results and encourage farmers' own experimentation. Here, participatory research refers primarily to farmer-researcher interactions of collaborative and collegiate types (Biggs 1989), with some consultative and one case of contractual on-farm research often occurring concurrently.

The FPR in Ikulwe began with characterization and diagnosis (C&D) activities. The initial C&D was not intended to give a complete understanding of the many interactions in the farming systems, recognizing that C&D would be an ongoing process (Fischler et al. 1996). Instead, the objective was to obtain a good overview of the farming systems and in-depth information of the soil resources. Farmers identified

14 soil types for which they had names, descriptions and preferred and non-preferred uses and management practices (Wortmann et al. 1998).

Participatory rural appraisal and research planning meetings followed the initial C&D. During the first research planning meetings, 20 constraints were identified. In small groups, farmers ranked subsets of five to seven problems using pairwise comparisons and counter methods (Beets 1990; the farmers preferred the counter to the pairwise methods). Initially, soil fertility was rated a low priority, possibly because farmers expected to receive inputs and anticipated short-term solutions to problems and low soil fertility was an accepted, long-standing and persistent problem. However, the farmers and researchers were not satisfied with the result and, after discussion, agreed that low soil fertility was a high-priority concern (Fischler 1997). Farmers then diagrammed and explored the causes of the priority problems and named soil erosion, intensive land use and overgrazing, compaction, little return of nutrients to the field and nutrient losses due to burning and leaching as the causes of low soil fertility. Possible solutions to the priority problems were identified and several selected for experimentation.

Farmers and researchers agreed on the need for genetic as well as management options. They opted to conduct trials on various topics. Trials for the first year evaluated (Wortmann et al. 1998): cassava varieties for resistance or tolerance to cassava mosaic virus; bean varieties for productivity under low-input conditions; planting density for control of groundnut rosette; hedgerows (alley cropping) with leguminous trees for soil improvement; composting; mulching; and green manures.

Other trials were initiated in subsequent years. Farmers worked with researchers to design the trials and assign responsibility for implementation. Trial designs varied from simple observation plots to replicated trials with several treatments. The trials were then implemented on farmers' fields with their management and farmers and researchers together evaluated the results. An extension agent and a farmer were selected to facilitate and supervise implementation, collect data and organize meetings of farmers with researchers (Fischler et al. 1996).

The FPR has continued to the present. During each dry season, research meetings attended by 20-45 farmers were held for 1 to 3 days to evaluate the research results of the previous season and to plan for the next season. Women accounted for 30 to 50% of the participants at these meetings. Farmers analysed the information from the trials through small group activities, often using diagramming and ranking exercises that facilitated the good participation of most farmers. Farmers then presented small group findings to the other farmers and researchers.

Experimentation on Green Manure/Cover Crops

Research activities on GMCCs in Ikulwe consisted of some experimentation designed by farmers and researchers and some farmers' own experimentation.

Trials Designed by Farmers with Researchers

Four major GMCC topics were addressed in trials designed by farmers and researchers together and implemented under farmer management. The effectiveness of *Crotalaria* fallow was compared to weedy fallow for improving soil productivity over a 3-year period in unreplicated trials conducted on eight farms. The production of *Crotalaria* green manure by intercropping with maize or bean was evaluated in Ikulwe over two seasons on five farms with two replications per farm. Sole-crop and intercrop production of *Mucuna* and *Lablab purpureus* (L.) Sweet and their effects on soil productivity and on weed suppression were evaluated over four seasons in unreplicated trials conducted on six farms. The effects of *Calliandra calothyrsus* Meissner hedgerow intercropping on the yield of a maize-bean rotation were evaluated on six farms with two replications per farm for 4-6 seasons (Wortmann et al. 1998; Fischler and Wortmann 1999). Highlights of the results include:

- Maize and bean yields were higher following sole crop of *Crotalaria* than following a weedy fallow (Table 3).
- Yield of the GMCC species was reduced 40-70% when intercropping with a food crop as compared to sole-crop production.
- Yields of food crops intercropped with *Crotalaria* were 61% for maize and 69% for bean compared to sole crop yields of 87% for maize and 85% for bean if *Crotalaria* was sown on the same day or about 3 weeks later than the food crop (Table 4).
- Maize grain yield response in the first season following sole-crop GMCC production ranged from zero to 240% in on-farm trials.
- *Mucuna* and *Lablab* were best for weed suppression and for control of soil erosion.
- Tillage and weeding requirements can be much less following GMCC legumes than when maize follows maize.
- Often, maize seed can be planted directly in the holes left from uprooting *Mucuna* and *Lablab*, reducing labour requirements the following season.
- Land productivity was not improved with alley cropping.

Table 3. Grain yields (kg ha^{-1}) of maize and beans grown as first and second crops after sole-cropped *Crotalaria* and weedy fallow (*Source*: Fischler and Wortmann 1999).

Treatment	1st subsequent crop[a]		2nd subsequent crop[a]	
	Maize	Beans	Maize	Beans
Crotalaria	3990 a	560 a	2630 a	740 a
Weedy fallow	2820 b	400 b	2150 a	660 a

[a] Means within a column followed by the same letter are not significantly different at the $P = 0.05$ level using LSD test.

Table 4. Grain yields of maize and beans (kg ha^{-1}) as affected by intercropping with *Crotalaria* (*Source*: Fischler 1997).

Treatment	Early intercropping		Late intercropping	
	Maize	Beans	Maize	Beans
Intercropped[a]	690	331	2433	807
Sole cropped[b]	1130	482	2791	942
LSD ($P = 0.05$)	373.7	120.3	NS	NS

[a] *Crotalaria* intercropped with either maize or bean.
[b] Sole-cropped maize or bean.

Farmers' Own Experimentation

An exciting outcome of the Ikulwe FPR was that farmers did much of their own experimentation with GMCCs (Wortmann et al. 1999), apparently stimulated by a number of factors. These included improved access to information on technical alternatives, availability of planting material and interactions with other FPR farmers, which encouraged experimentation and provided ideas.

Some of the farmers' own research followed the suggestions of a researcher. One case involved *Tephrosia* in the control of mole rats, a burrowing rodent that farmers recognize as a major crop pest. A researcher advised farmers that the roots of *Tephrosia* may be toxic to mole rats and that scattered plants in their fields may give control. Six farmers harvested seed (the plant is indigenous) and experimented. Researchers made irregular observations of this work and then conducted open-ended interviews after 3 years of work, the results of which showed effective control of the mole rats and significant adoption of *Tephrosia* by neighbouring farmers.

Much of the farmers' own experimentation was testing their ideas on options for integrating GMCCs into their cropping systems. Additional information was gained on intercrop production of legumes with maize. More important was the information gained on producing the legumes in association with coffee, banana, cassava and sweet potato because no formal research had been done on these cropping systems. The topic most addressed by farmers' own experimentation was intercrop production of *Canavalia* with banana, followed by maize-*Mucuna* intercropping but there was no farmers' own experimentation on sole-crop production of the legumes.

Researchers interviewed 19 of the farmer-researchers individually in an open-ended manner concerning 49 cases of experimentation, 35 of which were farmers' own experimentation. Farmers were asked to name the good and bad characteristics of each GMCC species. Improvement of soil fertility, followed by suppression of weeds, was the most frequently mentioned feature and these traits were mentioned with similar frequency for each species. Uprooting of the mature GMCC plants was easiest for *Canavalia*. *Lablab* was most preferred for forage production. The legumes were similarly mentioned as effective in reducing soil erosion. Farmers frequently expressed concern about the climbing tendency of *Mucuna* as well as *Lablab*. *Lablab* produced little, but edible, seed, while *Canavalia* and *Mucuna* produced much inedible seed. *Crotalaria* was frequently observed to be laborious to cultivate. *Canavalia*, *Crotalaria*, *Lablab* and *Mucuna* differed little from one another in intercrop compatibility with banana, maize and cassava. *Canavalia* was more compatible with sweet potato than were other legumes.

MOST PROMISING GREEN MANURE/COVER CROP OPTIONS

Information from on-station trials, formal on-farm trials and farmers' own experimentation was integrated to develop a decision guide on the use of GMCC legumes (Table 5).

Table 5. Guidelines for using *Canavalia*, *Crotalaria*, *Lablab* and *Mucuna* as green manure/cover crop species in central and eastern Uganda (*Source*: Fischler 1997; Wortmann et al. 1999).

If you want to	Plant......	Do not plant......
Produce in sole crop	*Mucuna* or *Lablab*	*Canavalia*
Intercrop with maize	*Canavalia*, or *Lablab* at very low plant density	*Mucuna*
Intercrop with newly planted banana or coffee	*Canavalia*	*Mucuna* or *Lablab*
Intercrop with established banana or coffee	*Canavalia* or *Mucuna* at low plant density	*Crotalaria*
Intercrop between sweet potato mounds	*Crotalaria* or *Canavalia*	*Mucuna* or *Lablab*
Intercrop with newly planted cassava	*Canavalia* or *Crotalaria* between rows of cassava	*Mucuna* or *Lablab*
Intercrop with established cassava	*Canavalia* or *Mucuna* at low density	*Crotalaria*
Produce fodder	*Lablab* or *Mucuna*	*Canavalia* or *Crotalaria*
Suppress weeds	*Mucuna* or *Lablab*	*Crotalaria* or *Canavalia*
Reduce nematodes	*Mucuna*	*Canavalia*
Produce durable mulch	*Crotalaria* and *Canavalia* (allow to mature)	*Lablab* or *Mucuna*

The options identified as most promising were:

- Sole-crop or intercrop production of *Mucuna* to improve soil productivity and suppress weeds;
- Intercrop production of *Mucuna* or *Lablab* at low plant densities with banana, especially where the forage is desired for feeding cows;
- Intercrop production of *Crotalaria* with bean, sowing *Crotalaria* 3 weeks after bean;
- Intercrop production of *Canavalia* or *Crotalaria* with sweet potato, planting both on the same day;
- Intercrop production of *Canavalia* with maize, sowing *Canavalia* 3 weeks after maize;
- Intercrop production of *Canavalia* or *Crotalaria* with sweet potato;
- Sole-crop production of *Mucuna* (or *Crotalaria grahamiana* Wight & Arn.) to suppress root-knot nematodes (*Meloidogyne* spp.) (CIAT 1999); and
- Production of scattered plants of *Tephrosia* in fields of annual crops to control mole rats (*Tephrosia* does, however, result in higher levels of root-knot nematode infection in subsequent crops).

GREEN MANURE/COVER CROPS AND INTEGRATED NUTRIENT MANAGEMENT

Soil productivity cannot be sustained by using GMCCs alone; researchers, extension staff and farmers need to use integrated nutrient management strategies with GMCCs as one component of the strategy. Producing short-term GMCC fallow on 20% of the annual crop land and intercrop production of legumes under 50% of the banana crop may be sufficient to reduce the N deficit for annual crop land to 30 kg ha^{-1} y^{-1} and to achieve a positive nutrient balance under banana (Wortmann and Kaizzi 1998). This assumes that the legumes are not harvested for forage and that the legume biomass is managed to minimize N losses to volatilization (Wortmann and Kaizzi 2000). Furthermore, legumes do not give more than a short-term improvement in the supply of nutrients other than N.

Smallholder farmers often have other nutrient-containing materials that might be used, although their supply may be small. Some farmers can afford to invest a little money in fertilizer. Naturally occurring vegetation that otherwise may be of little value, such as leaves and branches of *Lantana camara* L. or *Tithonia diversifolia* (Hemsl.) A. Gray, may be nutrient sources for nearby cultivated fields (Kaizzi and Wortmann 2001). A challenge is to use these diverse resources efficiently.

An approach to research and extension for integrated nutrient management in Uganda develops a conceptual framework as the starting point (Wortmann and Ssali 2001). Using information from various sources and the knowledge of researchers, a conceptual framework is developed for principal crops produced on a major soil type in an AEZ. The conceptual framework is often in the form of a tentative decision guide that considers the farmer's purchasing power, substitution values of organic materials for fertilizers and residual and cropping system effects. Optimal rates or combinations of nutrient sources are then estimated. This is followed by experimentation on numerous farms at two or more locations per major AEZ to gain additional information or to verify the estimates. As researchers become confident of their estimates, the decision guide is refined and provided to extension staff.

DISSEMINATION OF GREEN MANURE/COVER CROP TECHNOLOGY

Researchers facilitated the dissemination of GMCC technology in a number of ways as described below.

- Informal Seed Exchanges: Farmer participatory research stimulated interest amongst neighbours who requested seed from the FPR farmers, most of whom gave seed to other farmers. Of those farmers who had grown a GMCC species previously, 27% gave to 1-3 farmers; 26% gave to 4-6 farmers; 4% gave to 8-10 farmers; 10% gave to 10-18 farmers and 12% gave to more than 20 farmers. Only 21% did not give seed to other farmers
- Printed Materials: Leaflets on *Mucuna*, *Canavalia* and *Tephrosia* as well as a decision guide on the use of GMCC legumes (Table 5), were prepared, printed and disseminated. Several articles were printed in English language and vernacular newspapers.
- Farmer Group Visits to Ikulwe and Agricultural Show: The Ikulwe farmers invited groups of farmers from other communities to see their research results. Researchers also encouraged organizations to take groups of farmers to Ikulwe and gave the Ikulwe FPR committee the equivalent of about US$10 per group of eight or more visitors to help cover the expenses of hosting the groups. The visitors were able to see farms of FPR participants, discuss FPR findings and share perspectives. At the end of the first year, the committee's book of visitors listed 180. FPR farmers joined researchers, in at least one case, at a 3-day agricultural show, to inform others of the benefits of using GMCCs.
- Government and Non-governmental Extension: Numerous organizations obtained information and seed from researchers to use in their extension activities elsewhere. Some received training and technical support. A few took groups of farmers to Ikulwe. Often, their efforts were not well focused on GMCC technology and consisted of little more than giving seed to farmers. Others were more effective, complementing seed and information dissemination with farmer training and follow-up.

- Farmer Experimentation Mini-kits: In 1999, researchers prepared 3000 mini-kits to be distributed through extension collaborators to farmers. The kits typically consisted of seed of three GMCC legumes, information on each species and a decision guide on the selection of species for different situations. The expectation was that at least some farmers would have a good learning experience and see opportunities to integrate one or more of the species into their cropping systems. Some extension collaborators merely gave kits to selected farmers, while others provided training and allowed community members to select the recipients of the kits. The effectiveness of the mini-kit approach has not been assessed but it is an inexpensive means of enabling many farmers to try GMCC options.
- Stockists of Agricultural Inputs: Stockists were provided with 50-g packets of *Tephrosia* seed to sell to their customers. Stockists also received simple instructive posters on *Tephrosia* management photocopied onto A3 paper as well as leaflets. *Tephrosia* was planted to control mole rats but it also derives much of its N from the atmosphere (Wortmann and Kaizzi 2000). Demand exceeded seed supply but early adopters soon began supplying others with seed. Researchers planned to eventually promote *Tephrosia* as an improved fallow species, taking advantage of farmers' familiarity with the species. This program was put on hold, however, as it became apparent that *Tephrosia* resulted in increased infection by root knot nematodes in subsequent crops.

GREEN MANURE/COVER CROP TECHNOLOGY ADOPTION IN EASTERN AND CENTRAL UGANDA

Adoption of GMCC technology has not been assessed well but appears to have been slow. In Ikulwe, adoption was assessed 3 years after the start of activities in late 1995 by interviewing 22 farmers. Farmers who had participated in experimentation with a GMCC species previously and had it as a crop during the second season of 1995 were considered to have adopted. Adoption was poor (5%) for *Crotalaria*, intermediate for *Mucuna* (43%) and *Lablab* (45%) and high (62%) for *Canavalia* (Wortmann et al. 1998). The reasons given for discontinuation were: labour constraints (17 farmers), lack of benefit (7), land constraints (1), lack of seed (6) and other problems (6). The labour constraint was associated primarily with *Crotalaria*. It is noteworthy that insufficient land was seldom mentioned as an obstacle to adoption.

A banana-*Mucuna*-dairy system and the *Tephrosia* system for mole rat control have been most adopted. *Canavalia* produced by intercropping with food crops is agronomically effective but adoption has been slow. The use of *Mucuna* for managing perennial weeds has recently been promoted in another AEZ (M. Versteeg, personal communication, 1999).

Mucuna is used as forage in the banana-*Mucuna*-dairy system to supplement the diet of dairy cows. The system has been promoted primarily since 1998. Researchers provided printed information and *Mucuna* seed to several NGOs. Different extension approaches have been used but farmer training and visits to practicing farmers have usually been part of the promotion process. Where promoted, in at least four districts of the Lake Victoria Crescent AEZ, dairy farmers are using the system. In two of these districts, an NGO has a program to assist farmers to acquire dairy heifers, provided they first plant an area of Napier grass and a legume, such as *Mucuna*. In any case, adoption appears to be driven by the need for an improved fodder. Farmers find the system easy to adopt because it is undemanding to manage and *Mucuna* produces a lot of seed. The system has its weaknesses; while improving the cows' diet, it probably does not contribute much to soil N status. Owners of dairy cows are few and this system is inappropriate for poorer farmers unable to keep dairy cows. Farmers desire a species that produces acceptable forage for small ruminants and non-ruminants and produces palatable seed.

The *Tephrosia* system for mole rat control has been promoted and adopted in several districts of the Lake Victoria Crescent AEZ, the southern and eastern Lake Kyoga Basin and the south-west highlands of Kabale district. Because of its low cost and ease of use, the system spread from farmer to farmer and promotion through input suppliers was effective. Increased infection in subsequent crops by rootknot nematodes is a concern. *Tephrosia* obtains much N from the atmosphere but the plant density is low in this system and it may not contribute much nitrogen.

Intercrop production of *Canavalia* green manure is easily managed and effective in improving subsequent soil productivity. It has not been much promoted because most extension staff agree that improvement of soil productivity is insufficient reason for farmers to adopt a GMCC species. Traditional beliefs that the plant has mystical powers may constrain adoption of *Canavalia*.

Mucuna can be effective in the suppression of weeds in addition to improving soil productivity and is being promoted for perennial weed suppression in the Northern Moist Farmland AEZ of Uganda. Elevation and annual rainfall are similar to those of the Lake Victoria Crescent AEZ but the distribution of rainfall differs, with a long rainy period in a weakly bi-modal pattern followed by a long dry season. The soils are typically sandy loams in this AEZ, with less available water holding capacity than in Ikulwe also contributing to more severe effects from the dry season. The system being promoted involves establishment of *Mucuna* by relay sowing into a maize crop and allowing the *Mucuna* to grow following maize harvest until soil moisture is depleted during the following dry season. The expectation is that weed density will be low when the next crop is sown. A significant contribution to soil productivity is anticipated. The practice of cut-and-carry for feeding dairy cows is uncommon in this AEZ but herds of Zebu cows are likely to graze the mature *Mucuna* during the dry season. Even when mature, the *Mucuna* is likely to be more nutritious than the dominant grasses during the dry season.

CONCLUSIONS

Researcher-farmer collaboration improved the efficiency of developing GMCC systems. The research process further benefited when collaborating farmers experimented on their own to seek better ways of integrating GMCC legumes into their cropping systems. Returns to labour are more important than returns to land for many small-scale farmers. Early benefits, in addition to soil fertility management, are important for adoption; in fact, the alternative use of the legume is generally more important to extension staff and farmers than is the improvement of soil productivity. Farmer-to-farmer diffusion of GMCC technology has a role because a few farmers were observed to be highly effective in dissemination and most farmers at least gave seed to neighbours. There has been considerable adoption of a *Mucuna*-banana-dairy system and a *Tephrosia* system for mole rat control. *Canavalia* for soil improvement remains promising but has not received promotion. The effects of recent promotion of *Mucuna* for weed suppression as well as the improvement of soil productivity need to be assessed.

REFERENCES

Beets C W (1990) Raising and sustaining productivity of smallholder farming systems in the tropics. AgBe Publishing, Alkmaar, The Netherlands. 738 p.

Bekunda M A, Woomer P L (1996) Organic resource management in banana-based cropping systems of the Lake Victoria basin. Agric. Ecosyst. Environ. 59:171-180.

Biggs S D (1989) Resource-poor farmer participation in research: A synthesis of experiences from nine national research systems. International Service for National Agricultural Research (ISNAR), The Hague, The Netherlands. 337 p.

CIAT – Centro Internacional de Agricultura Tropical (1999) Meeting demand for beans for sub-Saharan Africa in sustainable ways. CIAT Annual Report, Cali, Colombia. p. 57-58.

Dreschsel P, Steiner K G, Hagedorn F (1996) A review on the potential of improved fallows and green manure in Rwanda. Agrofor. Sys. 33: 109-136.

Fischler M. (1997) Legume green manures in the management of maize-bean cropping systems in Eastern Africa with special reference to *Crotalaria* (*C. ochroleuca* G. Don.). Ph.D. thesis No. 12099, Swiss Federal Institute of Technology (ETHZ), Zurich, Switzerland. 211 p.

Fischler M, Wortmann C S (1999) Green manure research in eastern Uganda - a participatory approach. Agrofor. Syst. 47:123-138.

Fischler M, David S, Farley C, Ugen M, Wortmann C S (1996). Applying farmer participatory research methods to planning agricultural research. Experiences from Eastern Africa. J. Farm. Syst. Res. Exten. 6:37-48.

Fischler M, Wortmann C S, Feil B (1999) *Crotalaria* (*C. ochroleuca* G. Don.) as a green manure in maize-bean cropping systems in Uganda. Field Crops Res. 61:97-107.

Gachene C K K, Jarvis N J, Linner H, Mbuvi J P (1997) Soil erosion effects on soil properties in a highland area of Central Kenya. Soil Sci. Soc. Am. J. 61:559-564.

Harrop J F (1970) Soil. *In*: Jameson J D (Ed.) Agriculture in Uganda. Oxford University Press, UK. p. 43-71.

Kaizzi C K, Wortmann C S. (2001) Plant materials for soil fertility management in sub-humid tropical areas: Nutrient release and crop response. Agron. J. 93(4): 929-935.

Lupanga J J, Matee A Z, Mvena Z S K, Weber J J (1987) Social economic factors influencing the adoption of *Crotalaria ochroleuca* in farming communities around Peramiho. A preliminary study. *In:* Proceedings of a writers' workshop, Peramiho, Songea, Tanzania. Benedictine Publication, Ndanda Peramiho. p. 48-65.

Martin W S, Biggs C E J (1937) Experiments on the maintenance of soil fertility in Uganda. EAAJ 2: 371-378.

Raquet K (1990) Green manuring with fast-growing shrub fallow in the tropical highlands of Rwanda. *In*: Kotschi J (Ed.) Ecofarming practices for tropical smallholdings. Trop. Agroecol. 5:55-80.

Rupper G (1987) Cultivation of *marejea* (*Crotalaria ochroleuca*). Experience of Peramiho. *In*: Proceedings of a writers' workshop, Peramiho, Songea, Tanzania. Benedictine Publication, Ndanda Peramiho. p. 9-12.

Smaling E M A, Braun A R (1996) Soil fertility research in sub-Saharan Africa: New dimensions, new challenge. Comm. Soil Sci. Plant Anal. 27: 365-386.

Ssekabembe C K (1994) *Ficus natalensis* in banana plantations: An overview of research needs. *In*: Rwakaikara M C, Bekunda S M A, Tenywa J S (Eds.) Enhancing farmers efforts to combat soil degradation: A challenge to soil science technology, Kampala, Uganda. p. 261-268.

Wortmann C S, Eledu C A (1999) Uganda's agroecological zones: A guide for planners and policy makers. Centro Internacional de Agricultura Tropical (CIAT) Kampala, Uganda.

Wortmann C S, Kaizzi C K (1998) Nutrient balances and expected effects of alternative practices in farming systems of Uganda. *In*: Smaling E M A (Ed.) Nutrient balances as indicators of productivity and sustainability in sub-Saharan African agriculture. Agric. Ecosys. Environ. 71:115-130.

Wortmann C S, Kaizzi C K (2000) Tree legumes in medium-term fallows: N_2 fixation, nitrate recovery and effects on subsequent crops. African Crop Sci. J. 8:263-272.

Wortmann C S, Kirungu B (1999) Adoption of soil improving and forage legumes by small-holder farmers in Africa. *In*: Stür W W, Horne P M, Hacker J B, Kerridge P C (Eds.) Working with farmers: The key to adoption of forage technologies. Australian Centre for International Agricultural Research (ACIAR), AU. 32-53. (ACIAR Publication PR095.) p. 140-148.

Wortmann C S, Ssali H (2001) Integrated nutrient management in Eastern Africa: An approach to adaptive research and technology dissemination. J. Altern. Agric. 16:161-167.

Wortmann C S, Isabirye M, Musa S (1994) *Crotalaria ochroleuca* as a green manure crop in Uganda. Afr. Crop Sci. J. 2:55-61.

Wortmann C S, Fischler M, Alifugani F, Kaizzi C K (1998) Accomplishments of participatory research for systems improvement in Iganga District, 1993-1997. Occasional Paper No. 27, Centro Internacional de Agricultura Tropical (CIAT), Kampala, Uganda.

Wortmann C S, Kaizzi C K, Fischler M (1999) Farmers' experimentation on green manure/cover crops: A component of participatory research for improvement of Ugandan farming systems. *In*: Fujisaka S (Ed.) Systems and farmer participatory research: Developments in natural resource management. Centro Internacional de Agricultura Tropical (CIAT), Cali, Colombia. p. 118-127.

Wortmann C S, McIntyre B D, Kaizzi C K (2000) Annual soil improving legumes: Agronomic effectiveness, nutrient uptake, nitrogen fixation and water use. Field Crops Res. 68:75-83.

Chapter 10

Integrating *Mucuna* in the Maize-Based Systems of Southern Benin

M. Eilittä[1], A. Etèka[2] and R. J. Carsky[3]

[1]*Rockefeller Foundation-funded Post-doctoral Fellow*
[2]*Consultant*
[3] *International Institute of Tropical Agriculture (IITA), Nigeria*

SUMMARY

Resource-conserving technologies are urgently needed in West Africa and the Republic of Benin, where population density is high and soil fertility is fast decreasing. In the late 1980s, a development-oriented research project, Recherche Appliquée en Milieu Réel (Applied Research On-farm) of the Beninoise government's Institut National des Recherches Agricoles du Bénin (National Agricultural Research Institute), the International Institute of Tropical Agriculture and the Royal Tropical Institute of the Netherlands initiated on-farm demonstrations and trials with *Mucuna pruriens* (L.) DC. Farmer adoption followed as a response to its ability to suppress *Imperata cylindrica* (L.) Raeusch. and to maintain soil fertility in the maize (*Zea mays* L.)-based systems of the region. As a result, several organizations initiated extension efforts. They included the government's extension organization, Centre d'Action Régionale pour le Développement and Sasakawa-Global 2000. Farmers of the region utilized two maize-*Mucuna* systems: sole-cropped *Mucuna* short fallow (particularly in areas with severe *Imperata* infestation) and maize-*Mucuna* relay intercropping (in most fields). The relatively scant on-farm data on *Mucuna* biomass production and impacts on maize yield, and the impressions of those familiar with the systems, suggest that *Mucuna* grows well in the region, exerts positive impacts on maize yield and reduces *Imperata* infestation. Many farmers (3000 in 1993 and 10 000 in 1996) began testing *Mucuna* in their fields. Strong support from the extension organizations included free seed and a market to sell seed as well as linking *Mucuna* to a maize-production technology package. More recently, estimated adoption and utilization rates have been lower and by

M. Eilittä et al. (eds.), Green Manure/Cover Crop Systems of Smallholder Farmers, 237-261.

2002, sustained adoption rates for *Mucuna* in the region are clearly low. Because of the good knowledge base and biophysical performance of *Mucuna* in the area, niches for its inclusion in the region's cropping systems should be found.

INTRODUCTION

Resource-conserving technologies are urgently needed in West Africa. The region has a high population density that is quickly increasing and has a rapidly deteriorating resource quality. Lacking economic possibilities outside of farming, most of the region's inhabitants typically continue to practice subsistence farming with few external inputs. Such characteristics also apply to Benin, a small West African country of six million inhabitants, located between Nigeria and Togo, and especially to its southern part where most of the country's inhabitants live.

In the second half of the 1980s, researchers working with Recherche Appliquée en Milieu Réel (RAMR, Applied Research On-farm) became interested in *Mucuna pruriens* (L.) DC. On-farm demonstrations gave way to further trials and spontaneous farmer adoption. *Mucuna* seemed to both suppress *Imperata cylindrica* (L.) Raeusch., an important weed in the area, and to maintain soil fertility in the maize (*Zea mays* L.)-based systems of the region. This prompted a range of organizations to promote *Mucuna*, including the government's extension organization, Centre d'Action Régionale pour le Développement (CARDER, Regional Action Centre for Development) as well as a large non-governmental organization (NGO), Sasakawa-Global 2000 (SG2000) and some smaller organizations. The strongest of these extension efforts have now waned and future commitment to *Mucuna* extension in the region is unsure. Seemingly, *Mucuna* remains the most adopted cover cropping system in West Africa (Carsky et al. 2001).

Compared to many other promoted *Mucuna* systems worldwide, research in the region has been relatively extensive. In particular, the involvement of the International Institute of Tropical Agriculture (IITA) in these efforts, as well as with *Mucuna* work in general, has resulted in a number of publications on the topic (Carsky et al. 1998). Finally, although several studies have been conducted on *Mucuna* adoption in the region, the methods employed in them make it difficult to accurately estimate the number of *Mucuna* adopters or utilizers in the southern region. Two areas of research have received less emphasis, impeding the accurate characterization of the *Mucuna* experience in Benin. First, few on-farm observational studies have been conducted on farmer management of the system as well as the system's performance under farmer management. Second, the diffusion efforts have been little studied at a detailed level.

In the following, the Beninoise context is first described, after which attention turns to the maize-*Mucuna* systems in the region.

REGIONAL CONTEXT

Biophysical Environment

Southern Benin consists of three provinces: Mono, Atlantique and Ouémé (Figure 1). The average annual temperature is 27 °C, with the highest average monthly temperature (36 °C) in March and lowest (22 °C) in July (Houngnandan 2000). The area belongs to the subhumid zone with a relatively bimodal rain distribution. Rainfall decreases to the north and varies from 1000 to 1300 mm. The bimodal rainfall pattern allows for two growing seasons: the main one from April to July and the second, minor season, from September to November. The minor season is risky for maize cultivation because of erratic rainfall (Mutsaers 1991).

Upland soils in the area, called *terres de barre*, are classified as *sols ferralytiques* in the French system, while in the Food and Agriculture Organization (FAO) classification they are considered Acrisols (with low base saturation) or Lixisols (with moderate saturation). They have a sandy topsoil and clayey subsoil and have low organic C (0.8 ± 0.4%), K (0.15 ± 0.05 meq 100 g^{-1}) and cation-exchange capacity (5.6 ± 1.1 meq 100 g^{-1}) (Versteeg et al. 1998b, citing Kater, unpublished). Table 1 presents soil characteristics of 24 farmers' fields in four villages in southern Benin.

Table 1. Average values (range) in some soil parameters measured in farmers' fields in the villages of Zouzouvou, Eglime and Tchi and in the Niaouli research station in southern Benin (*Source*: Houngnandan et al. 2001).

Soil parameter	Site			
	Zouzouvou	Eglime	Tchi	Niaouli
pH ($H_2$0)	6.31 (5.61-6.78)	5.87 (4.55-6.25)	6.59 (6.16-7.40)	4.73 (4.35-5.11)
Total organic carbon (%)	0.62 (0.34-1.05)	0.90 (0.62-1.25)	1.19 (1.13-1.21)	0.29 (0.28-0.30)
Extractable P (µg g^{-1})	12.20 (3.58-44.40)	5.03 (0.79-13.20)	0.42 (0.38-0.54)	23.01 (16.48-29.53)
Available NO3 (µg g^{-1})	0.71 (0.21-1.47)	1.12 (0.52-1.85)	1.80 (1.08-3.39)	0.70 (0.68-0.72)
Available NH4 (µg g^{-1})	0.71 (0.23-1.48)	0.81 (0.56-1.02)	2.26 (0.98-2.39)	0.57 (0.46-0.68)
No. of fields	12	16	4	2

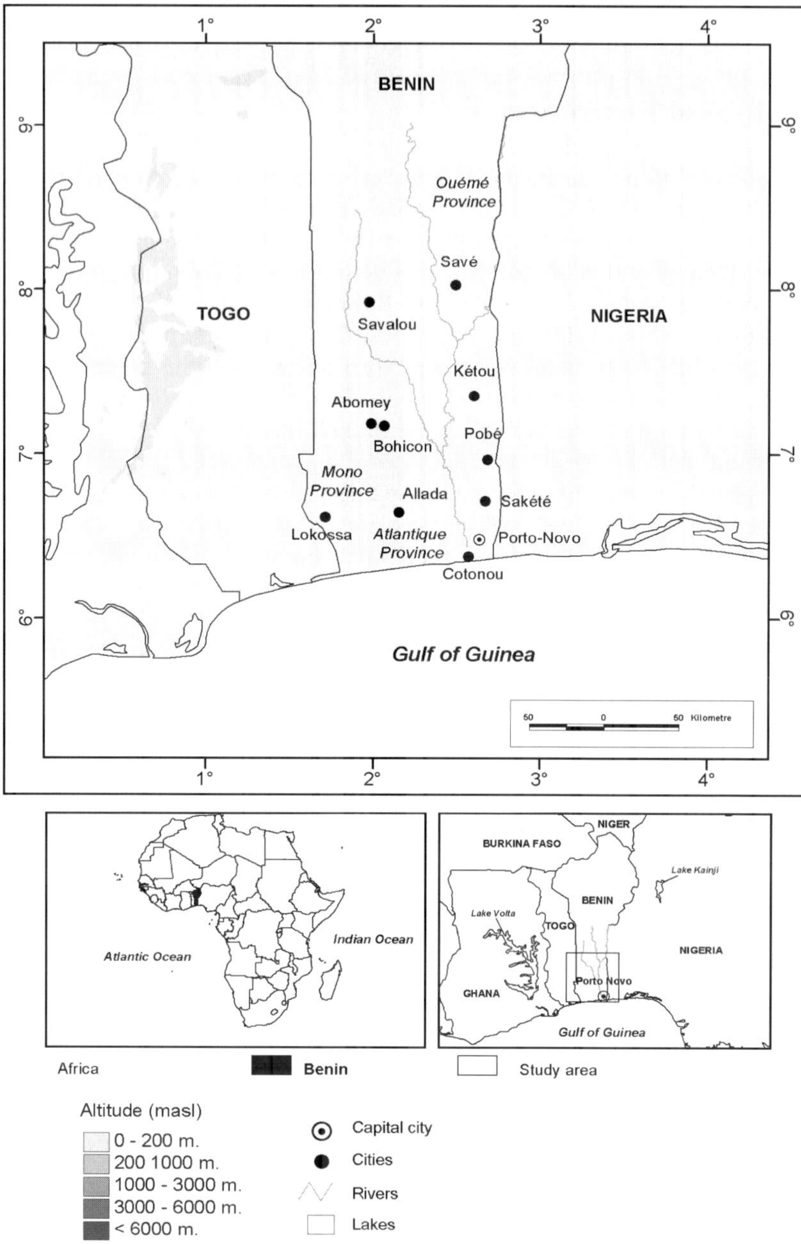

Figure 1. Location of the study area, southern Benin.

Cropping Systems

The major cropping system found in southern Benin is an oil palm (*Elaeis guineensis* Jacq.) – maize-based system, with oil palm, maize, cassava (*Manihot esculenta* Crantz), cowpea (*Vigna unguiculata* [L.] Walp.) and groundnut (*Arachis hypogaea* L.) as the major crops. Intercropping is the norm, with maize/cassava a common mixture in the main season and maize/cassava and cowpea/cassava common in the second season. In the Houngnandan (2000) survey, 72% of the fields had two crop mixtures, 22% had three crop mixtures and no field had only one crop. Oil palms, citrus (*Citrus* sp. L.) and other trees are found scattered in the fields. Fire is typically used to clear fallowed land, while hand hoeing is sometimes done. Very little fertilizer is used, except in areas where cotton (*Gossypium hirsutum* L.) is cultivated. Cotton is grown in some areas, although it is more commonly cultivated farther north.

The traditional oil palm fallow lasts about 15 years and renders several products—firewood, animal feed, palm oil and palm wine—the latter of which can be distilled into Sodabi liquor (Kang et al. 1991). Because of land pressure, fallow periods have declined and fallowing is becoming less common. The Houngnandan (2000) survey showed that an average of 61% of farmers practiced fallow but an average oil palm fallow lasted only 9 years. In the southern provinces, fallowing with *Acacia auriculiformis* Benth. produces both fuel wood and material for traditional fishing practices and has become increasingly popular (Versteeg et al. 1998b; A. Floquet, personal communication, 1999; J. Totongnon, personal communication, 1999; Houngnandan 2000). Only a small percentage of farmers, averaging 8% in the Houngnandan (2000) survey, supplement fallows with non-crop organic materials such as *Mucuna* mulch or home waste. According to the same survey, the practice of returning crop residues to the field for soil fertility maintenance is widespread, with 86% of farmers recycling cowpea residues and 67% recycling maize residues.

Productivity of the major food crop, maize, is very low. Typically, yields of 500 kg ha^{-1} are achieved in depleted fields without fertilizer (Versteeg et al. 1998b). Major agricultural problems in the area include weeds, insect pests and low soil fertility. Of the weeds, *Imperata* is particularly problematic (Vissoh et al. 1998). Livestock in the area consists of small ruminants (goats and sheep) as well as poultry.

Socio-economic Environment

Southern Benin, populated by several ethnic groups, has very high population densities. With an average density of about 220 inhabitants per km^2, the three southern provinces cover only about 10% of the country's area but contain about 70% of Benin's population of six million (Honlonkou and Manyong 1999). In the low-rising plateaus of the south, population densities of up to 350 inhabitants per km^2 are found (Vissoh et al. 1998, citing Manyong et al. 1996). Such high densities contrast with average population densities in the central (25-50 people per km^2) and northern (fewer than 25 people per km^2) regions of Benin.

As a consequence of high population density, average farm sizes are small, less than 1 ha (Vissoh et al. 1998). In the Houngnandan (2000) survey, the average farm size was 0.65, 1.42 and 0.63 ha in three villages in Mono province. Of the surveyed 232 farmers, 77% owned their fields, while the remaining 23% used fields that were rented, borrowed or sharecropped. Another researcher estimates that about half of all land in southern Benin is rented and that often tenants are young people with more labour availability (A. Floquet, personal communication, 1999).

Both men and women carry out farming activities in the region. The Houngnandan (2000) survey showed that 24% of farmers were female. Typically, the fields of women farmers are smaller and their labour constraints are higher. In addition, both young and old people are engaged in farming activities, indicating relatively poor economic prospects outside of farming. In the above-mentioned survey, 61% of farmers were less than 45 years old, while only 11% of farmers were over 60 years old.

Poverty is widespread in the region. Villages typically have primary schools but lack electricity, clinics and other infrastructure (including irrigation) and are generally accessible by dirt roads or paths. Small local markets exist but transportation of goods and people to larger markets is often difficult. In some villages, farmers have access to credit through cooperatives.

STRUCTURE AND FUNCTIONING OF THE *MUCUNA* SYSTEMS

Maize-*Mucuna* Systems and their Management

Two *Mucuna* types, one with black seed (var. *utilis*), another with white seeds (var. *cochinchinensis*), have been promoted in the region. In northern Benin, *Mucuna* matures only if planted with the rains and is therefore sole cropped. In the south, however, both the sole-cropped and relay-intercropped *Mucuna* systems allow for seed production.

Research and extension efforts have focused on two maize-*Mucuna* systems. The systems' main features have been described in a technical recommendation pamphlet developed by the SG2000 (Carsky et al. 1998; Vissoh et al. 1998) as follows:

(1) Sole-cropped *Mucuna* short fallow: For fields severely infested with *Imperata*, the recommendation has been to grow *Mucuna* sole-cropped over the major and minor cropping cycle. (Seemingly, the system has been practised also in highly degraded fields where maize can no longer be grown.) The technical advice given through CARDER has been to plant *Mucuna*, spaced 0.4 m by 0.8 m, at the beginning of the rains (typically in April) after field slashing with one seed per hole. If

necessary, weeds were to be slashed again after 1 month. *Mucuna* would then flower (in November) and produce seed (in December) and senesce (in December-January). For the dry season, the technical recommendation advised that a fire-break should be constructed around the plot.

(2) Maize-*Mucuna* relay-intercropping: *Mucuna* planting between maize rows has been recommended at 45 days after planting (DAP) maize (typically in May-June), at a density of 0.8 m by 0.8 m. Maize is harvested in July-August. No maize production can take place during the second, minor season, and *Mucuna* stays in the field through flowering (November), seed production (December) and senescence (December-January).

Due to lack of on-farm data, the share of farmers using these two systems has not been quantified; seemingly the great majority of *Mucuna* was grown under intercropped conditions (J. Totongnon, personal communication, 1999). Moreover, no data are available that describe farmer management in a detailed and quantitative manner. According to researchers and extension workers, farmers have typically followed the management recommendations described above, although some later *Mucuna* planting dates have been reported (J. Totongnon, personal communication, 1999; P. Vissoh, personal communication, 1999). Some farmers also apparently have not harvested seed, letting *Mucuna* reseed itself and controlling its growth during weeding operations (Nouatin 1999). In the south, residues are normally left as surface mulch, because most of the time no mounding or ploughing takes place. However, bush fires can burn the mulch.

System Productivity

Mucuna Biomass Production

Mucuna biomass production under farmer management has been poorly quantified. In general, the impressions of several researchers and extension workers suggest that in most fields, *Mucuna* biomass production is good and that *Mucuna* grows well, covering the ground rapidly.[1] Even at 45 DAP, maize growth in some high-fertility fields has been so good that *Mucuna* has been reported to compete with maize.

The following research results describe *Mucuna* productivity in the area:

- Sanginga et al. (1996) measured *Mucuna* biomass production in farmer-managed fields at 12 weeks after planting (WAP). The sampling purposefully included a range of *Mucuna* productivity (P. Houngnandan, personal information, 1999), so one cannot estimate typical variability from the research. In the same study, the authors found an average contribution of 167 kg ha^{-1} atmospheric N but, more disconcertingly, that 7 out of 34 farmer fields had no nodulation.

- Houngnandan et al. (2000) conducted a researcher-managed, on-farm experiment in three villages and 15 fields in 1995 and 1996. In the three villages they found an average biomass production of 2.2, 3.2 and 3.3 t ha^{-1} at 12 WAP and 4.3, 5.3 and 4.7 t ha^{-1} at 20 WAP. *Rhizobia* inoculation increased shoot dry weight and N accumulation; such effect was not consistent but varied by field and site. In a greenhouse trial using soils from two farmers' fields where *Mucuna* did not grow well, biomass production decreased by 74% and 66% for one soil and 58% and 28% for the other soil when N and P were absent in the complete fertilizer treatments.
- Importantly, Carsky and Etèka (2001) document the good persistence of *Mucuna* mulch under the dry-season conditions of southern Benin. In the region, new growth is sufficient (because of the sporadic rains during the dry season) to compensate for the simultaneous disappearance of mulch. In drier regions with fewer or no sporadic rains during the dry season, monthly *Mucuna* mulch disappearance rate of about 1 t ha^{-1} has been documented due to wind and termites.

***Mucuna*'s Impact on Maize Yield and *Imperata cylindrica* Infestation**

Most researchers and extension workers also agree that *Mucuna*'s impact on the following year's maize yield is relatively high and consistent and that it persists for both main and minor maize crops, whether *Mucuna* was grown under relay-intercropped or sole-cropped conditions. Moreover, *Mucuna* has been shown to greatly reduce *Imperata* infestation. It is thought that *Mucuna* only suppresses but does not eliminate the weed. Although relatively much on-farm data on yield impacts are available, most of the work has been conducted under researcher-managed experimental conditions:

- Versteeg et al. (1998b), in reporting results from 15 farmer-managed on-farm trials, found on average a 70% higher maize yield after intercropped *Mucuna* than after monoculture maize.
- Totongnon (1998) measured maize yield in 1997 and 1998 in eight relatively fertile farmers' fields after they had been cultivated with *Mucuna* in 1996. In 1997, yields after *Mucuna* were significantly higher (main season 1707 kg ha^{-1} and minor, 1110 kg ha^{-1}) than yields in plots with no prior *Mucuna* cultivation (main season 800 kg ha^{-1} and minor, 846 kg ha^{-1}). There was no residual impact of *Mucuna* in 1998.
- In an on-station experiment in Niaouli, Houngnandan (2000) found a 94% yield increase (from 372 to 720 kg ha^{-1}) of main season maize after a sole-crop of *Mucuna* had been planted in early July of the

previous year. He also found much higher yield impact of *Mucuna* when it was incorporated (1.24 t ha^{-1}) than when applied as mulch (0.87 t ha^{-1}).

- In the communities studied by the University of Hohenheim, yield increases in both major and minor season maize from the previous year's relay-intercropped *Mucuna* have been 30-50% (A. Floquet, personal communication, 1999).
- Dovonou (1994) reports that 1-year *Mucuna* fallow reduced *Imperata* shoot density from 270 to 32 shoots per m^2; Udensi et al. (1999) and Chikoye and Ekeleme (2000) have confirmed this impact. The latter documented that different *Mucuna* accessions (and associated growth duration) have varying capacities to suppress the weed. Economic analysis of *Mucuna* for *Imperata* control has also been done (Chikoye et al. 2002).

Costs and Benefits of *Mucuna* Use

Relatively few economic studies on the system have been conducted. Manyong et al. (1998, cited in Vissoh et al. 1998) found positive returns in the second year of *Mucuna* cultivation at both farmer and regional level. However, in the event that *Mucuna* seed can be sold, the system was shown to be economically beneficial from the first year onwards. The following *ex ante* benefit-cost ratios over 8 years of utilizing the system were obtained: 1.24 for the system with *Mucuna*, 0.62 for the system without *Mucuna*. If the seed were sold, benefit-cost ratio was up to 3.56. The system showed a declining yield trend over time, indicating that additional external inputs are needed to maintain productivity. The authors estimated that, if used throughout the Mono province, adoption of *Mucuna* would mean a saving of 6.5 million kg of N y^{-1}, or US$1.85 million.

Factors Impacting the Performance of the Systems

There seems to be general agreement that agronomically, *Mucuna* is well adapted to southern Benin and that in most farmer environments of the region the maize-*Mucuna* system is relatively productive, with yields higher than those obtained without *Mucuna*.

The most important positive factors impacting the performance of the system are that:

- Climatic factors typically favour *Mucuna* biomass production, apparently enabling rapid growth in most fields. In fields with high soil fertility, competition with *Mucuna* has been reported even when it has been planted 45 days after maize (C. Akakpo, personal communication, 1999). Occasional poor performance of *Mucuna* has been reported due to a dry spell (J. Totongnon, personal communication, 1999).

- The soils of southern Benin typically favour *Mucuna* production and allow for clear yield impacts on the following year's maize crops.
- Farmer management of the system has seemed to follow recommendations rather closely, enabling high biomass production of *Mucuna* and positive impacts on the following year's maize crop. On occasion, farmers seemingly have not collected *Mucuna* seed, causing self-reseeding (A. Floquet, personal communication, 1999).
- High maize yield increases are seemingly able to counteract the loss of minor season maize production, making the system economically more attractive than maize production without *Mucuna*. As discussed later, however, economic gains that can be realized only after 1 year may mean relatively little in conditions of extreme poverty.

Relatively few factors seem to negatively affect the system's agronomic and economic performance and productivity:

- In extremely poor soils, *Mucuna* growth has been poor.
- *Mucuna* planting date has been mentioned as a factor that at times has negatively affected the system's performance. While late *Mucuna* planting has caused low biomasses, early *Mucuna* planting (even at recommended date) has reportedly caused competition with maize in very fertile fields.
- Near-complete discontinuation of subsidized *Mucuna* seed markets had clear negative impacts on the system's economic performance. Such seed markets had made the system demonstrably profitable from the first year onwards.
- Uncontrolled fires that destroy the *Mucuna* mulch take place in the area (Vissoh et al. 1998), greatly impacting the system's performance.

As discussed in the following section, despite the apparently good agronomic and economic performance of the system, several factors have strong negative impact on the system's adoption potential.

ADOPTION OF THE SYSTEM

In the late 1990s, the success of *Mucuna* in Benin was relatively widely reported in literature. Surprisingly rapid rates of adoption and/or testing by farmers were reported and generally explained by *Mucuna*'s ability to control *Imperata*, a relatively common weed in the southern provinces, as well as by its ability to restore soil fertility.

More recent reports and the observations of several researchers familiar with the current situation suggest that either *Mucuna* is relatively rarely used in the system or the absolute number of utilizers/adopters is relatively low. Many individuals familiar with the *Mucuna* system have the impression that rather than being consistently used by farmers in the region it is used occasionally, whenever *Imperata* and/or soil fertility problems in a field are very serious. Others consider that in absolute numbers, *Mucuna* adoption has remained low and that this, as much as utilization patterns, accounts for the comparatively few fields that can be seen under *Mucuna*.

The Focus and Context of Adoption

Although several studies have been conducted on *Mucuna* adoption in the region, the context of such adoption has been discussed relatively little in the literature and requires clarification. *Mucuna* work took place in the context of the research efforts of RAMR as well as the extension efforts of CARDER and SG2000. Across these three organizations, another line of work was pursued concurrently with *Mucuna*—the promotion of a package of an improved maize variety and inorganic fertilizer as a way to improve maize production of the smallholders in the region. Seemingly, at the village and farm level, the *Mucuna* technology was linked to the improved maize + fertilizer package, in that the same organizations were promoting the two packages concurrently and often to the same farmers. In the case of SG2000, such promotion efforts also included inexpensive credit for the purchase of the fertilizers.

History of Diffusion Efforts

Several organizations have been involved with the *Mucuna* diffusion efforts, from either research or extension perspectives; such efforts have been recounted in a number of articles, including Versteeg et al. (1998a; b). The RAMR project played a key role in the initial development of the technology, and in its continued on-farm testing. Technical advice was provided by IITA and the Royal Tropical Institute of the Netherlands also aided in the early development efforts. The Beninoise extension agency, CARDER, first brought the technology into large-scale extension. The agency was later aided in these efforts by SG2000 along with some smaller NGOs, which provided additional funding for field operations and facilitated *Mucuna* extension efforts through seed purchases.

The *Mucuna* efforts in southern Benin began in the village of Zouzouvou, Mono Province in 1986-87, where a demonstration plot was established within a project focusing on increased food production and improved soil fertility. Farmer visits to such demonstration sites, which were located often on school grounds, were encouraged. Farmers reportedly became impressed by *Mucuna*'s ability to suppress *Imperata* and in 1988, 15 of them requested *Mucuna* seed to test in their fields. Such testing was monitored with results indicating *Imperata* suppression, reduced need for weeding and increased maize yield with *Mucuna*. In the following year, additional farmers started to use *Mucuna*. In 1990, CARDER began pre-extension tests in 12 more villages in the Mono Province, involving 180 farmers. In 1991, trials were expanded to other southern provinces.

In 1992, SG2000 initiated its efforts on *Mucuna* in the region. Such efforts were meant to complement, not compete with, CARDER's efforts and, for example, staff from the Ministry of Rural Development were located within SG2000 (P. Vissoh, personal communication, 1999). In concrete terms, SG2000 facilitated the CARDER efforts in three ways:

(1) It provided field allowances and vehicles to CARDER field-level extension agents and technicians involved in the *Mucuna* promotion efforts.
(2) It purchased *Mucuna* seed from farmers and distributed it free of charge to new farmers involved in extension efforts; after the first cropping cycle, farmers were supposed to return the same quantity of seed to the organization.
(3) Training village-level staff in the technologies was promoted. A technical bulletin on the establishment of *Mucuna* was developed as a training tool.

The work was conducted through a Training-and-Visit system by establishing demonstration plots in farmers' fields to give farmers an opportunity to evaluate and decide whether to adopt the technology. Reportedly, the demonstration efforts of CARDER-SG2000 were coupled with spontaneous farmer adoption of the *Mucuna* systems.

Until 1995, such demonstration plots were sited in individual farmers' fields, with a typical size of 5000 m^2. Starting in 1995, an effort was made to reach a larger number of farmers rapidly and two modifications were made to the system. First, the size of the demonstration plots was reduced to 500 m^2; second, instead of being managed by an individual farmer, each plot became the responsibility of a group of farmers, averaging 10 in number. Moreover, at that time, SG2000 stopped paying field allowances directly to the CARDER staff and no longer provided equipment for them (P. Vissoh, personal communication, 1999).

In addition to the earlier mentioned organizations, others in the area were carrying out diffusion efforts on *Mucuna*. These others included two NGOs (Centre Régional pour le Développement et la Santé [CREDESA], Projet de Développement de l'Elevage dans le Borgou Est [PDEBE]) and a University of Hohenheim project conducting research in six communities in southern and central Benin. Their participation, however, has been limited relative to the other organizations.

An important aspect of the research efforts in Benin was the post-harvest processing to reduce the L-dopa content of *Mucuna* flour used to make *pâte*, a staple in Benin. Such research efforts were relatively successful, reducing the L-dopa content to about one-tenth of that found in seed (Versteeg et al. 1998a). These efforts have not resulted in food utilization of *Mucuna* to any significant degree. In northern Benin, some utilization of *Mucuna* hay as cattle feed has been reported (Yai 1998) but the project was suspended after a 2-year period. In the south, at least one farmer, having utilized *Mucuna* in his citrus orchard, obtained good results with weed suppression and productivity; the system also reduced labour requirements (Totongnon and Lame 2000).

In 1996, SG2000 discontinued work with *Mucuna* in Benin. This greatly impacted the ability of CARDER to continue extension efforts (P. Vissoh, personal communication, 1999) and put an end to most of the seed purchases (J. Totongnon, personal communication, 1999). Another organization, the University of Hohenheim project, stopped its field activities in 1998.

From the initial research efforts of RAMR, much emphasis was placed on interaction with farmers and their constant feedback to researchers. The latter saw the process as an interactive one in which farmers were able to focus research activities on several occasions, greatly increasing the efficiency of agricultural research (Versteeg et al. 1998b). For example, the farmers directed efforts to *Mucuna*'s ability to control *Imperata*. In the later work, funded and partially directed by SG2000, good contacts with farmers were considered essential to success (Galiba et al. 1998).

Clearly, SG2000 enabled CARDER to work more effectively with *Mucuna* extension but certain problems persisted in the organization, such as getting equipment to the village-level workers on time and their motivation and monitoring (J. Totongnon, P. Vissoh and A. Floquet, personal communication, 1999). An SG2000 staff member considered that the extension efforts were relatively more successful when the project was in touch with individual farmers instead of farmer groups and when it worked with larger plots from which farmers were able to pay back the seed (P. Vissoh, personal communication, 1999). More detailed adoption studies are needed to evaluate the impact of the temporary seed markets created by SG2000 on *Mucuna* utilization. However, it seems clear that with only 4 years of extension efforts it is difficult to achieve a lasting impact.

Adoption/Utilization Rates at Different Periods

In the following, *Mucuna* utilization at various periods is discussed, as reported in the literature.

Between 1986 and 1991

Most of *Mucuna* planting in the early phases of technology development and diffusion efforts took place in a few communities in the Mono Province. It should be noted that the government's extension and pre-extension efforts focused on demonstration plots that were located on farmers' fields. Exact quantification of the demonstration plots and spontaneous farmer adoption has not been conducted.

The progress in number of users as reported in the literature is:

- 1986-1987: Project demonstration plots were initiated in Mono Province.
- 1988: Farmer-managed trials with 20 farmers were conducted.
- 1989: 103 farmers were testing *Mucuna*; result of spontaneous adoption (Vissoh et al. 1998, quoting Versteeg and Koudokpon 1990).
- 1990: CARDER conducted pre-extension tests with 180 farmers located in 12 more villages in Mono (Versteeg et al. 1998b).
- 1991: 500 farmers were testing *Mucuna* in the three southern provinces as result of CARDER's efforts.

Between 1992 and 1996

A large increase in *Mucuna* planting occurred between 1992 and 1996 when SG2000 became actively involved in the work. Relatively high rates of farmer use of *Mucuna* have been cited for this period. In 1993 and 1996, 3000 and 10 000 farmers in all of Benin were testing *Mucuna* in their fields (Vissoh et al. 1998). As discussed earlier, this testing took place in demonstration plots of individual farmers (until 1994) or farmer groups (from 1995). What were the resulting rates of *Mucuna* adoption? Manyong et al. (1996), in surveying *Mucuna* use in the original Mono province in 1994, found a high rate of adoption (24%) in the villages where RAMR had originally worked. Galiba et al. (1998) found that 74% of participating farmers continued *Mucuna* use for at least 3 successive years, while 83% used it for 2 years. The remaining farmers were either using *Mucuna* infrequently (in exhausted or *Imperata*-invaded plots) or had abandoned it. Moreover, a spontaneous diffusion ratio of seven new farmers for every farmer reached by the extension efforts was reported (Galiba et al. 1998).

Houngnandan (2000) conducted a survey on farmer utilization of soil amendments in June-August 1996. A formal questionnaire was completed for 232 randomly sampled farmers residing in three villages that had been a part of the RAMR project. According to the survey, 61% (148) of the farmers fallowed their fields and of these, an average of 26.8% (38) utilized a *Mucuna* fallow of typically 1 year; such farmers would represent 16.4% of the population in the villages. The share of farmers using planted fallows of various kinds varied greatly by village. The author concludes that, while farmers had good knowledge of *Mucuna*, they did not grow it every year for a variety of reasons, such as a pressing need for food production and the cited poor growth of *Mucuna* in exhausted fields. Instead, *Mucuna* is cultivated when the need arises, especially for *Imperata* suppression.

Between 1997 and 2002

More recently, estimated adoption rates have been lower, which – if true – may be explained as a result of more accurate surveys, their broader coverage (i.e. villages covered where research and extension efforts had not been active) or discontinuation of *Mucuna* use. Three reports have been completed with information on the recent adoption patterns of *Mucuna*.

Honlonkou and Manyong (1999) and Honlonkou et al. (1999) surveyed 10 purposively selected villages between November 1997 and March 1998. All households in each village were stratified into four groups: users of *Mucuna* only, users of mineral fertilizer only, users of both *Mucuna* and mineral fertilizer and users of neither technology. Villages were chosen that were known to include households in all four categories, that is, it was known that some *Mucuna* users were in each village (A. Honlonkou, personal communication, 1999). The sample of households surveyed within a village (totalling 40) reflected the proportion of the four groups in that village. The researchers found relatively low adoption rates for *Mucuna*. Of the 580 farmers surveyed, 7% utilized *Mucuna*. The authors point out that if this adoption rate was extrapolated to southern Benin, there would have been 14 000 *Mucuna* adopters in late 1997-early 1998.[2] The average area in *Mucuna* cultivation was 0.4 ha per farmer, accounting for 2% of the farmland. In analysing past adoption rates in the 10 surveyed communities, the authors concluded that adoption never exceeded 8% of farmers and the area planted to *Mucuna* never exceeded 3% of land area in these villages. Such adoption rate constitutes a significant drop from the earlier relatively high adoption estimates of Manyong. This study found a decrease in adoption rate (11%) in 1997, after the discontinuation of the SG2000 efforts. Moreover, in 1997, the rate of abandonment (4.6%) was for the first time greater than the rate of new adoption (3.4%). Figure 2 depicts the *Mucuna* adoption pattern in the southern provinces according to this study.

Another survey conducted in four of the six provinces in Benin (Alohou and Hounyovi 1999) included only one southern province, Ouémé. Two villages were sampled in each of two *secteurs* (administrative unit where CARDER specialists are located). In these four villages, farmer adoption rates were relatively high (at 26%, 17%, 13% and 12%). In the two *secteurs*, *Mucuna* was the third or fourth most utilized soil-conserving technology after inorganic fertilizer, residue burning and in one village, rotation.

Finally, in 1998, the University of Hohenheim project surveyed farmers who had tested various resource-conserving technologies with the project in 1994-96 (Doppler et al. 1996). For the *Mucuna* relay-cropping system, adoption rate was about 20%, while its rejection rate was 35% and a further 44% of farmers considered that it would need more assessment. Interestingly, only 8% of farmers had adopted the sole-cropped *Mucuna* system for *Imperata* control and a full 56% had rejected it; the prevalence of bush fires in the area explains this. Reasons for rejecting the relay-cropping *Mucuna* system included its unsuitability for fields with oil palm, the difficulty of land preparation and the disappearance of the market for seed.

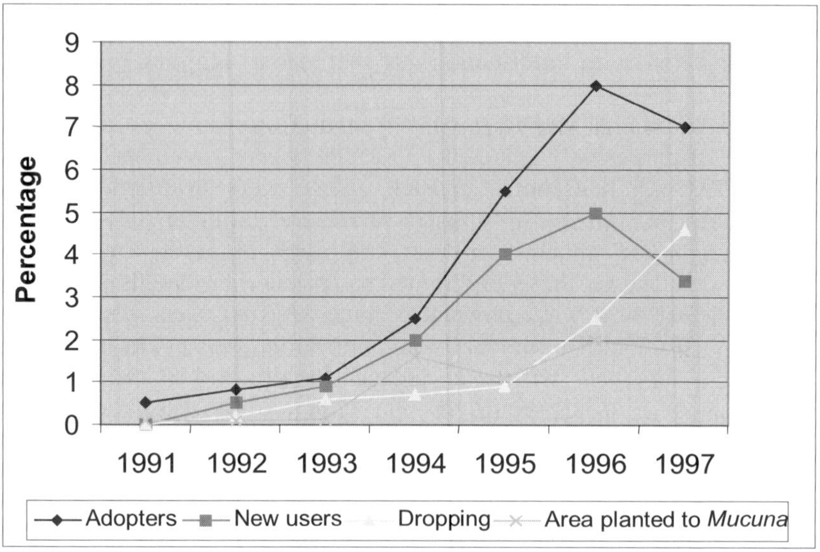

Figure 2. Estimation of *Mucuna* adoption rates (%) in southern Benin from 1991 to 1997 (*Source*: Honlonkou et al. 1999).

In summary, it is difficult to estimate the extent of *Mucuna* adoption in southern Benin for several reasons. First, there is a lack of adoption studies in randomly chosen villages in the south. Several studies either utilized villages where RAMR was active or where *Mucuna* was known to be used. Second, no quantification has been conducted of spontaneous adoption outside of CARDER or RAMR demonstration plots.

Extrapolation from these studies to southern Benin as a whole is therefore not possible, because it would assume uniform farmer-to-farmer diffusion both within and between villages. A researcher working in the region surveyed communities surrounding the study villages and reports surprisingly poor diffusion of soil fertility technologies to these surrounding communities; her estimations for adoption rate in the south were 1-2% (A. Floquet, personal communication, 1999). A long-time researcher at the government's extension agency estimated that in one-fifth of villages there is some *Mucuna* adoption and that the average adoption rate at the village level is low. *Mucuna*, according to this researcher, is the second most adopted resource-conserving technology after *A. auriculiformis* (J. Totongnon, personal communication, 1999). If such estimations are valid, adoption figures presented in many of the studies discussed above are more likely to represent adoption rates in communities with high, rather than average, adoption rates. In 2002, casual observations when travelling in southern Benin suggest that adoption is indeed very low. Yet interest in *Mucuna* continues in the area and a project funded by the Dutch government plans to integrate *Mucuna* in organic cotton growing systems. Because of the good knowledge base and biophysical performance of *Mucuna* in the region, future efforts might well find new and productive niches for it.

Farmer Views on the *Mucuna* Technology

Relatively few studies have focused on farmer perceptions of *Mucuna* technology in southern Benin. Honlonkou and Manyong (1999) and Honlonkou et al. (1999) surveyed farmer perceptions of *Mucuna* technology in late 1997-early 1998 against 10 criteria. They found that, on average, farmers were satisfied with *Mucuna* technology, which had a higher global satisfaction (79%) than had mineral fertilizers (49%). Appreciation criteria with particularly large percentage difference in the satisfaction score in favour of *Mucuna* over mineral fertilizers were (in decreasing order of importance): weed control, cost, preservation of soil fertility in the long run, ease of availability and long-term effect on soil fertility. The following appreciation criteria had smaller percentage difference (less than 15%) in satisfaction score (in decreasing order of importance): auto-diffusion, not worsening the weed problem, intensity of soil fertility improvement and willingness to use the technology again. Finally, on rented land, mineral fertilizers were ranked higher than *Mucuna*.

In another study, Alohou and Hounyovi (1999) asked farmers in four villages in southern Ouémé province to rank different soil-fertility technologies; the higher the rank, the more effective the technology in restoring soil fertility. Of the technologies ranked in two villages, *Mucuna* was among the highest ranking, along with fertilizer, alley cropping, animal manure and compost.

Totongnon (personal communication, 1999), in reporting on on-farm research in Atlantique province, noted that farmers were more interested in the cassava fallow than the maize-*Mucuna* system because it produced food during the minor season. Houngnandan (2000), working in Mono province, reported that farmers justified the occasional, rather than constant, use of *Mucuna* in several ways. In the village of Zouzouvou, farmers felt that land was urgently needed for food production. Farmers also reported that neither *Mucuna* nor maize grew in very exhausted fields.

Critical Factors Impacting the Adoption of the System

Factors Favouring Adoption

A number of factors can explain the increased farmer utilization of *Mucuna* in the 1990s as well as its seeming decline. The most important of these are the extension efforts funded by SG2000. Although some farmer interest in *Mucuna* clearly was initiated during the RAMR and CARDER trials and extension, *Mucuna* use greatly increased in response to SG2000 efforts. Although no detailed studies exist on the impact of these extension efforts, it seems that they fostered *Mucuna* adoption in many ways, as discussed below.

First, *Mucuna* was made more widely available to farmers throughout Benin: *Mucuna* seed was disseminated and CARDER became more effective through paid incentives (i.e. field allowances) and vehicles given to some district officials and village agents involved with *Mucuna* work. In a study by Manyong et al. (1996), access to external research and/or extension agents was a factor positively associated with *Mucuna* adoption.

Creation of profitable *Mucuna* seed markets was another way to foster adoption. Although CARDER had been buying some *Mucuna* seed from farmers, the quantities purchased and prices were low, at CFA25 kg^{-1} (US$0.07-0.08; US$1 = CFA300-350 until 1994; C. Akakpo, personal communication, 1999). In contrast, SG2000 purchased about 50 t of seed between 1992 and 1996 and at prices that were from 75 to CFA100 kg^{-1} (1992 US$0.21-0.24; 1993 US$0.28-0.32) to CFA100-150 kg^{-1} (starting 1994 US$0.143-0.214; US$1 = CFA700 from 1994). The presence of seed markets was a factor positively associated with adoption in a study by Manyong et al. (1996), while an adoption study of Honlonkou et al. (1999) concluded that the 'incentives created by Sasakawa Global 2000 for rapid diffusion of the technology could not be overemphasized.' However, the same authors maintain that the 11% reduction in adoption in 1 year after the discontinuation of *Mucuna* extension was small and that it could be concluded that *Mucuna* had been adopted mainly for reasons other than seed markets (i.e. weed control and soil fertility). A field-level agronomist with long experience in *Mucuna* work disagrees and maintains that the discontinuation of well-priced *Mucuna* markets acted as a disincentive for farmers to continue planting (C. Akakpo, personal communication, 1999).

A third means of fostering adoption was by linking *Mucuna* technology to the overall maize package promoted by CARDER and SG2000, which consisted mainly of improved maize variety and fertilizer, along with an accompanying credit.[3] It is not clear how common the notion was that the various components were linked but it seemingly had been relatively widespread at the field level (P. Vissoh, J. Totongnon, C. Akakpo, personal communication, 1999). The associated credit could have provided a further incentive to *Mucuna* adoption because it was quite attractive. In a Honlonkou et al. (1999) adoption study, access to informal but not formal credit explained *Mucuna* adoption.

Other factors favoured the adoption of *Mucuna* systems in southern Benin. Clearly *Imperata* infestation in the farmers' fields acted as an impetus to the initially observed farmer interest and spontaneous adoption of *Mucuna* (Versteeg et al. 1998b) and in later studies has been positively linked to adoption (Manyong et al. 1996; Honlonkou and Manyong 1999; Honlonkou et al. 1999). Deteriorating soil fertility in the region would seemingly be a great incentive for further *Mucuna* adoption and farmers have ranked *Mucuna* high because of its positive impacts on soil fertility (see above). The associated problem of ever-mounting land pressure in the area should guide farmers to intensifying land-use technologies as extensifying is no longer an option.

Factors with Negative Effect

Several factors in the area negatively affect *Mucuna*'s adoption potential in the region. First, the toxicity of *Mucuna* seed places the most serious limitation to its wider adoptability in the area, given the high pressure on land and widespread poverty. Allocating land to a crop whose products cannot be immediately consumed is often not possible, even if increased maize yield the following year would compensate such lost production (Totongnon et al. 2000). In an interview, a researcher reported stress among farmers with sacks of *Mucuna* seed during hungry times, many of whom had tried eating it and became ill (A. Floquet, personal communication, 1999).

Second, heavy land pressure in the area constrains fallowing generally unless fertility is so poor that no food production is possible. Although high land pressure itself could positively affect the adoption of technologies that intensify land use in comparison to traditional fallows, because *Mucuna* seed is unusable, its effect in southern Benin is seemingly mainly negative due to the loss of second season cultivation. In one study, small farm size was indicated as a factor constraining adoption in the southern Ouémé province (Alohou and Hounyovi 1999), while in another study (Honlonkou et al. 1999) very small farmlands were shown to be an obstacle to adoption and the higher number of male adopters of *Mucuna* was partially explained by their larger landholdings. Despite very low second-season maize yields in the area, maize cultivated and harvested during the second season is often urgently needed by the farming families (Akakpo et al. 2000; Totongnon et al. 2000). Even a production of 400 kg ha^{-1} is tolerated when it is essential for feeding the family (C. Akakpo, personal communication, 1999). Thus, ironically, increasing land pressure coupled with extreme poverty might be expected to worsen *Mucuna*'s prospects in Benin as farmers' planning horizons become extremely short and farmers have to devote all available land to food production. The adoption study of Manyong et al. (1996) suggested that the medium-scale farmers were more likely to adopt the *Mucuna* technology because they lacked land for traditional fallow but had sufficient land such that losing second-season maize production did not immediately threaten their families' food security.

Third, insecure landholding is common in the region and discourages longer-term investment. Secure tenure of land has been found to positively influence the adoption of *Mucuna* (Honlonkou et al. 1999) and a common perception is that this constraint affects especially women, who in the south typically do not own land but borrow fields from their husbands (J. Totongnon, personal communication, 1999). Moreover, women are often obliged to first weed their husbands' fields. This alone, with no consideration of women's other labour demands, places serious labour constraints on them, which may make it even harder for women farmers to adopt the technology.

Fourth, most projects associated with *Mucuna* work in the region have been discontinued. SG2000 formally stopped working on *Mucuna* in 1996, while the University of Hohenheim project, involved with research on soil conservation in the region for 15 years (with 4 years of on-farm research efforts), had finished its field activities by 1999. Decreasing interest in *Mucuna* after SG2000 stopped buying *Mucuna* seed is an indication of the artificial incentive that such seed markets provided.

Fifth, lack of focused efforts and funding to CARDER is a serious obstacle to further *Mucuna* work. If field-level extension workers do not have vehicles and are not motivated to reach farmers, the backbone of any such extension efforts is broken. Currently, the ability of the CARDER staff to reach farmers has seemingly greatly declined because of the discontinuation of SG2000 funding. The on-farm research-extension partnership of RAMR-CARDER during the mid-1990s apparently had offered a relatively effective way to further develop technologies to better suit farmers' needs.

Finally, a study on the process of *Mucuna* adoption in Benin found that, while the initial work in the region allowed interactive learning to take place between farmers and researchers, the approach that was adopted after SG2000 became involved was more top-down. This caused researchers and extension workers to miss out on opportunities to learn and to improve the technologies (Douthwaite et al. 2002).

Several additional field- and village-level constraints to further *Mucuna* adoption have been cited. These include bush fires (mentioned as the main reason for rejecting sole-cropped *Mucuna* for *Imperata* control in the study by the University of Hohenheim; Doppler et al. 1996), rats (C. Akakpo, personal communication, 1999), the difficulty in obtaining seed because of price speculation, the difficulty of incorporating *Mucuna* mulch (Alohou and Hounyovi 1999) and *Mucuna's* tendency to smother young palm trees in the field (Manyong et al. 1996).

In summary, although *Mucuna*'s prospects in southern Benin are affected by contradicting influences, its current prospects seem less promising than those in the mid-1990s, despite the seemingly good agronomic and economic performance at that time. Discontinuation of projects and funding allocated to *Mucuna* work after only a few years of active efforts already has had a negative impact. However, because of the knowledge base on *Mucuna* and its good biophysical performance in the region, it is possible that in the future, innovative niches for it will be found, such as the very recent interest in the incorporation of *Mucuna* in organic cotton-growing systems of the region.

RESEARCH NEEDED

The *Mucuna* experience in southern Benin points to a number of critical research issues concerning *Mucuna* and green manure/cover crops (GMCCs) generally.

Multiple Products

There is a general need to develop GMCCs with multiple products, especially those that satisfy the food demands of the poorest. In particular, studies conducted on post-harvest processing of *Mucuna* should be reviewed to determine whether processing might have the potential to reduce L-dopa content to a level that poses no harm to human beings and animals. In southern Benin, the combination of extreme poverty and high land pressure appears to discourage continual utilization of even GMCC systems that are agronomically and economically well functioning, if no immediate products are available from the system. These two conditions, extreme poverty coupled with high land pressure, will presumably become even more common in many other developing regions, including throughout coastal West Africa, in the near future.

Seed Purchases

The impact of seed purchases on adoption during the diffusion process should be reviewed. Purchases of GMCC seed from farmers are often necessary at the beginning of extension projects but such markets are typically artificial and therefore temporary. It is not known whether, in the long-term, such temporary markets have positive, neutral or negative impact on adoption.

Systems Approach to Research

It may have been worthwhile to include a number of other GMCCs in the research efforts. For example, in locations where *Mucuna* growth was poor, pigeon pea (*Cajanus cajan* [L.] Millsp.) may have provided sufficient biomass and an edible seed already in the first year.

CONCLUSIONS

Biophysically, *Mucuna* is well adapted to the conditions of southern Benin. It generates a large amount of biomass, exerts positive impact on maize yield and suppresses *Imperata*, a common weed in the area. Its current adoption rates are low, however, due to the lack of multiple uses, land constraints, insecure land tenure and discontinuation of most projects that researched and promoted it in the 1990s. Because *Mucuna* is well adapted to the region and the knowledge base on it is good, niches for its inclusion in the region's cropping systems should be found.

NOTES

1. A visiting researcher's trip report in September 1993 observes poor development and colour of *Mucuna*, presumably because of late and sporadic rains, which apparently had occurred also the year before with no lasting impact on the eventual biomass production (Buckles 1993).
2. As noted, however, all villages chosen were known to include *Mucuna* users; direct extrapolation from the study villages to the southern Benin level is therefore not possible.
3. CARDER's original package consisted of improved variety, fertilizer, *Mucuna* and, at times, pesticides (C. Akakpo, personal communication, 1999). In SG2000 work until 1995, all farmers participating in the *Mucuna* work also participated in maize work. Such a package was similar to that of CARDER but contained double the amount of fertilizer and credit for its purchase. The following package was applied to 0.5 ha: (a) improved seed, (b) fertilizer (200 kg ha^{-1} NPKSB, 14-23-14-5-1; and 100 kg ha^{-1} urea) and (c) a pesticide. From 1989 to 1991, farmers were given a 100% credit and they paid it back after harvest (no interest). From 1992, such conditions applied only to first-year farmers and others had to pay 50% on delivery. A village extension agent monitored and supervised cropping operations to make sure they were conducted correctly (P. Vissoh, personal communication, 1999).

REFERENCES

Akakpo C, Amadji F, Carsky R J (2000) Intégration du *Mucuna* dans les systèmes culturaux du sud Bénin. *In:* Carsky R J, Etèka A, Keatinge J D H, Manyong V M (Eds.) Cover crops for natural resource management in West Africa: Proceedings of a workshop organized by the International Institute of Tropical Agriculture (IITA) and Centre d'Information et d'Echanges sur les Plantes de Couverture en Afrique (CIEPCA), 26-29 October, 1999. IITA, CIEPCA, and Cornell International Institute for Food, Agriculture and Development of Cornell University (CIIFAD), USA. p. 175-184.

Alohou E, Hounyovi A (1999) Etude sur la vulgarisation et l'adoption des technologies de gestion de la fertilité des sols. Ministère du Développement Rural, Cotonou, Bénin.

Buckles D (1993) Trip report. Inter-Office Memo. Centro Internacional de Mejoramiento de Maíz y Trigo (CIMMYT), Mexico City, Mexico.

Carsky R J, Etèka A (2001) Persistence of *Mucuna pruriens* biomass during the dry season along an agroecological transect in Benin. *In:* Carsky R J, Etèka A, Keatinge J D H, Manyong V M (Eds.) Cover crops for natural resource management in West Africa: Proceedings of a workshop organized by the International Institute of Tropical Agriculture (IITA) and Centre d'Information et d'Echanges sur les Plantes de Couverture en Afrique (CIEPCA), 26-29 October 1999. IITA, CIEPCA, and the Cornell International Institute for Food, Agriculture, and Development of Cornell University (CIIFAD). p. 55-61.

Carsky R J, Tarawali S A, Becker M, Chikoye D, Tian G, Sanginga N (1998) *Mucuna* – Herbaceous cover legume with potential for multiple uses. Resource and Crop Management Research Monograph No. 25, International Institute of Tropical Agriculture (IITA), Ibadan, Nigeria. 52 p.

Carsky R J, Becker M, Hauser S (2001) *Mucuna* cover crop fallow systems: Potential and limitations. *In:* Tian G, Ishida F, Keatinge J D H (Eds.) Sustaining soil fertility in West Africa. Soil Science Society of America (SSSA) Special Publication No. 58. SSSA and American Society of Agronomy, Madison, Wisconsin. p. 111-135.

Chikoye D, Ekeleme F (2000) Response of *Imperata cylindrica* to smothering by different *Mucuna* accessions. *In:* Carsky R J, Etèka A C, Keatinge J D H, Manyong V M (Eds.) Cover crops for natural resource management in West Africa: Proceedings of a workshop organized by the International Institute of Tropical Agriculture (IITA) and Centre d'Information et d'Echanges sur les Plantes de Couverture en Afrique (CIEPCA), 26-29 October 1999. IITA, CIEPCA, and Cornell International Institute for Food, Agriculture, and Development of Cornell University (CIIFAD). p. 67-70.

Chikoye D, Manyong V M, Carsky R J, Ekeleme F, Gbehounou G, Ahanchede A (2002) Response of speargrass (*Imperata cylindrica*) to cover crops integrated with hand-weeding and chemical control in maize and cassava. Crop Prot. 21:145-156.

Doppler W, Floquet A, Bierschenk T (1996) Adoption of soil improving and agroforestry innovations in family farms in southern Benin. *In*: University of Hohenheim, Standortgemässe Landwirtschaft in Westafrika. SFB 308, Arbeits und Ergebnisbericht, Germany. p. 572-608.

Douthwaite B, Manyong V M, Keatinge J D H, Chianu J (2002) The adoption of alley farming and *Mucuna*: Lessons for research, development and extension. Agrofor. Syst. 56(3): 193-202.

Dovonou H (1994) Influence de la couverture du *Mucuna pruriens* variété *utilis* sur la densité du chiendent. *In:* Wolf J (Ed.) Systèmes agraires et agriculture durable en Afrique sub-Saharienne. Proceedings of a regional conference, 7-11 February 1994, International Foundation of Science, Stockholm, Sweden. p. 401-404.

Galiba M, Vissoh P, Dagbénonbakin G, Fagbohoun F (1998) Réaction et appréhensions paysannes liées à l'utilisation du pois mascate (*Mucuna pruriens* var. *utilis*). *In:* Buckles D, Etèka A, Osiname O, Galiba M, Galiano G (Eds.) Cover crops in West Africa: Contributing to sustainable agriculture. International Development Research Centre (IDRC)-International Institute of Tropical Agriculture (IITA)- Sasakawa-Global 2000 (SG2000), Canada. p. 55-65.

Honlonkou A N, Manyong V M (1999) Learning from farmers in a changing environment: The case of *Mucuna* fallow in southern Benin. Poster presented at the regional workshop on farmers and scientists in a changing environment: Assessing the impact of research in West Africa, 22-26 February 1999, Cotonou, Benin Republic.

Honlonkou A N, Manyong V M, N'Guessan Tchetche (1999) Farmers' perceptions and the dynamics of adoption of a resource management technology: The case of *Mucuna* fallow in southern Benin, West Africa. Int. For. Rev. 1(4): 228-235.

Houndékon V A, Manyong V M, Gogan A C, Versteeg M N (1998) Déterminants de l'adoption de *Mucuna* dans le département du Mono au Bénin. *In:* Buckles D, Etèka A, Osiname O, Galiba M, Galiano G (Eds.) Cover crops in West Africa: Contributing to sustainable agriculture. International Development Research Centre (IDRC)-International Institute of Tropical Agriculture (IITA)- Sasakawa-Global 2000 (SG2000), Canada. p. 45-53.

Houngnandan P (2000) Efficiency of the use of organic and inorganic nutrients in maize-based cropping systems in Benin. Ph.D. thesis, University of Gent, Gent, Belgium. 196 p.

Houngnandan P, Sanginga N, Woomer P L, Vanlauwe B, van Cleemput O (2000) Response of *Mucuna pruriens* to symbiotic nitrogen fixation by rhizobia following inoculation in farmers' fields in the derived savanna of Benin. Biol. Fertil. Soils 30: 558-565.

Houngnandan P, Sanginga N, Okogun J A, Vanlauwe B, van Cleemput O (2001) Assessment of soil factors limiting growth and establishment of *Mucuna* in farmers' fields in the derived savanna of Benin Republic. Biol. Fertil. Soils 33:416-422.

Kang B T, Versteeg M N, Osiname O, Gichuru M. (1991) Agroforestry in Africa's humid tropics: Three success stories. Agrofor. Today 3(2): 4-6.

Manyong V M, Houndékon V A, Gogan A C, Versteeg M N, van der Pol F (1996) Determinants of adoption for a resource management technology: The case of *Mucuna* in Benin Republic. *In*: Senwen Z, Yunlong W (Eds.) Advances in agricultural and biological environment engineering. Proceedings of International Conference on Agriculture and Biological Environment (ICABE), Beijing, 15-19 August 1996. China Agricultural University Press, Beijing, China. p. I-86-I-93.

Manyong V M, Houndékon V A, Gogan A C (1998) Factors affecting the adoption of a short-fallow improved technology: The case of *Mucuna* in the southwest of Benin Republic. International Institute of Tropical Agriculture (IITA), Ibadan, Nigeria.

Mutsaers H J W (1991) Opportunities for second cropping in southwestern Nigeria. RCMP Research Monograph No. 4. Resource and Crop Management Programme, International Institute of Tropical Agriculture (IITA), Ibadan, Nigeria. 28 p.

Nouatin G (1999) Farmer solution for the introduction of *Mucuna* into farming systems. Centre d'Information et d'Echanges sur les Plantes de Couverture en Afrique (CIEPCA) Newsletter 3:7.

Sanginga N, Ibewiro B, Houngnandan P, Vanlauwe B, Okogun J A, Akobundu I O, Versteeg M N (1996) Evaluation of symbiotic properties and nitrogen contribution of *Mucuna* to maize grown in the derived savanna of West Africa. Plant Soil 179: 119-129.

Totongnon J (1998) Rapport recherche-développement. Centre d'Action Régionale pour le Développement (CARDER), Atlantique, Benin. Unpublished report. 12 p.

Totongnon J, Lame B. 2000. Le *Mucuna* en culture fruitière: Cas de vergers d'agrumes et de papayers dans le Département de l'Atlantique, République du Bénin. *In:* Carsky R J, Etèka A, Keatinge J D H, Manyong V M (Eds.) Cover crops for natural resource management in West Africa. Proceedings of a workshop organized by International Institute of Tropical Agriculture (IITA) and Centre d'Information et d'Echanges sur les Plantes de Couverture en Afrique (CIEPCA), 26-29 October 1999. IITA, CIEPCA, and Cornell International Institute for Food, Agriculture, and Development of Cornell University (CIIFAD). p. 148-167.

Totongnon J K, Lame B, Zannou P (2000) Quelques performances et contraintes d'adoption du pois d'Angole (*Cajanus cajan*) et du pois mascate (*Mucuna pruriens* variété *utilis*) en milieu paysan dans le Départament de l'Atlantique, République du Bénin. *In*: Carsky R J, Etèka A C, Keatinge J D H, Manyong V M (Eds.) Cover crops for natural resource management in West Africa. Proceedings of a workshop organized by the International Institute of Tropical Agriculture (IITA) and the Centre for Cover Crops Information and Seed Exchange in Africa (CIEPCA), 26-29 October 1999. CIEPCA, IITA and the Cornell International Institute for Food, Agriculture and Development (CIIFAD), Ibadan, Nigeria. p. 109-119.

Udensi U E, Akobundu I O, Ayeni A O, Chikoye D (1999) Management of cogongrass (*Imperata cylindrica*) with velvetbean (*Mucuna pruriens* var. *utilis*) and herbicides. Weed Technol. 13:201-208.

Versteeg M N, Koudokpon V (1990) *Mucuna* helps control *Imperata* in southern Benin. West African Farming Systems Research Network (WAFSRN) Bull. No. 7. p. 7-8.

Versteeg M N, Amadji F, Etèka A, Houndékon V A, Manyong V M (1998a) Collaboration to increase the use of *Mucuna* in production systems in Benin. *In:* Buckles D, Etèka A, Osiname O, Galiba M, Galiano G (Eds.) Cover crops in West Africa: Contributing to sustainable agriculture. International Development Research Centre (IDRC)- International Institute of Tropical Agriculture (IITA)- Sasakawa-Global 2000 (SG2000), Canada. p. 33-43.

Versteeg M N, Amadji F, Etèka A, Gogan A C, Koudokpon V (1998b) Farmer adoptability of *Mucuna* fallowing and agroforestry technologies in the coastal savanna of Benin. Agric. Syst. 56 (3): 269-287.

Vissoh P, Manyong V M, Carsky R J, Osei-Bonsu P, Galiba M (1998) Experiences with *Mucuna* in West Africa. *In:* Buckles D, Etèka A, Osiname O, Galiba M, Galiano G. (Eds.) Cover crops in West Africa contributing to sustainable agriculture. International Development Research Centre (IDRC)- International Institute of Tropical Agriculture (IITA)- Sasakawa-Global 2000 (SG2000), Canada. p. 1-32.

Yai K (1998) Expériences du projet de développement de l'élevage du Borgou-est (PDEBE) sur les plantes de couverture. *In:* Buckles D, Etèka A, Osiname O, Galiba M, Galiano G (Eds.) Cover crops in West Africa: Contributing to sustainable agriculture. International Development Research Centre (IDRC)- International Institute of Tropical Agriculture (IITA)-Sasakawa Global 2000 (SG2000), Canada. 239 p.

Chapter 11

Sesbania rostrata in Rice-Based Farming Systems of Northern Thailand

P. Gypmantasiri, B. Limnirankul and C. Phothachareon

Multiple Cropping Centre, Chiang Mai University, Thailand

SUMMARY

Sesbania rostrata Bremek. & Oberm., a stem-nodulating species, has been the most successful green manure in increasing the yield of rainy-season lowland rice but has found little adoption in Asia, partly because its niches have not been sufficiently defined. In northern Thailand, rainfall is 1100-1300 mm. In the rain-fed areas of the upper north, only one crop of rice (*Oryza sativa* L.) can be planted (July to November-December), while under irrigation, rainy-season rice (from July-August to November-December) is followed by one to two dry-season crops that are harvested by May, leaving a period of 2-3 months before the next rice planting. Efforts to incorporate *Sesbania* in the rice-based systems of the upper north have focused on both systems.

On-station research efforts at the Multiple Cropping Centre of Chiang Mai University, initiated in 1993, have included an irrigated double-cropping rice-soybean (*Glycine max* [L.] Merr.) system, that is, rainy-season rice and dry-season soybean. *Sesbania,* broadcast in ploughed soil in mid-May and allowed to grow for 55 days before incorporation, performed well, with fresh weight of 10.6 to 21.9 t ha^{-1}, N accumulation of 78-123 kg ha^{-1} and, in the absence of fertilizer, a positive impact on following rice yield (4.64 vs. 3.93 t ha^{-1} for fertilized rice). On-farm efforts on a similar rice-soybean system and on a rain-fed single-cropping rice system have been ongoing since 1997 in Chiang Mai province. Efforts have included diffusion work and have had a participatory and wide focus—intensification of the rice system. On farmer fields in 1997, overall rice yields were low because of a long dry spell. Across fields, the average yield after *Sesbania* was slightly lower (3.25 t ha^{-1}) than for the fertilized rice but the effect was not consistent from one field to another. In 1998, impact of *Sesbania* varied by field. *Sesbania* biomass production

M. Eilittä et al. (eds.), Green Manure/Cover Crop Systems of Smallholder Farmers, 263-283.

varied (0.9-8.4 t DM ha^{-1}) both among villages (because of environmental factors, particularly soil type) and within villages (mainly because of different sowing dates of *Sesbania*).

Benefit-cost analysis indicating beneficial cost-benefit ratio (1.62) included only the direct economic benefits to the farmer (yield increase and savings in inorganic fertilizer) and added costs of seed purchase and preparation, land preparation prior to *Sesbania* planting and labour for broadcasting seed. The surveyed farmers affirmed these costs and benefits, verifying the importance of the need for additional land preparation before *Sesbania* planting as a main constraint for farmer adoption, which has varied by community. Of the surveyed farmers, 75% found that incorporation of *Sesbania* caused no problems or hardship, 80% experienced improved soil conditions and 68% increased rice yield. About 25% encountered problems, including that the system required more time and labour. Over 60% of the participating farmers were determined to continue using *Sesbania*. The future of rice-based *Sesbania* systems in northern Thailand is unsure and, among others, developments in the policy and economic spheres will impact future adoption. Ability to collaborate among government agencies (particularly in seed production) and to develop effective extension packages will be particularly influential. Future research should include niche identification, assessment of cost-effective ways to establish and stabilize *Sesbania* productivity and the development of integrated nutrient management practices.

INTRODUCTION

Sesbania rostrata Bremek. & Oberm., a stem-nodulating species, has been the most successful green manure in increasing the yield of rainy-season rice in the lowland ecosystem (Becker et al. 1995). Many studies have shown that *Sesbania* can accumulate high N (80-100 kg N ha^{-1}) in 45-60 days of growth and have concluded that the incorporation of *Sesbania* before rainy-season rice (*Oryza sativa* L.) can add large quantities of biologically fixed N (about 80%) to lowland rice production systems, thus improving soil fertility and increasing rice yield.

Despite good biophysical performance, *Sesbania* has not been greatly adopted in the rice-based systems of Asia, perhaps partly because its niches have not been sufficiently defined. Rice-based systems exhibit a great variability from the often relatively unintensive rain-fed systems with one crop a year to the extremely intensive irrigated systems where three rice crops a year are possible. The potential of *Sesbania* to contribute to soil fertility and yield improvement understandably differs by the intensity of cropping in these systems.

The rice-based systems in northern Thailand exhibit a similar variability in intensity. For almost a decade now, the Multiple Cropping Centre (MCC) of Chiang Mai University has been experimenting with

Sesbania in the lowland rice-based systems of the region, both in on-station and on-farm conditions. This paper describes the work conducted and discusses the experiences of the farmers and researchers in trying to incorporate *Sesbania* in the rice-based systems of the region.

RICE CULTIVATION IN NORTHERN THAILAND

The Environment

Topographically, northern Thailand is divided into upper and lower north. The upper north is a mainly mountainous region, composed of about 70% mountains, 10% lowland valleys and 20% uplands (Figure 1). The lower north is dominated by undulating uplands (30%) and lowlands (50%), which extend to the central plain, forming the important 'rice bowl' of Thailand. In the lower north, only about 20% of the land is classified as mountainous. The total annual average rainfall in the region is 1100-1300 mm, with a dry spell of 3 weeks commonly occurring between late June and mid-July. Most of the rainfall occurs between May and October, with August and September the rainiest months. The region's soils are sandy loams, sandy clay loams and clays. Typically, soil organic matter content is below 1.5%, P ranges from very low to low (10-30 ppm) and K is below 50 ppm.

Culturally, the upper north is diverse. A number of ethnic groups are settled on the highlands and ethnic Thais live in the lowlands and uplands. Only during the last three decades has agriculture in the highlands been transformed into a more commercialized enterprise specializing in high-value vegetable crops and subtemperate fruit trees, frequently an indication of successful opium crop replacement programs by various development agencies.

Ethnic Thais, who form the largest ethnic group in the region and the country, mainly inhabit the lowland valleys. In addition to ethnicity, the two regions can be distinguished by food (and associated food crop cultivation) patterns. In the upper north, glutinous rice is consumed as the staple, while in the lower north, non-glutinous rice prevails.

The 1998-99 household farm-income survey by the Office of Agricultural Economics (OAE 2000) reported that the net cash income of the average farm household in the north was Baht 68 202 (US$1568; US$1 = 43.5 Baht), of which the net farm cash income was only 35% (Baht 24 190 or US$556). Non-farm cash income, constituting 65% (Baht 44 012 or US$1012), is particularly important for farmers in the rain-fed areas where only single cropping is normally carried out per year.

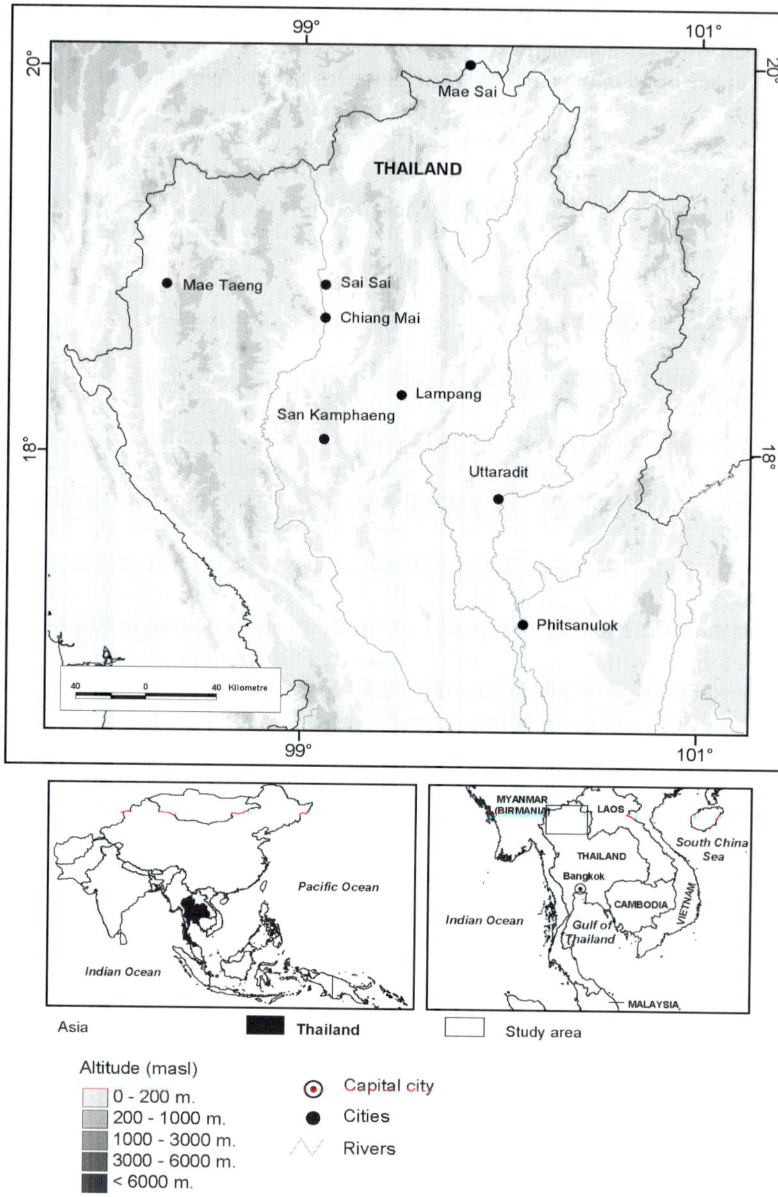

Figure 1. Location of the study area, northern Thailand.

The agricultural services and infrastructure are adequate, particularly for rice production. The Bank of Agriculture and Agricultural Cooperatives (BAAC) and Government Savings Bank (GSB) are the main credit institutions. The extension services administered by the Department of Agricultural Extension (DOAE) of the Ministry of Agriculture and Cooperatives (MOAC) have basic structure at the subdistrict (i.e. *tambon*) level since the MOAC adopted the Training and Visit system in the mid-1970s with partial financing by the World Bank. The farm inputs are available at the district and subdistrict levels, while the DOAE also sets up seed multiplication centres to provide seeds of a number of food crops, such as rice, soybean (*Glycine max* [L.] Merr.), mungbean (*Vigna radiata* [L.] R. Wilczek) and vegetables. The forage legume seeds are available from the Department of Livestock (DOL), while seed of green manure/cover crops (GMCCs) is available only from the Department of Land Development (DLD).

Family labour is important for agricultural production in the region. In the upper north, where rice is grown for subsistence, the traditional labour exchange system still exists. In contrast, in the lower north, where farm size is larger, farm mechanization and the rice broadcasting technique have greatly reduced labour constraints. The available family labourers, averaging three persons per household in the region, share almost equal responsibility in farming. Women in particular have always played an important role in farming in the region (Shinawatra 1994).

THE RICE SYSTEMS

The rice-growing areas in northern Thailand occupy approximately 2 million ha or 22% of national rice areas. The upper north, consisting of nine provinces, includes about 0.57 million ha in rice cultivation, while the lower north, covering eight provinces, has 1.43 million ha of area planted in rice. The rice-growing environments in the north, and particularly in Chiang Mai, are favourable and average rainy-season yields are 2.69 t ha^{-1} in the north and 3.21 t ha^{-1} in Chiang Mai (Table 1), substantially higher than the national average. The average dry-season rice yields are much higher, averaging 3.90 t ha^{-1} in Chiang Mai. In both regions, only about 30% of the rice area is under irrigation. The remaining 70% of the lowland rice is rain fed and allows for single cropping only. In the single-cropping systems, the land remains fallow for about 7 months after the rice harvest.

Table 1. Average rice yield (t ha^{-1}) in Thailand, 1994-97 (*Source*: OAE 2000).

Area	Average rice yield (t ha^{-1})	
	Rainy season	Dry season
Thailand	2.14 ± 0.06	4.36 ± 0.21
North	2.69 ± 0.18	4.43 ± 0.18
Chiang Mai	3.21 ± 0.27	3.89 ± 0.79

Most farmers in the lower north, particularly those located in the irrigated areas, have large farm holdings, ranging from 4 to 16 ha and grow rice commercially, practicing triple rice cropping throughout the year. The rice-growing season is not limited by cool temperature as in the upper north, so the modern high-yielding varieties (HYVs), developed and released by the Department of Agriculture (DOA), are well adapted to the region and cultivated under the broadcasting system. Rice cropping throughout the whole year is possible through the use of medium-maturing varieties that are photoperiod insensitive and high yielding, through the heavy application of mineral fertilizers and through the utilization of combine harvesters. The dominant rice variety is a non-glutinous Chainat 1, which is grown for domestic and export markets.

There are only few opportunities for *Sesbania* to be incorporated in the continuous rice monoculture system of the lower north, unless farmers are concerned about regenerating soil fertility and reducing dependency on mineral fertilizers.

In contrast, the rice systems in the upper north are mainly subsistence or semi-commercial. In this region, 80% of rice is planted to glutinous rice, particularly the RD 6 variety, for household consumption. The small-scale rice farmers of this region still prefer the tall plant types of local varieties because of the multiple uses that rice straw has in mushroom culture, as mulch in cash crop production and as livestock feed. Those farmers that are located in the rain-fed lowlands own larger holdings and are able to allocate a portion of their land to growing non-glutinous rice as a cash crop. Under these circumstances, farmers select high-quality rice varieties (i.e. fragrant jasmine rice) such as KDML 105 and RD 15 for planting as a cash crop. The rain-fed lowlands enable the planting of only one crop, rice, which in this system is typically planted in June and harvested in October-November, followed by a 7-month fallow (Figure 2).

The irrigated lowland rice system of the upper north is more diversified than the rain-fed system. In the irrigated system, double or triple cropping is practiced with short-maturing cash crops, such as soybean, potato (*Solanum tuberosum* L.), onion (*Allium cepa* L.), garlic (*Allium sativum* L.), tomato (*Lycopersicon esculentum* Mill.) and tobacco (*Nicotiana tabacum* L.), all planted after the rainy-season rice. At times, a dry-season rice is also planted. In this system, rainy-season rice is normally planted in late July-early August and harvested in November-December and the dry-season crops are planted immediately thereafter (Figure 2). Typically, dry-season crops are harvested by May, leaving a period of 2-3 months before the next rice planting. The Chiang Mai Valley, for instance, is considered to have the most diversified and intensified land-use system in the north because of its favourable environmental conditions, good irrigation facilities and infrastructure and market opportunities. These additional crops, while providing cash to farmers, also supply large amounts of residual fertilizer to the succeeding dry-season rice, which enables high yields of up to 6 t ha^{-1} without additional fertilizer applications.

These irrigated lowland rice cropping systems of the upper north are resource and labour intensive. Because the farm holdings are fragmented with an average farm size of 0.8 ha, farm machinery is of small type, such as the two-wheeled tractor, an important asset to farmers. If needed, heavy farm machinery can be rented. Farm mechanization has speeded up fieldwork and most importantly, has enabled timely planting. In land preparation, the two-wheeled tractor has replaced animal traction almost completely.

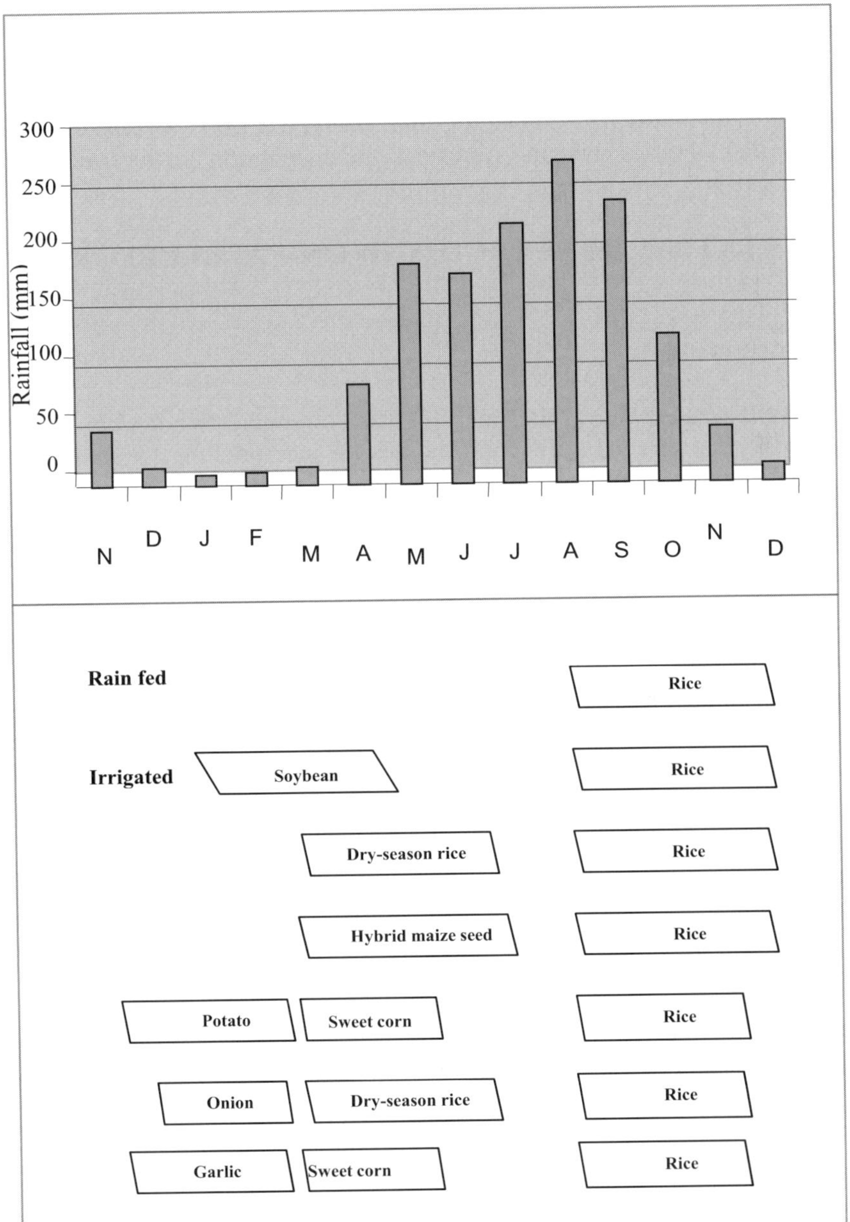

Figure 2. Monthly rainfall (mm) and cropping systems in the rain-fed and irrigated lowlands of Chiang Mai Valley.

The degree of commercialization mainly depends on farm size and water availability. Rain-fed lowland farmers with larger holdings are able to allocate land for planting quality rice to earn cash. Irrigated lowland farmers with smallholdings only fulfil their household rice requirement from the rainy-season rice crop and invest in cash cropping during the dry season. In both systems, any rice productivity improvement with available farm resources or the reduction of dependence on external inputs will contribute significantly to household food security and income stability.

In the following, on-station and on-farm efforts to incorporate *Sesbania* in the rice cropping systems of the upper north are described.

ON-STATION RESEARCH AND ON-FARM EFFORTS

Introduction

As discussed above, the intensive triple-cropping rice system of the lower north offers very few opportunities for incorporating *Sesbania* as a green manure. Cropping in this system is continuous and strongly dependent on applications of inorganic fertilizers. *Sesbania* requires a minimum of 30 days to grow sufficiently to make any kind of contribution to the system; preferably, it should grow at least 45 days. Additional time is required for land preparation before *Sesbania* planting and the incorporation of *Sesbania* before rice planting but in these systems, even a free 30-day period is difficult to find, let alone 45 days. In contrast, in the systems in the upper north, the fields are typically left fallow either during December-June (in the rain-fed single-cropping system) or April-August (in the irrigated double-cropping system). Rainfall typically starts in April and is normally sufficient between May and July to support the growth of *Sesbania*. Even in the irrigated triple-cropping rice-cash crop systems some room is available prior to rainy-season rice crop planting, particularly if photoperiod-sensitive rice varieties are utilized that can be planted as late as in late July-early August. These rice systems of the upper north therefore offer an opportunity to incorporate *Sesbania* as an early rainy-season crop, planted in May and incorporated in July, prior to the planting of the rainy-season rice crop.

Consequently, most of the efforts to incorporate *Sesbania* in the rice systems of northern Thailand have focused on the single-cropping rain-fed and double-cropping irrigated systems. The efforts have taken place both on station and, more recently, on farm. On-station research efforts at the MCC of Chiang Mai University started in 1993 and continue to date. The MCC on-farm work in the Chiang Mai Valley has been ongoing since 1997 when experimentation was initiated on five farms in the San Sai district. The Chiang Mai Valley is representative of the lowland agricultural systems of the upper north, where rice is the main crop in the rainy season and typically diverse cash crops are cultivated during the dry season. This on-farm work has grown greatly in importance during the last few years. An additional effort by the MCC includes collaboration with the DOA on the extension of organic rice systems to farmers in Phayao province. As a result, *Sesbania*, together with compost and crop residues, has been promoted to farmers of the region for soil fertility maintenance. That work was initiated only in 2001 and is not reported here.

In addition to MCC, Lamphun's DLD, located near Chiang Mai, has initiated the promotion of *Sesbania* together with other GMCCs, such as sunn hemp (*Crotalaria juncea* L.), mungbean and cowpea (*Vigna unguiculata* [L.] Walp.), as pre-rice GMCCs in Lamphun province. A few successful cases have been reported, particularly with the use of sunn hemp. However, farmers still depend on the DLD Lamphun office for seed supply.

On-Station Research

On-station research has focused on the irrigated double-cropping rice-soybean system, that is, rice in the rainy season (July/August-December) and soybean in the dry season (December/January-April). Both *Sesbania* performance and its impact on rice and soybean yield were evaluated.

Sesbania was broadcast with a seeding rate of 20 kg ha^{-1} in ploughed soil in mid-May when early rains provided enough moisture for seed germination. The crop was allowed to grow for 55 days, after which it was incorporated into the soil. Rice seedlings of 25 days were transplanted at 25 × 25 cm spacing with a single plant per hill. No additional mineral fertilizer was applied in the plots with previous *Sesbania*. The conventional plots without *Sesbania* were broadcast with 16-20-10 at 156 kg ha^{-1} 20 days after transplanting and top-dressed with urea at booting stage at 63 kg ha^{-1}. These fertilizer applications concur with the recommendations of DOA's Good Agricultural Practice (GAP) Program. The field was flood irrigated with about 10 cm of water above the ground. Incidence of pests and diseases was insignificant, so no control was carried out. Herbicide was used to control weed infestation. Water was drained about 20 days before rice harvest. Crop sampling was taken from a 2 m^2 plot for yield determination. Each year, two to three high-quality improved rice varieties were used to compare with the standard high-quality variety, KDML105 or RD6. After harvest, the rice plants were mowed to ground level so that the stubble would form a light mulch. Soybean was hill-planted at 40 × 25 cm with 4-5 seeds per hill. The stands were thinned to three plants per hill 7 days after emergence. The other management practices for soybean also reflected those of farmers, such as the use of pre-and post-emergence herbicides, three insecticide sprayings and four irrigations. Soybean residues were removed from the field, a common farmer practice, either for use in mushroom culture or because soybeans are taken to the homestead for threshing.

Sesbania performed well in on-station conditions. The plant height ranged from 1.1 to 2.1 m and fresh weight ranged from 10.6 to 21.9 t ha^{-1} on sandy clay loam soil. The N content ranged from 2.6 to 3.5%, giving N accumulation from 78 to 123 kg ha^{-1} (Table 2).

Table 2. Fresh and dry weight, N content and N accumulation of *Sesbania rostrata* in sandy clay loam soil, Chiang Mai University, 1993-98 (*Source*: Gypmantasiri et al. 2000).

Year	Fresh weight (t ha^{-1})	Dry weight (t ha^{-1})	N (%)	N accumulation (kg ha^{-1})
1993	10.64	2.54	3.06	77.7
1994	16.01	3.41	2.59	88.3
1995	17.33	2.80	3.54	99.1
1996	16.48	2.91	3.06	89.0
1997	18.78	3.50	2.80	98.0
1998	21.88	3.68	3.34	122.9

The effect of *Sesbania* on subsequent rice yield averaged across varieties (Table 3) was positive, increasing it on average by about 18%. With *Sesbania*-incorporated plots, the average yield across varieties over 6 years was 4.64 t ha^{-1}, while the non-*Sesbania* plots provided average yield of 3.93 t ha^{-1}. The yield increase was between 13 and 19% in all years except one, when it was 26%. *Sesbania*'s effect was quite consistent for all varieties evaluated in the study.

Table 3. Dry biomass of *Sesbania rostrata*, incorporated before rice transplanting, and grain yield of rice and soybean in irrigated lowland system at Chiang Mai University, 1993-98 (*Source*: Gypmantasiri et al. 2000).

Year	*Sesbania* biomass (t ha^{-1})	Grain yield (t ha^{-1})			
		Rice		Soybean	
		Without *Sesbania*	With *Sesbania*	Without *Sesbania*	With *Sesbania*
1993	2.54	3.94	4.44	n.d.	n.d.
1994	3.41	3.89	4.89	2.60	2.56
1995	2.80	3.95	4.69	2.40	2.44
1996	2.91	3.59	4.23	2.55	2.44
1997	3.50	3.85	4.57	1.94	2.00
1998	3.68	4.36	5.00	2.56	2.45
Mean	3.14	3.93	4.64	2.41	2.38
s	0.45	0.25	0.29	0.27	0.22
CV	14.30	6.40	6.30	11.20	9.20

This on-station study did not show any clear indication of a carry-over effect of *Sesbania* on soybean yield (Table 3). The mean soybean yield in the *Sesbania*-incorporated treatment was 2.38 t ha^{-1} over the 5-year period, while plots with no prior *Sesbania* yielded 2.41 t ha^{-1}.

On-farm Efforts

The on-farm efforts of the MCC of Chiang Mai University initiated in 1997 and are ongoing. As mentioned earlier, on-farm research mainly focused on the double-cropping irrigated systems (i.e. rainy-season rice followed by soybean); in addition, some work was conducted on the single-cropping rice system of the rain-fed lowlands. All on-farm work took place in Chiang Mai Valley, located in Chiang Mai province.

Researchers worked with either single farmers or with farmer groups, depending on the community, beginning in 1997 with six farmers in San Sai district. In 1998, a rain-fed lowland site in San Kamphaeng district was added and 11 farmers from the two districts joined the testing program. In 1999, one irrigated lowland site in Mae Taeng district was added. The number of participating farmers in the on-farm trials was increasing.

Several methods were used to work with farmers. The general approach was participatory, with researchers acting as facilitators in the process. Farmer meetings were typically conducted in the evenings to encourage farmer participation. Farmer facilitators were identified. In the meetings, work plans for the upcoming season were developed and the results of the past season assessed. Fields were monitored at least once in 2 weeks, typically with groups of farmers. At the end of the season, crop cutting was done with farmers to estimate the yield. Chiang Mai University students assisted in the on-farm research by taking notes, arranging meetings and by helping with the fieldwork.

The efforts have gone beyond *Sesbania*-rice systems and beyond research. In fact, the focus of the efforts has been the intensification of the rice-soybean system and the work has included the introduction of new rice varieties. Diffusion efforts have also been important. Community-level seed production of *Sesbania* and rice seed was also initiated; this production needs to rely on individual farmers rather than farmer groups because of lack of communal land.[1] In addition, farmer field days were arranged both in the villages and on station, typically at a time when *Sesbania* was ready to be incorporated into the soil.

In 1997, overall rice yields were low due to a long dry spell. The average rice yield following *Sesbania* was slightly lower (3.25 t ha^{-1}) than in plots without *Sesbania* (3.40 t ha^{-1}). In two fields, rice yield following *Sesbania* was higher (by 28 and 12%) than without it. In three fields, plots without *Sesbania* yielded more (by 18, 29 and 30%), while in one field, the yields were about the same. In the rice plots with no *Sesbania*, farmers applied inorganic fertilizer (16-20-0) at a rate ranging from 63 to 156 kg ha^{-1}; no fertilizer was applied in the *Sesbania* plots (Table 4).

Table 4. Average rice yield (t ha^{-1}) with and without *Sesbania rostrata* in farmer fields in three districts of Chiang Mai Province, 1997-99 (*Source*: Gypmantasiri et al. 2000).

System	Rice yield[a]		
	San Sai	San Kamphaeng	Mae Taeng
With *Sesbania*			
1997	3.25 ± 0.70 (6)	-	-
1998	4.23 ± 1.09 (2)	5.29 ± 0.10 (3)	-
1999	4.11 ± 0.47 (16)	4.84 ± 0.99 (8)	4.39 ± 0.67 (16)
Without *Sesbania*			
1997	3.40 ± 0.43 (6)	-	-
1998	3.71 ± 0.76 (2)	5.27 ± 0.99 (3)	-
1999	4.89 ± 0.26 (2)	4.46 ± 1.22 (3)	3.27 ± 0.36 (5)

[a] The numbers in parenthesis refer to number of farmers participating in the study.

In 1998, a rain-fed lowland site at San Kamphaeng was added in the on-farm trial. Eleven farmers from two districts joined the testing program. Yield samples were obtained from five farmers because the remaining fields were either damaged by rodents or harvested by farmers before sampling could be carried out. In the non-*Sesbania* plots, farmers continued to apply inorganic fertilizer (16-20-0) at a rate ranging from 63 to 156 kg ha^{-1}. In two fields in the San Sai district and in one field in San Kamphaeng district, yields following *Sesbania* were clearly higher than without *Sesbania*. In one field of San Kamphaeng district, yields in the two plots were about the same and in another field, yield in the non-*Sesbania* plot was higher.

Sesbania biomass production varied considerably among and within villages (Table 5). In the year 2000, average biomass yield (4.7 t ha^{-1}) was highest in San Sai, with considerable variability (from 1.4 to 8.4 t ha^{-1}). Lowest average biomass was measured in San Kamphaeng with average biomass below 2 t ha^{-1}. Variability within communities was mainly due to different sowing dates of *Sesbania* (ranging from 15 May to 9 June in 1999), giving the dry biomass yield ranging from 150 to 4000 kg ha^{-1}. Environmental factors, particularly soil type, explained much of the variability in biomass production among the different communities. During the on-farm research process, farmers have learnt the importance of optimal sowing dates for *Sesbania*. At present, farmers with double cropping systems, such as rice-soybean, have managed to broadcast *Sesbania* seed in early to mid-May when sufficient water is available.

Table 5. Biomass production of *Sesbania* (t DM ha^{-1}) in farmers' fields in three districts in the Chiang Mai Province, 2000 (*Source*: Gypmantasiri et al. 2000).

District	Average (range)	No. of farmers
San Sai	4.70 +/- 0.23 (1.4-8.4)	15
San Kamphaeng	1.79 +/- 0.10 (0.9-4.0)	7
Mae Taeng	3.04 +/- 1.86 (0.9-7.0)	15

In conclusion, on-station trials indicate a good possibility of obtaining increased rice yields after *Sesbania* incorporation. Increased yield is not consistent in on-farm conditions, which is hardly surprising.

EVALUATING SESBANIA IN RICE SYSTEMS OF NORTHERN THAILAND

Benefits, Costs and Limitations

While the *Sesbania*-rice and *Sesbania*-rice-soybean systems typically give a clear but modest rice yield increase, they also have a number of costs and limitations. These factors are discussed below, both from farmer and researcher perspectives.

Benefits and costs of the *Sesbania* system were assessed by surveying 53 farmers who had cultivated *Sesbania* with rice in the year 2000 (Gypmantasiri et al. 2000). The monetary benefits and costs are discussed below, along with farmer perceptions.

Based on the survey, the yield increment from the *Sesbania*-incorporated unfertilized rice field was estimated at 20% or 800 kg ha^{-1}, giving economic benefits of US$132 ha^{-1} (year 2000 farm-gate prices of Baht 7 kg^{-1} or US$0.16) (Table 6). In addition, based on farmer surveys, the inorganic N fertilizer use was estimated to decrease by US$2 ha^{-1}.

Table 6. Partial budget based on a survey of 53 farmers for the *Sesbania rostrata*-rice system in the irrigated rice-based farming systems in Chiang Mai Province[a] (*Source*: Gypmantasiri et al. 2000).

Item		Baht rai^{-1}	US$ ha^{-1}
Costs			
Ploughing		420	60
Sesbania seed		90	13
Additional labour for *Sesbania*		62	9
	Total costs	572	82
Benefits			
Reduction in inorganic N		14	2
Increased rice yield		917	131
	Total benefits	931	133
	Benefits-costs	359	51

[a] 1 rai = 1600 m^2; US$1 = 43.5 Baht.

A number of other benefits result from the cultivation of *Sesbania*. Field observations confirm that the *Sesbania* system also results in less weed pressure in the subsequent rice crop. The 55- to 60-day *Sesbania* crop provides complete ground cover and suppresses all weed growth. *Sesbania* incorporated in the field also provides better soil physical conditions for rice transplanting and improves seedling stands. After puddling, the sandy loam soil can form a moderately hard layer if the soil particles are allowed to settle for a few days before transplanting, which can make transplanting difficult and seedling stands less stable. However, only yield increase and savings in the inorganic fertilizer, that is, direct economic benefits to the farmer, have been included in the calculations of economic benefits, estimated at US\$133 ha^{-1}.

Farmers confirm these benefits. The information on farmer perceptions was gathered through group discussions and formal surveys. Farmers emphasized the following benefits:

- Soil fertility improvement: Some farmers had reduced the quantities of inorganic fertilizer to half.
- Ease of land preparation: *Sesbania*'s stems are brittle and the land preparation is typically no more difficult than without *Sesbania*.[2]
- No delay in rice transplanting: Because *Sesbania* quickly accumulates sufficient biomass, rice transplanting typically is not delayed to a harmful degree, especially if farmers use photosensitive varieties with optimal planting dates ranging from late July to early August.
- Easier establishment of rice seedlings: Farmers noted that *Sesbania* residues improved soil conditions for transplanting.
- Fewer weeds in the subsequent rice crop: In the intensively cropped areas (i.e. double- and triple-cropping) weed incidence is typically severe and herbicide use common. Farmers have observed less severe weed incidence in the rice field following *Sesbania* cropping, which has allowed them to decrease the use of herbicides.

The added costs of incorporating *Sesbania* as a pre-rice green manure crop include: seed purchase and preparation, land preparation prior to *Sesbania* planting and labour use for broadcasting seed. Since land needs to be prepared anyway prior to rice planting, ploughing *Sesbania* into the soil does not represent an additional cost. The seed is available from the DLD at Baht 30 kg^{-1} (US\$0.70). With the recommended seeding rate of 20 kg ha^{-1}, seeding cost is US\$13 ha^{-1}. Seed scarification either by hot water treatment or by sulphuric acid is needed to improve germination. To achieve better seed-soil contact, ploughing is essential. Broadcasting seeds in unploughed, weed-infested soil usually results in poor germination. Where weed infestation is heavy, a big tractor is preferred to a small two-wheeled tractor for land preparation. Cost of ploughing is estimated to be Baht 420 rai^{-1} (i.e. US\$60 ha^{-1}). The labour use for broadcasting seed is not a constraint because one family member can complete it within a few hours in a typical rice field. In all, additional labour for management is estimated at Baht 62 rai^{-1} (US\$9 ha^{-1}). A two-wheeled tractor can incorporate *Sesbania* plants into the rice field and its cost does not differ significantly from that of the normal land preparation prior to rice planting. Based on the farmer surveys, the total added costs incurred by adopting the *Sesbania* system are estimated at Baht 572 rai^{-1} (US\$82 ha^{-1}).

Farmers affirm these costs. In the surveys and group discussions, the farmers mentioned the following costs and limitations:

- Higher cost of land preparation: Before *Sesbania* planting, ploughing is necessary for uniform and vigorous seedling establishment. This seems to be the main constraint for farmer adoption.
- Difficulty of obtaining seed: The availability of seed is an important constraint to the promotion of green manures in the lowland rice-based farming systems. At present, *Sesbania* seed is only produced by the DLD regional office at Khon Kaen, which is situated relatively far from Chiang Mai province in north-east Thailand.[3] The farmers considered it high priced. So far, government research organizations in the country have not reported *Sesbania* germplasm collection and evaluation. There is potential to engage farmers in the production of *Sesbania* and other GMCC seed as contracted seed growers in the lowland and upland areas during the rainy season. The crop, harvested in November, will not compete with rice harvest for labour.
- Early rainy-season dry spells: The prolonged dry spell in early June could damage or retard seedling growth during *Sesbania* establishment even though it is known to be drought tolerant.
- Unfamiliarity with *Sesbania*: For example, information on the need for, and methods of, seed scarification should be provided to farmers.
- Insect pressure: Blister beetles that feed on flowers could damage seed production. A spraying program is necessary.
- Poor adaptability to highland environments with elevations above 1000 masl: In these environments, the 55- to 60-day crop generated much less biomass than in the lowland environment.

With additional benefits of US\$133ha^{-1} and costs of US\$82 ha^{-1}, the cost-benefit ratio is beneficial (1.62), indicating the economic potential of the rice-*Sesbania* system. Similarly, research in the north-east region of Thailand by DLD and the International Network on Soil Fertility and Sustainable Rice Farming (INSURF) during 1988-91 demonstrated highest income in a rice system where both *Sesbania* and inorganic NPK fertilizer were used (Arunin et al. 1994).

Farmer Interest in Testing *Sesbania*

Another measuring stick for evaluating the rice-based *Sesbania* systems is the farmers' interest in continuing to utilize them. Varying degrees of farmer adoption have been observed. Farmers in San Sai district, with good irrigation facilities, have managed both *Sesbania* and rice crops better and the initial group of farmers almost entirely has continued cultivating *Sesbania* with rice. In San Sai, three farmers who worked closely together have started *Sesbania* seed production, which has enabled them to use *Sesbania* on all rice fields since the 2001 planting season without requesting seed from the MCC or the DLD. In contrast, the rain-fed lowland site in San Kamphaeng has not performed well because of poor *Sesbania* establishment caused by water shortages. Farmers who initially participated are less committed and new individuals have joined the testing.

Farmers of the new site at Mae Taeng, with moderately good irrigation facilities, show the most enthusiasm over *Sesbania*. The growth of *Sesbania* in 2000 was less vigorous than at the San Sai site because of the clayey soils, yet farmers are seemingly satisfied with the visual effects of *Sesbania* on rice growth and development. Farmers immediately started producing their own seed by planting *Sesbania* along the banks of irrigation canals or on the upper slope near the homestead. Since *Sesbania* significantly increased the rice yield, farmers have initiated community rice seed production to add more economic benefit from the yield increment. But, most importantly, the group sees the provision of quality rice seed to the neighbouring farmers as a community service. In the near future, the community rice seed production will be accredited by the subdistrict organization (*tambon*), which means that it will be made known to all *tambon* members of the Mae Taeng district. This can offer good marketing prospects for the community.

Farmers participating in the on-farm experimentation were surveyed after the 1999-2000 growing season about their experiences with the rice-*Sesbania* system (Table 7); about 75% of these farmers commented that incorporating *Sesbania* in rice farming did not cause any problems or hardship. Almost 80% observed that soil conditions had been improved and 68% experienced an increment in rice yield. However, about 25% complained that the process required more time and labour and added cost to production, and that incorporating *Sesbania* into soil made land preparation for rice transplanting less convenient.

Table 7. Advantages and disadvantages of *Sesbania* cultivation as perceived by farmers in three districts of Chiang Mai province, 1999-2000 (n = 53; *Source*: Farmer interviews, 1999-2000; Gypmantasiri et al. 2000).

Farmer perception	% farmers
Advantages:	
Soil conditions improve	79.2
Rice yield increases	67.9
Fertilizer use decreases	5.7
Disadvantages:	
More time, more labour and more cost	24.5
Land preparation for rice transplanting more difficult	18.9
No problems associated with the use of *Sesbania*	75.5

Farmers who did not obtain any yield increase in rice realized that it was due to poor establishment of *Sesbania*, which resulted in low biomass. Other yield-reducing factors identified by farmers were damage from rodents and diseases.

Over 60% of the participating farmers were determined to continue using *Sesbania* as a green manure on their own initiative. The main reasons cited were that they observed better soil conditions in the plots with prior *Sesbania*, that they wanted to reduce the use of mineral fertilizers and that they had obtained higher rice yield. Those who wanted to stop testing *Sesbania* pointed out that they had limited land and would rather invest in economic cash crops. Another reason for abandoning *Sesbania* was the added cash input for land preparation.

It seems that farmers adopt *Sesbania* if they can observe an increase in rice yield in their own plots and if their farmland has no opportunity cost. Other important factors inducing adoption include the receipt of accurate information on the system and a close relationship between farmers and researchers, which helps provide confidence in technology development and diffusion.

THE WAY FORWARD

Factors Currently Fostering Adoption

While the future of rice-based *Sesbania* systems in northern Thailand is unsure, several developments in the policy and economic spheres could increase its adoption. Some of these are outlined below.

Increased Interest in Sustainable Agriculture

Especially since the economic meltdown of 2nd July 1997, both government and non-governmental organizations have promoted sustainable agriculture as an alternative approach to economic recovery. The integrated farming system is seen as a way to develop a self-reliant household economy. Recycling of on-farm resources and the utilization of land-use technologies are essential to achieve a land-use system that is self-reliant and food security-oriented. The GMCC systems, if compatible with the farmers' circumstances, are one of the 'best bets' to achieve this. In addition, the DLD has also launched a program of compost technology, GMCC systems and soil conditioners to improve soil productivity.

Increased Fertilizer Price

The prices of commonly used inorganic fertilizers such as 16-20-0, 15-15-15, 16-16-16 and urea have increased over 40% since July 1997, causing farmers to look for solutions in integrated nutrient management.

More Stable Future Prices for High-quality Rice

The country is now promoting the production of high-quality non-glutinous rice, KDML 105. Despite price fluctuations over seasons, since 1997 the farm-gate price of KDML 105 has always remained above Baht 6 kg^{-1} (US$ 0.16 kg^{-1}). The price advantage of high-quality rice, together with a 20% yield increment by *Sesbania*, are expected to provide incentives for farmers to adopt. Moreover, with improved land availability resulting from increased yields of the subsistence glutinous rice in *Sesbania* systems, farmers could set aside another area for non-glutinous KDML 105 as a cash crop.

Land Tenure

The rice farmers in Chiang Mai Province have an average farm size of 0.8 ha and about 25% of farm households rent land for cultivation. Such conditions would seemingly be limitations to the incorporation of GMCCs in the systems. However, in at least two of the districts, the costs for rent are fixed and renting periods are relatively long (3 years) and renewable, which encourages farmers to invest in the GMCC systems.

Key Factors Impacting Future Adoption

Developments in arenas given below are particularly influential for the future adoption of *Sesbania*-rice systems.

Collaboration of Government Agencies

Government agency collaboration in seed production is particularly important. In Thailand, the improvement of food crops and non-food crops is under the DOA. Seed multiplication and extension programs are under the mandate of the DOAE, except for some specified crops, including forage crops (under the responsibility of the DOL) and soil-improving legumes (responsibility of the DLD). The DLD's *Sesbania* seed multiplication program takes place mainly on station, so the quantity available is limited. This type of departmental structure and sectorial responsibility for agricultural development is not appropriate for the promotion of GMCC systems. Close collaboration between the two departments is needed to ensure that soil productivity is considered along with food crop production and that seed production needs are met in the future. On-farm production of seed offers an effective avenue to improve the availability of *Sesbania* seed.[4] If access to seed is improved, it is expected that the system will be better adopted.

Development of Effective Extension Packages

In 1999, the DOA launched the GAP Program, whose aim is to design and promote the use of good agricultural practice for improving farmer yields. The recommended cultural practices of each crop are reviewed and booklets are published and made available to extension agents and farmers. In the past, the DOAE successfully promoted the use of modern high-yielding rice and maize varieties in the lowlands through a seed exchange program, in which farmers wanting HYV seed just exchanged the seed of a local variety for a modern HYV in a 1:1 ratio. It has also promoted the production of soybean by providing, as one package at subsidized cost, both soybean seed and inoculum. In the current program, however, the topic of GMCCs is not emphasized. It should be possible that in the GAP Program, rice seed and *Sesbania* could be promoted as one resource-conserving rice production technology.

On-farm Testing of 'Best Bet' GMCC Systems

So far, the *Sesbania*-rice system is regarded as the most promising rice-based GMCC system. However, the system requires tillage and sufficient soil moisture for good seedling establishment and the minimum time needed for substantial biomass accumulation is 55-60 days. Clear niches for *Sesbania* therefore need to be defined through on-farm testing in diverse environments.

Reaching Out to the Whole Farming Community

Farmers utilizing the *Sesbania* system were prepared to take risks after the MCC experimental results were described to them. Other farmers may have fewer resources and are less likely to participate in new ventures. To reach out to more farmers, recommendation domains within the target area should be defined after a preliminary survey and prior to the on-farm testing.

Research Issues

Positive future developments with the rice-*Sesbania* systems also depend on effective research on the following issues:

- Using the farming systems approach to identify environmental niches where *Sesbania* offers the greatest potential: rain-fed lowland, partially irrigated, double-cropping and triple-cropping areas.
- Identifying cost-effective methods of establishing and stabilizing *Sesbania* productivity in the pre-rice system.

- Developing integrated and balanced nutrient management practices to increase and stabilize rice yield.

CONCLUSIONS

The on-station studies and participatory on-farm experimentation with *Sesbania* in rice farming systems of the upper north have indicated its potential to enhance soil fertility and increase rice yields. The lowland niches where *Sesbania* is expected to be effective are the irrigated rice-based double-cropping systems, particularly the rice-soybean system, and the rain-fed systems where single cropping of rice is practiced. The intensive triple-cropping areas have less flexibility for inclusion of *Sesbania* before rice transplanting.

Large variability in *Sesbania* growth and biomass production within the community is caused mainly by differences in sowing dates. Variability among communities is likely caused by variations in soil types. The species shows better growth and higher biomass yield in the sandy clay loam than in the heavy clay soils.

Despite the relatively low benefits obtained from the on-farm analysis, farmers who observed the positive increase in rice yield from *Sesbania* have adopted the system and continue using *Sesbania* as a green manure crop in rice farming, thereby either reducing or discontinuing their use of inorganic fertilizers.

The participatory approach has proved to be an effective way to diffuse the *Sesbania* technology, not only because it has fostered the development of a technology that works in farmers' fields, but also because it has enabled relationships based on mutual trust between farmers and researchers.

NOTES

1. Interestingly, while researchers were advocating seed production in a portion of the field, farmers found it unacceptable because of the high land pressure. Instead, farmers produced seed on the banks of the irrigation canals.
2. Note that some farmers consider *Sesbania* incorporation difficult (Table 7); this discrepancy may be because of differences in soil types.
3. DLD also has had a farmer seed production scheme since 1998.
4. Seed yield of a currently available photosensitive variety was 1.38 t ha^{-1}. The crop was planted in June/July and harvested in November. The only production constraint was blister beetle damage on flowers. Application of insecticide was required. With the present *Sesbania* seed price set at Baht 30 kg^{-1} (US$0.81) by DLD, contracting farmers to produce seed is a promising option.

REFERENCES

Arunin S, Pongwichina P, Aragon E L (1994) Integrated nutrient management strategies: the INSURF experience in northeast Thailand. *In:* Ladha J K, Garrity D P (Eds.) Green manure production systems for Asian ricelands. International Rice Research Institute (IRRI), Philippines.

Becker M, Lasha J K, Ali M (1995) Green manure technology: Potential, usage, and limitations. A case study for lowland rice. Plant Soil 174: 181-194.

Gypmantasiri P, Thong-Ngam K, Limnirankul B, Phothachareon C, Phuangmanee J, Lodkeow S, Phuchareon N (2000) Farmer participation in sustainable agriculture development: Improving rice production efficiency through the use of green manure. A report to the Thailand Research Fund. Thailand Research Fund, Chiang Mai, Thailand (In Thai)

OAE – Office of Agricultural Economics (2000) Agricultural statistics of Thailand, crop year 1998-99. Agricultural Statistics No. 10/2000. Office of Agricultural Statistics, Ministry of Agriculture and Cooperatives, Bangkok, Thailand.

Shinawatra B (1994) Research on roles of women in agriculture in Thailand. Paper presented at the conference on Roles of Women in Farming Systems, Had Yai, Thailand, 8th March 1994. (In Thai)

Chapter 12

Survey of Green Manure/Cover Crop Systems of Smallholder Farmers in the Tropics

M. Flores Barahona[1] and M. Janssen[2]

[1]Centro Internacional de Información sobre Cultivos de Cobertura (CIDICCO), Honduras
[2]Consultant

SUMMARY

Experiences with green manure/cover crops (GMCCs) are mounting throughout the tropical regions. However, information is inadequate on the GMCC systems practiced by farmers. With the financial support of The Rockefeller Foundation and as a component of the GMCC Exploration Project, in 1998 the International Centre for Information on Cover Crops (CIDICCO, the Spanish acronym) initiated a process of documenting GMCC systems utilized by smallholder tropical farmers. This chapter describes that effort, the Catalogue of GMCC Systems Used by Smallholders of the Tropics, and characterizes the information on the 72 GMCC systems collected by the year 2000. The largest share of the systems is from Africa (51% of all systems), followed by Latin America (42%). The systems include a total of 27 main crops, with maize (*Zea mays* L.) predominant (over 66% of the systems). Thirty-six GMCCs are utilized in the surveyed systems, with *Mucuna* sp. Adans as the most frequent (25% of the systems), followed by cowpea (*Vigna unguiculata* [L.] Walp) and *Canavalia ensiformis* (L.) DC. In these systems, simultaneous intercropping arrangements were the most common (36% of the systems), followed by rotational systems (13%). Although many of the surveyed systems are traditional (44%), recently adopted systems form a slight majority (56%) and about one-third of all systems (38%) had been developed through research and followed up with research support. There was little exact information on adoption but about half of the systems were described as well adopted, a further 35% as moderately so and adoption in most was described as increasing (42%) or stable (32%). About one-half of the systems surveyed include both subsistence and cash

M. Eilittä et al. (eds.), Green Manure/Cover Crop Systems of Smallholder Farmers, 285-301.

crops and one-third only subsistence crops. Of the GMCCs, in one half of the systems either nothing is removed or only the seeds are removed. In the large majority of the systems, women play an active role. In addition to the survey, an interactive Web site has been developed to harbour the catalogue.

INTRODUCTION

Experiences with green manure/cover crops (GMCCs) are mounting throughout the tropical regions. This is the result of the interest and efforts of numerous research institutions, extension agencies and most importantly, the work of thousands of innovative farmers. In addition to many other advantages, GMCCs are viewed as a low-cost method to restore soil organic matter and prevent rapid decline in soil fertility, a key cause of food insecurity of tropical smallholders (Chapter 6 this volume; Becker and Johnson 1998).

However, information is lacking on the GMCC systems practiced by farmers in tropical regions. Although there are a number of databases on GMCC species[1], the diversity of systems integrating GMCCs has not been mapped nor has the information been systematized. In 1998, the International Centre for Information on Cover Crops (CIDICCO, the Spanish acronym)[2] initiated a process of documenting the GMCC systems that smallholder farmers utilize in the tropics. This was done with the financial support of The Rockefeller Foundation and as a component of the GMCC Exploration Project. By the year 2000, this project, the Catalogue of GMCC Systems Used by Smallholders of the Tropics, had collected detailed information on 72 GMCC systems.

This chapter will first describe the project and its two components, the survey and the development of a Web site. Then the 72 surveyed systems are characterized. Finally, some lessons emerging from the preliminary analysis of the surveys as well as an ongoing effort to continue the surveying process will be outlined.

THE PROCESS

Background and Development of the Survey

The process initiated when a group of individuals with experience in research and extension of GMCCs formed a working group to lead the GMCC Exploration Project (see Introduction, this volume). One of the first actions of the working group was to begin to gather information on GMCC systems utilized by smallholder farmers in the tropical regions, with the purpose of sharing these findings with a wide audience of agricultural development workers, extension agents and researchers. It was decided that a minimum requirement for a system to be surveyed was that farmers at least had tested it. In other words, a system could be surveyed regardless of its degree of adoption, that is, whether it was in its initial stage of adoption or a well-established practice.

For the purpose of this project, the working group adopted the following definition for GMCCs:

> 'The use of leguminous or non-leguminous plants (e.g. natural or improved fallow with several species, such as *Tithonia diversifolia* (Hemsl.) A. Gray, as ground cover and canopy in various temporal and spatial configurations used in crop or animal production systems. The purpose of using these species can be to improve one or more of the following: human or animal diet, erosion control, availability and recycling of N, P and other nutrients, soil moisture, temperature regulation and weed and pest control.'

A 12-page questionnaire was designed to gather relevant information on GMCC systems, including their agronomic, ecological and socio-economic context. After several revisions, the final version of the questionnaire was sent to over 150 individuals around the world. The major sections of the survey are: GMCC System Description, Summary of Key Aspects of the GMCC System, Use and Adoption by Farmers, Critical Factors, Information Available and Regional Information.

Each section consists of 4 to 11 main questions, some of which ask for detailed, quantified data on the system, while others solicit the respondent's impressions of certain attributes of the system. The survey is available from the following Web sites: www.cidicco.hn/proyecto.htm (for the Spanish version) and www.cidicco.hn/current_projects.htm (for the English version).

The Survey

Respondents were diverse and consisted of researchers who had helped develop a particular system, individuals who had participated in diffusion efforts of a system or others who had closely followed a system and its performance. They were affiliated with a wide array of institutions: international centres, national research organizations, donor agencies, non-governmental organizations (NGOs) and academic institutions. To ensure that relevant individuals were contacted and important GMCC systems were covered, the working group elaborated a list of potential systems to be documented and a directory of persons and institutions known to be researching or promoting GMCCs in Africa, Asia and Latin America.

During the surveying process, E-mail played a critical role in reaching people dispersed over 24 countries and three continents. Earlier, this would not have been possible, given the poorly developed electronic facilities available in most developing countries. The initial doubts regarding its efficacy for obtaining information slowly dissolved as the completed survey forms began to arrive. Less than a dozen surveys came via normal mail.

While E-mail assisted with the logistical aspects, some key contacts greatly facilitated the process of surveying. In particular, the project would not have been possible without the contributions of Joseph Mureithi of the Kenya Agricultural Research Institute (KARI) and Gloria Melendez of the University of Costa Rica. They acted as CIDICCO's liaisons and were able to collect information on a large number of systems in their regions. Despite efforts to obtain information on GMCC systems in Asia, information was received on only five systems. As discussed later, the richness of experiences with GMCCs in Asia needs to be documented in the second round of surveying.

Even though the Working Group first believed that the completion of 200 surveys would be possible, with the passing of time it became obvious that it would be necessary to set a more realistic goal of 80. Altogether, 82 surveys were received but 10 were discarded because of incomplete or inconsistent information. In the second section of this report, these 72 completed surveys are characterized.

The Development of a Web Site

All the cases collected so far are available through an interactive Web site developed by CIDICCO. This site runs out of the CIDICCO Web page (www.cidicco.hn) and is being regularly updated to include new cases from several colleagues. Among other features, the site contains all the cases collected, relevant articles and GMCC research summaries, a gallery of GMCC photographs and a few short videos of the systems described. Links to related sites are also included, particularly that of the Soil Health Portal (http://mulch.mannlib.cornell.edu/) and of CIDICCO for the Spanish version. The Soil Health Portal site runs out of Cornell's Main Library server.

Initially, cases are available directly from a world map, which highlights the number of cases collected from a particular country as the cursor moves over the map. Another option is to run a search by country, by main crop and by cover crop. A more complete database capable of performing cross searches will be added soon. Meanwhile, the user will be able to read the entire case and to print it. The site also contains an electronic version of the questionnaire to complete and send to CIDICCO with new cases.

Finally, the first edition of the Catalogue was launched only in English as most of the cases were originally written in that language. Efforts are underway to translate the whole site to Spanish and French.

CHARACTERIZING SURVEYED SYSTEMS

It should be noted that because only a small percentage of GMCC systems utilized by tropical smallholders has been surveyed, the systems described here are not necessarily representative of either general trends or of trends in a particular country or continent. Many more systems have not been documented and ongoing efforts are targeting those.

In the following, the location and structure of these surveyed systems are reviewed, after which some of their socio-economic characteristics, including adoption rates and products, are described. Appendix I lists the common and Latin names of the main crops and Appendix II those of GMCCs.

Location and Structure of the Surveyed Systems

Of the 72 completed surveys, the largest share is from Africa (51% of all systems), followed by Latin America (42%) and, as a distant third, Asia (7%). As mentioned earlier, the relative abundance of systems from Africa and Mesoamerica mainly reflects the productivity of the liaison persons in those regions. This is also clearly seen in the distribution of the systems by country, as the largest share of systems is from Kenya (19 systems or 26% of all) and Costa Rica (11 systems, 15%). Each of the following countries has five systems (i.e. 7% of the systems each): Thailand, Malawi, Mexico, Guatemala and Benin, while Honduras has 4 (6%). In 11 countries, only one or two systems were surveyed.[3] Altogether, the GMCC systems surveyed are located in 19 countries.

The systems include 27 main crops (Table 1). Maize (*Zea mays* L.) is by far the predominant main crop as over two-thirds of the systems include it, attesting to its importance as a staple food, its adaptability to intercropping situations and its consequent popularity in GMCC systems. Other main crops include cassava (*Manihot esculenta* Crantz), which is the main crop in one eighth of the systems) and common beans (*Phaseolus vulgaris* L., main crop in only 7% of the systems). Pumpkin (*Cucurbita* sp. L., 6%), sorghum (*Sorghum bicolor* [L.] Moench) and coffee (*Coffea arabica* L.) follow with 4% each. Most main crops (21 in total) have only one or two systems. Interestingly, all but one system reported from Africa are maize based, while in Latin America, only 11 of the 30 systems (37%) include maize. Whether this reflects a wider diversity of main crops used with GMCCs in Latin America will need to be analyzed later, when information on more systems has been collected.

Table 1. Number and share of the main crops in the surveyed systems.

Number of systems	% of systems (each)	Main crop[a]
49	68	Maize
9	13	Cassava
5	7	Common beans
4	6	Pumpkin
3	4	Coffee, sorghum
2	3	Sweet potato, Napier grass, cocoyam, cowpea, potato, plantain, cotton, groundnut
1	1	Rice, mango, sugar cane, tobacco, onion, black pepper, ginger, various herbs, *palmito*, soybean, wax gourd, millet, *ipecacuanha*

[a] Full Latin names are given in Appendix I.

Thirty-six GMCCs are utilized in the surveyed systems, considerably more than the number of the main crops reported (27). *Mucuna* sp. Adans is the most frequent GMCC and is reported in one fourth of the systems (Table 2). The dominance of certain crops is less dramatic with the GMCCs than with the main crops. Cowpea (*Vigna unguiculata* [L.] Walp.) and *Canavalia* (*Canavalia ensiformis* [L.] DC.) are the second-most common GMCCs, each present in one eighth of all systems. Other common GMCCs include pigeon pea (*Cajanus cajan* [L.] Millsp.), groundnuts (*Arachis hypogaea* L.), *Tephrosia vogelii* Hook. f. and miscellaneous fallow species.

Table 2. Number and share of green manure/cover crops (GMCCs) in the surveyed systems.

Number of systems	% of systems (each)	GMCCs[a]
18	25	*Mucuna*
9	12.5	Cowpea, *Canavalia*
6	8.3	Pigeon pea
5	6.9	Groundnut, *Tephrosia*, fallow species
3	4.2	*Leucaena*, *Crotalaria*
2	2.8	Lima bean, mung bean, perennial peanut, *Lablab*
1	1.4	*Tithonia*, soybean, *Gliricidia*, *tarwi*, tree leaf mulch, *Acacia*, *choreque*, *Brachiaria*, black oats, sunn hemp, rice bean, *Lithocarpus*, broad bean, rye grass, radish, butterfly pea, *Desmodium*, *Stylosanthes*, *Sesbania*, *Schima*

[a] Full Latin names are given in Appendix II.

Of all the 36 GMCCs, only pigeon pea and cowpea are reported in all three continents. Another 11 GMCCs are reported in two continents:

- Black oats (*Avena strigosa* Schreb.),
- *Brachiaria plantaginea* (Link.) Hitchc.,
- Broad bean (*Vicia faba* L.),
- *Canavalia*,
- *Lathyrus nigrivalvis* Burkart,
- Common bean,
- *Crotalaria ochroleuca* G. Don,
- Lima bean (*Phaseolus lunatus* L.),
- *Mucuna* spp.,
- *T. vogelii* and
- Mung bean (*Vigna radiata* [L.] R. Wilczek).

The remaining 24 crops are reported only in one of the three continents. Of the GMCCs in the survey, for the most part (i.e. 63%) only one system represents each GMCC and most GMCCs therefore are found only in one country. This may indicate a need to exchange experiences and seed from one country to another, although it may simply be caused by the limited number of systems. In systems from Africa, *Mucuna* was included in only 16% but in those from Latin America, *Mucuna* is included in 40%.

Interestingly, several countries have more GMCCs than surveyed systems, indicating the use of several GMCCs in one system. For example, five countries (Burkina Faso, Guinea, Zambia, Zimbabwe and Nicaragua) are represented by only one system, which has two GMCCs. Similarly, in Senegal, Brazil and Thailand, the number of GMCCs surpasses the number of systems. In absolute numbers, the greatest diversity of GMCCs in the surveyed systems is found in the countries from which most surveys were received, Kenya and Costa Rica.

The relative abundance of GMCCs compared to main crops means that several main crops are combined with a wide variety of GMCCs. For example, the 49 maize systems are associated with 21 different GMCCs but particularly with *Mucuna* (12 systems) and cowpea (8 systems) (Figure 1). There are eight different GMCC-maize systems with only one example (i.e. system) per combination. Interestingly, in some cases common beans, peanut, soybean (*Glycine max* [L.] Merr.) and cowpea are reported as a main crop, while in other cases they are used as GMCCs.

Over one quarter of the systems include more than one main crop indicating prevalence of intercropping (Figure 2). Interestingly, a similar share of the systems includes more than one GMCC (Figure 3). However, having more than two main crops was more common than having more than two GMCCs.

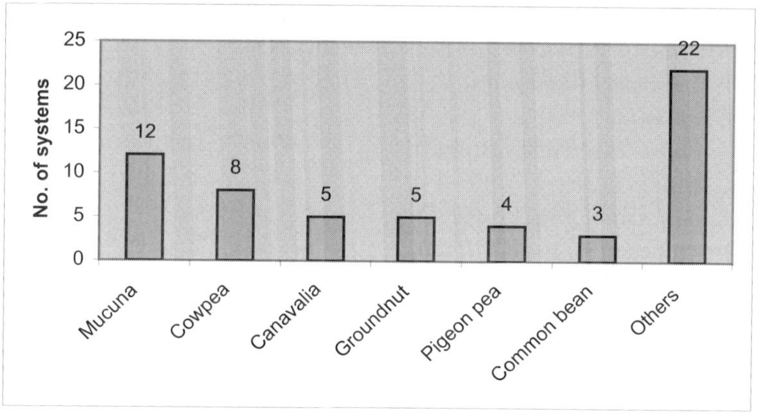

Figure 1. Number of maize systems (n =49) with different green manure/cover crops (n = 59, as some systems contain more than one).

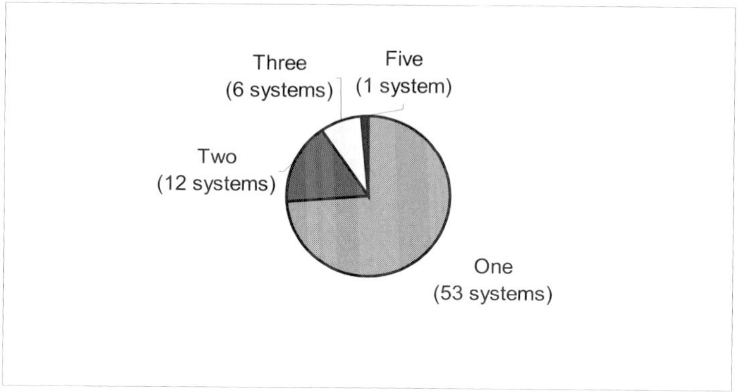

Figure 2. Number of main crops in the surveyed systems (n = 72).

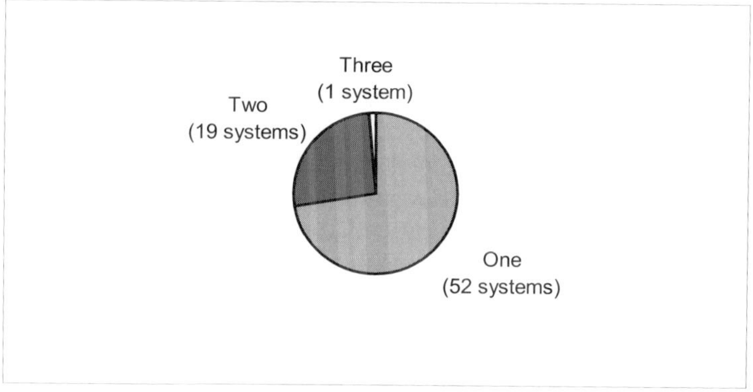

Figure 3. Number of green manure/cover crops in the surveyed systems (n = 72).

The survey form included various alternatives for temporal and spatial arrangements of the GMCCs and the answers indicate a wide variability in the systems covered. Clearly, simultaneous intercropping arrangements (i.e. planting of the main crop and GMCCs at the same time) are the most common, with 36% of the systems exhibiting only this arrangement (Figure 4). Rotational systems, where the GMCC is planted immediately after harvesting the main crop, are also quite common (13% of the systems), while there is one system with only relay intercropping (i.e. where the GMCC is planted after planting the main crop but before it is harvested). A full 46% of the systems surveyed have two different temporal/spatial arrangements. For example, within one system, one GMCC may be intercropped with the main crop and another one in rotation, after the main crop is harvested.

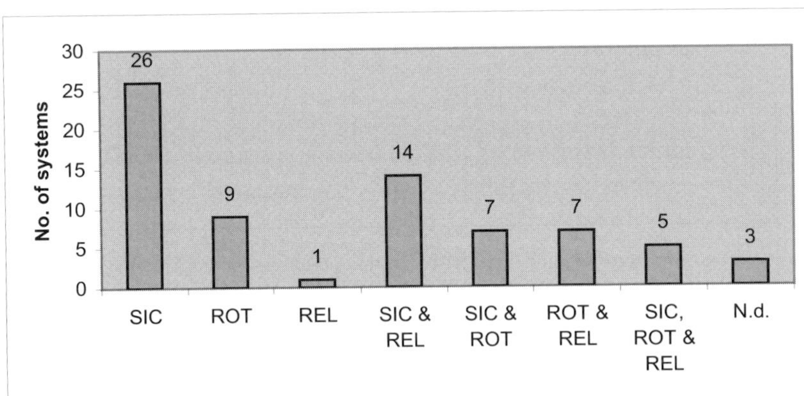

Figure 4. Spatial and temporal arrangements in the surveyed systems (n = 72). (Note: SIC = simultaneous intercropping, ROT = rotation, REL = relay cropping, N.d .= no data, not reported in surveys.)

Socio-economic Factors

Although many of the surveyed systems are traditional (44%), a slight majority (56%) was reported as having been recently adopted by farmers. However, it should be noted that several systems are traditional in an area but have been introduced into another area where they are (correctly) classified as recent systems.

Another issue that was assessed through the surveys was the involvement of research in the development and improvement of GMCC systems (Figure 5). More than one-third of the systems surveyed (38%) has been developed through research and is being monitored and improved through research follow-up. Slightly fewer systems are classified as traditional in this question (33%) and less than one-sixth of the systems (15%) are characterized as having been developed by research but without any further follow-up. Notably, a considerable number of respondents mention the need for research to improve the systems. This was the case both in researcher-developed and in traditional systems. In the case of researcher-developed systems, research can play a role in recommending agronomic and management modifications, which will increase the likelihood of adoption by farmers. In the case of the traditional systems, research can also indicate technical adaptations that will help farmers to enhance the present performance level.

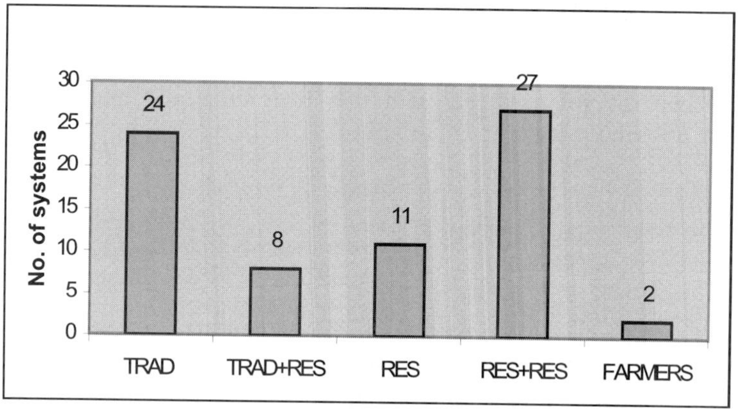

Figure 5. Role of farmers and researchers in the development and improvement of the surveyed systems (n = 72). (Note: TRAD = traditional system, TRAD+RES = traditional with research follow-up, RES = researcher developed; RES+RES = researcher developed with follow-up; FARMERS = recently developed by farmers.)

Survey questions regarding adoption of the systems often did not yield exact figures because in many cases no formal adoption studies have been conducted. About half of the systems are described as being well adopted and a further 35% are characterized as having been adopted to a moderate extent. Ten respondents did not provide information on adoption rates, in some cases because the system had been introduced recently and its adoption could not be estimated as yet. Despite these limitations, for most systems adoption is characterized as either increasing (42% of systems) or as stable (32%). A decreasing trend in adoption is observed for only 17% of the systems.

Adoption is often closely related to land tenure. Farmers tend to adopt long-term soil improving practices more easily when they own land. In some cases, landowners let their cattle graze on crop residues on land rented to others, thus making it hard for those renting to make any long-term investments in soil improvement. Unfortunately, the surveys do not provide information that is specific enough to establish the impact of land tenure on adoption.

The quantity and type of products removed from the system is an important consideration for GMCC systems. From an agronomic viewpoint, typically the more products that are removed from the system, the less benefit there is to the soil. For the farmers' short-term benefits, the situation may be the reverse. About one-half of the systems surveyed include both subsistence and cash crops, one third of the systems include only subsistence crops and 15% include only cash crops. This underlies the importance of considering the multiple interests of farmers who expect to obtain not only soil fertility benefits but also food, feed and/or cash income to meet their needs.

The survey also solicited information on the products of the main crop and of the GMCCs that are removed from the field. In most systems (65%), only the edible portion (i.e. grains, beans, tubers, fruit or nuts) is removed from the main crop. In one-fourth of the systems surveyed, main crop residues are utilized in one way or another. For example, cows may graze on the field after harvesting or the main crop residues are harvested manually and fed to the animals. In only 7% of the systems surveyed are other, secondary products removed from the fields, such as young leaves for food, or the whole plants are harvested as in the case of common beans. Only one of the systems surveyed has as its principal aim to provide improved pastures and silage.

Of the GMCCs, in one-half of the systems either nothing is removed or only the seeds are removed, leaving the residues on the soil surface to be incorporated or mulched (Figure 6). Remarkably, in slightly less than one-third of the systems (31%), no product is removed from the GMCCs. In one-fourth of the systems surveyed, foliage or foliage and seed are removed. If the GMCC grain is edible, typically almost all harvest is used for food and only a small part for seed. The non-edible grains are generally harvested for use as seed or for selling to earn cash. Often, the GMCC grains are harvested with pods and threshed later.[4] In some cases the pods are fed later to the cattle.

In the large majority of the systems, women play an active role (Figure 7). Women's participation ranges from occasional help with the harvest to doing most of the farming work.

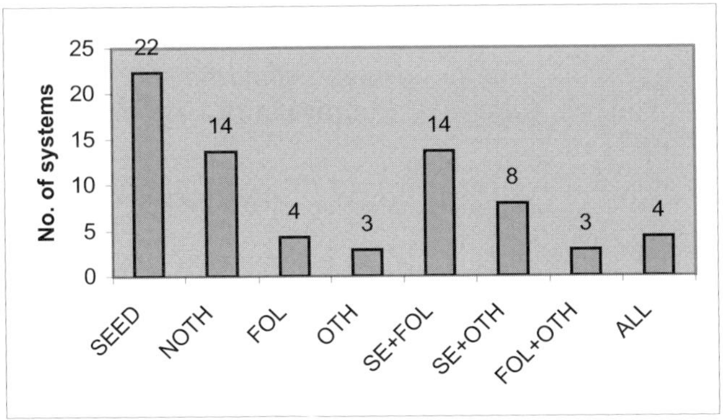

Figure 6. Utilized plant part of the green manure/cover crops in the surveyed systems (n = 72; SE = seed, NOTH = nothing, FOL = foliage and OTH = other).

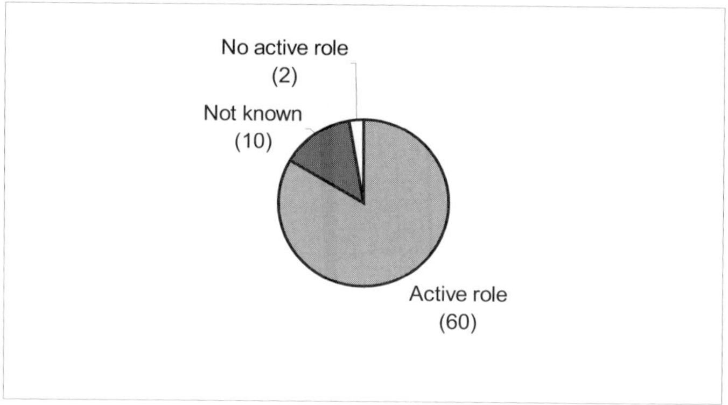

Figure 7. Involvement of women in the management of surveyed systems (n = 72).

EMERGING ISSUES

The completed surveys have certain shortcomings, which should be taken into account. For example, most of them lack an estimate on the number of people utilizing a given system. They often also lack data on agronomic factors (such as biomass production) and economic performance because these have not been properly documented. This is particularly the case with systems that are considered traditional, which are often practiced by a large number of farmers. In contrast, more information is available on recently introduced GMCC systems, which have often been promoted by NGOs.

As presented earlier, some main features of the surveyed systems include:

- Maize is the predominant food crop with which a diversity of GMCCs is grown in different temporal and spatial arrangements.
- There is a greater diversity of GMCCs than main crops in the surveyed systems. Not surprisingly, some popular GMCCs, such as *Mucuna*, cowpea, *Canavalia* and pigeon pea are included in most of the systems.
- Several species have been adapted to ecoregions different from their origin.
- Simultaneous intercropping is the predominant single arrangement in the surveyed GMCC systems. This may be related, among other things, to restrictions in land availability. Most commonly, several spatial and temporal arrangements can be found in any one system.

No analysis of the data in the surveys has been conducted. Observations on the information include:

- Reasons for using GMCCs are strongly tied to their multiple purposes such as food, animal feed, source of cash income, reduction of production costs, weed control, soil restoration, shade for perennials and mulching for moisture conservation.
- Use of GMCCs to address poor soil fertility and erosion is generally viewed as a complement to, and not a substitute for, inorganic fertilizers. Where high price of fertilizers is not a major constraint, GMCCs used together with fertilizers result in economic benefit.
- While reliable information about adoption (and abandonment) is scarce, adoption seems more likely when modifications are made to systems that farmers are familiar with, as compared to the introduction of completely new GMCC systems.
- Unreliability and variability of rainfall are major limitations to the use of GMCCs.
- Lack or limited availability of GMCC seeds is another constraint to further diffusion of GMCC systems.
- Research needs to 'escort' extension by NGOs or government agencies. Dissemination of practices with insufficient research follow-up has led to abandonment or stagnation due to, for example, emerging diseases or competition with other farming activities.
- Farmer-developed systems have been promoted without due consideration of their suitability to the local farming system. Identification of the appropriate niches is important.
- Insufficient understanding of traditional GMCC systems is a hindrance to taking advantage of the knowledge and understanding that they embrace.
- Gender roles in GMCC systems need clarification. Participation of women is often mentioned in the questionnaires but specifics are lacking.

FUTURE EFFORTS

Although information on a relatively large number of systems (72 in all) was collected during the first phase of the project, it is estimated that this number is but a small sample of the richness of the GMCC systems that can be found in tropical countries. Recent discussions in two GMCC electronic discussion lists, MULCH-L (in English) and COBERAGRI-L (in Spanish) are illustrative of this diversity.[5] For example, in a message sent to COBERAGRI-L, Roland Bunch, a well-known development consultant, reported that he personally has observed 98 different GMCC systems with 37 different species in 23 countries. A researcher from Venezuela responded by wanting to know whether the *Crotalaria juncea*/sorghum rotation utilized in his country was included in the list. It was not and, interestingly, that particular system is not included in this survey either.

Most GMCC systems have therefore not been adequately documented and the information that is available on them only circulates within a region where they are practiced. These factors limit the potential contribution of GMCCs to the improvement of smallholders' agriculture in other regions.

Therefore, there is not only a need but also an excellent opportunity to considerably expand the number of systems to be included in this ongoing effort. Based on the experience gained during the first phase and the identification of new contacts, it is estimated that information on 200 more systems can be obtained through a second phase of surveying. An additional component of this ongoing work includes the completion of the Web site. As of September 2002, 20 new cases have been received and at least 30 more cases are in the process of completion by contacts in the Philippines, Cuba and Brazil.

This new surveying effort is being carried out systematically, by targeting certain underrepresented regions. For example, geographically it is well known that considerable experience with GMCCs exists in South and South-east Asia, namely India, Indonesia, Thailand, Vietnam and the Philippines. However, the present collection of systems contains only a handful of systems from Thailand. Moreover, although only two systems from Brazil were collected during the first phase, in the southern states of Santa Catarina and Parana alone, more than a dozen species of legume GMCCs have been adapted to several agricultural systems (Monegat 1991; Chapter 1, this volume). In addition, agro-ecologically more emphasis has been given in the past to the research and promotion of GMCCs in the humid tropics. The current systems reflect this bias. The arid and semi-arid areas of Africa and Latin America present a challenge to the use of GMCCs. However, efforts carried out by international research institutes[6] as well as experiences gained by many national researchers and non-governmental groups are worthy of being documented and disseminated to other similar regions.[7] In addition, GMCCs can benefit high altitude zones (1500 - >3000 masl) (Pilbeam et al. 1999), another neglected agro-ecological zone. Nepal in Asia and the Andean region in South America are areas that would benefit from a broader range of GMCC options. So far, however, the inventory includes only two systems from the Andean region.

Additionally, particular emphasis will be given to the collection of information on systems practiced by women. As is well known, women play a major role in subsistence agriculture in several countries of Africa (Quisumbing et al. 1995). Systems utilized by women may present important lessons in regard to, for example, low labour demand as a consideration in increasing further adoption of GMCCs. Additional emphasis is also given to systems that include animal interactions. Animals are a crucial component of all agricultural systems, especially in Africa's drier agro-ecological zones. Animal components in GMCC systems bring about new sets of issues that need to be considered.

NOTES

1. For example, LEXSYS, International Institute of Tropical Agriculture (IITA), The Legume Network (www.nmw.ac.uk/gctefocus3/networks/legumes.htm) and Purdue University's database (www.hort.purdue.edu/newcrop).
2. www.cidicco.hn
3. Countries with two systems were Senegal and Brazil. Countries with one system were Zimbabwe, Zambia, Uganda, Nicaragua, Guinea, Ecuador, Congo, Burkina Faso and Bolivia.
4. In Figure 6 these cases are being reported as 'seed and other products removed.'
5. For Mulch-L (in English), contact Lucy Fisher lhf2@ cornell.edu and for Coberagri-L (in Spanish), contact milton.flores@cidicco.hn. Additionally, EVEC-L is available in French, contact: Robert Carsky r.carsky@cgiar.org.
6. For example, the IITA, International Centre for Research in Agroforestry (ICRAF) and International Crops Research Institute for the Semi-Arid Tropics (ICRISAT).
7. Some 60 researchers across West Africa, affiliated with CoverCropNet, are working in a regional network helping to refine the role of cover crops in sustainable agriculture. For additional information, visit http://www.cgiar.org/spipm/ccropnet/crop/aims.

REFERENCES

Becker M, Johnson A (1998) The role of legume fallow in intensified upland rice-based cropping systems in West Africa. Nutr. Cycl. Agro-ecosyst. 6:1-11.

Monegat C (1991) Plantas de cobertura do solo: Cararacterísticas e manejo em pequenas propriedades. Centro Internacional de Información sobre Cultivos de Cobertura (CIDICCO), Tegucigalpa, Honduras. 337 p.

Pilbeam C J, Tripathi B P, Munakarmy R C, Gregory P J (1999) Economic benefits of integrated nutrient management in three major cropping systems in the mid-hills of Nepal. Mountain Res. Dev. 19(4): 333-344.

Quisumbing A R, Brown L R, Feldstein H S, Haddad L, Peña C (1995) Women: The key to food security. Food policy report. International Food Policy Research Institute (IFPRI), Washington D.C. 22 p.

APPENDIX I. COMMON AND LATIN NAMES OF THE MAIN CROPS IN THE SURVEYED SYSTEMS.

Common name	Genus	Species
Black pepper	*Piper*	*nigrum* L.
Cassava	*Manihot*	*esculenta* Crantz
Cocoyam	*Xanthosoma*	Schott *colocasia* Schott
Coffee	*Coffea*	*arabica* L.
Common bean	*Phaseolus*	*vulgaris* L.
Cotton	*Gossypium*	*hirsutum* L.
Cowpea	*Vigna*	*unguiculata* (L.) Walp.
Ginger	*Zingiber*	*officinale* Roscoe
Groundnut	*Arachis*	*hypogaea* L.
Herbs	Various species	
Ipecacuanha	*Psychotria*	*ipecacuanha* (Brot.) Stokes
Maize	*Zea*	*mays* L.
Mango	*Mangifera*	*indica* L.
Millet	*Panicum*	*miliaceum* L.
Napier grass	*Pennisetum*	*purpureum* Schumach.
Onion	*Allium*	*cepa* L.
Palmito	*Bactris*	*gasipaes* Kunth
Plantain	*Musa*	*x paradisiaca* L. (pro sp.)
Potato	*Solanum*	*tuberosum* L.
Pumpkin	*Cucurbita*	sp. L.
Rice	*Oryza*	*sativa* L.
Sorghum	*Sorghum*	*bicolor* (L.) Moench
Soybean	*Glycine*	*max* (L.) Merr.
Sugar cane	*Saccharum*	*officinarum* L.
Sweet potato	*Ipomoea*	*batatas* (L.) Lam.
Taro	*Colocasia*	*esculenta* (L.) Schott
Tobacco	*Tabacum*	*nicotianum* Berchtold & Opiz.
Wax gourd	*Benincasa*	*hispida* (Thunb.) Cogn.

APPENDIX II. COMMON AND LATIN NAMES OF THE GREEN MANURE/COVER CROPS IN THE SURVEYED SYSTEMS.

Common name	Genus	Species
Acacia	*Acacia*	spp. Mill.
Black oats	*Avena*	*strigosa* Schreb.
Brachiaria	*Brachiaria*	*plantaginea* (Link) Hitch.
Broad bean	*Vicia*	*faba* L.
Butterfly peas	*Clitoria*	*ternatea* L.
Canavalia	*Canavalia*	*ensiformis* (L.) DC.
Choreque	*Lathyrus*	*nigrivalvis* Burkart
Common bean	*Phaseolus*	*vulgaris* L.
Cowpea	*Vigna*	*unguiculata* (L.) Walp.
Crotalaria	*Crotalaria*	*ochroleuca* G. Don
Desmodium	*Desmodium*	*uncinatum* (Jacq.) DC.
Fallow	Various species	
Gliricidia	*Gliricidia*	*sepium* (Jacq.) Walp.
Groundnut	*Arachis*	*hypogaea* L.
Lablab	*Lablab*	*purpureus* (L.) Sweet
Leucaena	*Leucaena*	*leucocephala* (Lam.) De Wit
Lima bean	*Phaseolus*	*lunatus* L.
Lithocarpus	*Lithocarpus*	*sootepensis* (Craib) A. Camus
Mung bean	*Vigna*	*radiata* (L.) R. Wilczek
Perennial peanut	*Arachis*	*pintoi* Krapov. & W.C. Gregory
Mucuna	*Mucuna*	sp. Adans
Pigeon pea	*Cajanus*	*cajan* (L.) Millsp.
Radish (Napo)	*Raphanus*	*sativus* L.
Rice bean	*Vigna*	*umbellata* (Thunb.) Ohwi & Ohashi
Rye grass	*Lolium*	*multiflorum* Lam.
Sesbania	*Sesbania*	*sesban* (L.) Merr.
Schima	*Schima*	Reinw. ex Blume
Soybean	*Glycine*	*max* (L.) Merr.
Stylosanthes	*Stylosanthes*	*hamata* (L.) Taubert
Sunn hemp	*Crotalaria*	*juncea* L.
Tarwi	*Lupinus*	*mutabilis* Sweet
Tephrosia	*Tephrosia*	*vogelii* Hook. f.
Tithonia	*Tithonia*	*diversifolia* (Hemsl.) A. Gray
Tree leaf mulch	Various species	

Learning From the Case Studies

The Editors

INTRODUCTION

This chapter reviews the case studies from the point of view of two issues that are important regarding GMCC systems: actual and potential adoption and impact. We address these issues by first reviewing historic adoption of these systems and then considering four themes, each important when farm-level adoption and impact of GMCC systems is considered: food security, smallholder economy, natural resources and diffusion factors.

Drawing firm lessons from these cases as a group is difficult, however, for a number of reasons. As described in the Introduction, the cases are relatively few in number but highly diverse, covering many GMCCs, systems, main crops and various biophysical, socio-economic and policy environments. While such diversity may help us learn more widely about factors affecting impact and adoption, it also makes drawing of definitive conclusions difficult. The case studies also have differing emphases on diffusion and research and varying amount of information available on them. Their analysis is made even harder because the experiences with, and therefore also lessons from, GMCC systems are jumbled together with experiences regarding project management, agricultural extension approaches, research methodologies and others. While all are relevant to GMCC research and development, issues specific to GMCC systems are harder to discern. Finally, these case studies describe moving targets, since the systems described, their circumstances, the associated extension efforts and their adoption are not constant. Key factors enabling good performance or farmer acceptance of a system may have changed before this book is published and at times the potential impact of such factors or their combination will become evident only after circumstances have changed.

What follows thus explores some of the main issues affecting adoption and impact of the experiences with GMCC systems in this volume.

M. Eilittä et al. (eds.), Green Manure/Cover Crop Systems of Smallholder Farmers, 303-347.

TRENDS IN ADOPTION

The GMCC systems described vary greatly in the extent of diffusion efforts (when conducted) and in their cultivation and adoption. Table 1 presents an overview of the adoption of the GMCC systems; when available, results of various studies are presented. In the following, adoption is used as shorthand for the cultivation of the GMCC system by farmers, whether they are testing a technology or cultivating it over the long term.

Of the cases presented, GMCC adoption is by far the greatest in the state of Santa Catarina, Brazil (Chapter 1), where use of GMCCs and conservation tillage has increased enormously in the past 20 years. In 1987, only 112 000 ha (5%) was under GMCCs but by 1997, the figure had risen to 793 000 ha—or about 44% of the land area. Similar increases in cropland under reduced or zero tillage have been reported and GMCC and reduced tillage adoption are still on the increase. Several other cases describe GMCC systems that have been adopted at a relatively large scale, including:

- *Fríjol tapado*: almost 25% of Costa Rica's beans (*Phaseolus vulgaris* L.), a staple food in 1998, were produced under the system (Chapter 2);
- Maize (*Zea mays* L.) – *Mucuna pruriens* (L.) DC. in northern Honduras: in 1997, adoption was by 45% of the hillside farmers (Chapter 3);
- The improved fallow systems in eastern Zambia: 20 0000 farmers have planted improved fallows (Chapter 5); and
- Maize-*Mucuna* in southern Benin: 10 000 farmers were testing the technology in 1996 (Chapter 10).

Although no adoption studies were conducted in the Ruvuma region of southern Tanzania (Chapter 7), from the purchases of *marejea* (*Crotalaria ochroleuca* G. Don) seed it is clear that its cultivation was extensive in the 1980s.

As is also evident from Table 1, for most of the systems presented in this volume, the status of adoption is relatively low and for several its trend is downward. Although no formal studies were conducted in Uganda (Chapter 9), adoption of most systems is judged slow. Similarly, the current adoption of planted forages in coastal Kenya (Chapter 8) is low, mainly because dairying is an expensive enterprise and animal diseases cause high cattle mortalities. In northern Thailand (Chapter 11), only a few dozen farmers are utilizing the *Sesbania rostrata* Bremek. & Oberm. technology, principally an indication of the small scale and recency of this effort by a research institute.

Table 1. Past and recent adoption, adoption trends and factors favouring and limiting adoption of the green manure/cover crop (GMCC) systems in this volume.

Chapter	Recent trend	Past adoption figures	Recent adoption	Factors favouring adoption	Factors limiting adoption
1. GMCCs-Cons. Till., Brazil	↑	- 1987: 112 000 ha (5% cropland)	- 1997: 793 000 ha, or 44% cropland	- Lower labour use and cost for ploughing and harrowing - Endogenous development of appropriate technology - Private sector involvement - Support of agric. research, national and local policies and extension - Farmer awareness - Funding by external development organizations	- Availability of seed, information on management, appropriate equipment - Conflicting production structure/work calendar/objectives/farmer needs - No immediate economic benefit - Not convinced of need for conservation technologies
2. *Tapado*, Costa Rica	↓	- Pre-Columbian, widely used - 1980: 80% of Costa Rica's bean production	- Late 1990s: 24% of Costa Rica's bean production	- Farmer importance on food security - Requires mainly labour but only little, leaving time for other activities - Minimal use of external inputs, labour and capital - Low risk - Provision of cash	- Globalization + market liberalization - Rural-to-urban migration - Increasing land pressure - Competition with more profitable enterprises - Increased climate risk - Low use of inputs due to low capital availability
3. Maize-*Mucuna*, Honduras	↓	- 1970s: Introduced, slow expansion - 1980s-90s: +5% growth y^{-1} - Early 1990s: 65% adoption - 1990s: Disabandonment, 4.5% y^{-1}	- 1997: 45% adoption	- Fertilizer effect - Ease of land preparation - Moisture conservation - Weed and erosion control - Secure land tenure - Importance of maize	- *Rottboellia* weed infestation - Importance of reseeding forgotten - Good market access

Continued.

Table 1. (Continued)

Chapter	Recent trend	Past adoption figures	Recent adoption	Factors favouring adoption	Factors limiting adoption
4. Maize-*Mucuna*, Mexico	↓	- 1992: 150 farmers cultivated - 1993: 723 farmers planted - 1994: 1457 farmers received seed	- Early 2000: Adoption limited in most communities	- Secure land tenure - Importance of maize production, soil fertility problem, increased input prices	- Importance of pests and diseases as a crop production constraint - Lack of food/feed uses and markets for *Mucuna* - Low maize profitability and yields - Conflicting government policies
5. Improved fallows, Zambia	↑	- 1992: on-farm efforts - 1996: 50% planted 2nd fallow - 1999: 7% had planted - 2001: 71% of those who planted 1996-97 had adopted	- Altogether, 20 000 farmers have planted	- Yield and labour benefit - Soil fertility decline; low availability of fertilizer	- Pests and diseases (*Sesbania*) - Livestock grazing (pigeon pea) - Expensive seed and difficult establishment (*Gliricidia*) - Poor germination, fire, drought, weeds, time lag to receive benefits - Lack of land, time, strength to manage
6. Best-bet legumes, Malawi	?	- Grain legumes widely used but over-reliance on maize	- No data	- Rotational (groundnut, soybean): superior in caloric and economic net benefits - Intercropped (pigeon pea, *Tephrosia*): temporal complementarity in resource use - *Mucuna*: food use, soil fertility benefit	- Seed cost (groundnut), fluctuating market price (soybean), residue incorporation difficult (both) - Poor seed yield and livestock damage (pigeon pea) - No food use (*Tephrosia*) - *Mucuna*: L-dopa and processing needs

Continued.

Table 1. (Continued)

Chapter	Recent trend	Past adoption figures	Recent adoption	Factors favouring adoption	Factors limiting adoption
7. *Marejea*, Tanzania	↓	- 1940s: Known in region - 1970s: Large-scale production; no data - 1987 and 1988: 200 t of seed distributed outside the region - Early 1990s: Adoption slowed	- Little or no adoption	- Seed purchases and access to fertilizers - Social pressure - Mass media and politicians - Large land size, poor soil fertility and positive maize yield impact	- Discontinuation of seed purchases and fertilizer exchange - Discontinuation of strong focus on *marejea* by the mission project - Improved availability of fertilizers
8. Forages, Kenya	↓	- 1970s: screening - 1980s: extension - 1987: focus on hedgerows - 1992: 80 farmers had planted *Clitoria*	- Slow and decreasing adoption	- High price of milk and profitability of dairying - Yield benefit from legume integration - Facilitative efforts by extension - Iinstitutional support	- Higher labour use - Cattle diseases - Expense and capital intensity of dairying - Poor infrastructure and long distances limit dairying - Preference of food crops in labour/land allocation
9. GMCCs, Uganda	Limited	- 1990s: farmer groups experimenting	- Slow adoption	- Some systems satisfy need for improved fodder - *Tephrosia* controls mole rats	- Labour constraints - Lack of benefits - Land scarcity - Lack of seed

Continued.

Table 1. (Continued)

Chapter	Recent trend	Past adoption figures	Recent adoption	Factors favouring adoption	Factors limiting adoption
10. Maize-*Mucuna*, Benin	↓	- 1987: 20 testing - 1990: 180 testing - 1991: 500 testing - 1993: 3000 testing - 1996: 10 000 testing - 1996: 16% using in project villages	- 1997-98: 7% in villages with known use - 1997: rate for abandonment> adoption	- Access to and markets for *Mucuna* seed - Linkage to maize package - *Imperata* infestation - Deteriorating soil fertility - Land pressure, extensification not an option	- Lack of processing methods for *Mucuna* - Occupies land in lieu of needed food crop - Lack of constant market for *Mucuna* seed - Land pressure, food crops needed - Insecure landholding - Termination of projects - Limited funding for extension - Top-down extension
11. Rice-*Sesbania,* Thailand	Limited	- 1997: 7 farmers in one district - 1998: 11 farmers in 2 sites in 2 districts - 1999: 3 sites in 3 districts, increased no. of farmers	- Limited farmer experimentation continues	- Improved soil - Increased rice yield - High price of and decreased fertilizer use - Land tenure and access - Increased interest in sustainable agriculture and high-quality rice	- Greater use of time and labour - Higher cost - Difficulty of land preparation

Note: In the table, terminology used in each case study, which in some cases differentiates between 'testing' and 'adoption', has typically been kept.

For several of the systems, these present adoption levels are much lower than earlier ones. In some cases, the decrease occurred as a project working in the area was discontinued. In southern Benin in 1997, when 1 year after Sasakawa-Global 2000, a nongovernmental organization (NGO), discontinued its efforts on *Mucuna*, an annual abandonment rate of 4.6% was not offset by a 3.4% new adoption rate (Chapter 10). In Benin, current adoption rates, based on impressions of those familiar with the system, remain low. Similarly, in southern Veracruz, Mexico (Chapter 4), in 1993 over 700 farmers were estimated to have planted *Mucuna* in all communities where one of the NGOs worked. Thereafter, emphasis on *Mucuna* diffusion decreased, cultivation waxed and waned but currently, few farmers are cultivating *Mucuna*. In the Ruvuma region of southern Tanzania (Chapter 7), *marejea* cultivation quickly decreased when its seed exchange with inorganic fertilizers was discontinued and at the moment, there is either no or very little *marejea* cultivation. Similarly, in coastal Kenya, the number of farmers planting Napier grass (*Pennisetum purpureum* Schumach.) has declined since 1993 (Chapter 8).

Decreasing adoption is evident not only in those systems that have been the targets of development efforts. Both of the systems with spontaneous farmer adoption, *fríjol tapado* in Costa Rica (Chapter 2) and maize-*Mucuna* in northern Honduras (Chapter 3), also have seen their adoption decrease in the past decade. In 1980, *fríjol tapado* was practiced by 80% of Costa Rica's bean producers and 47% of the country's bean production resulted from this system. In Honduras, a survey conducted in 1997 reported a discontinuation of the maize-*Mucuna* system by about 45% of the farmers (annual disadoption rate of 4.5%) who were practicing it in the early 1990s.

Similarly, low adoption has been reported for other GMCC systems. In Rwanda, little adoption was seen after 15 years of efforts on improved fallows and green manures (Drechsel et al. 1996). Adoption of green manures in lowland rice (*Oryza sativa* L.) systems has dramatically declined since the 1970s (Becker et al. 1995) and despite considerable support, pre-rice green manures have not been adopted (Chettri et al. 2003). In general, adoption of legume-based GMCC technologies in Africa is poor (Sumberg 2002; Morse and McNamara 2003) and, somewhat provocatively, Boonman (1999) states that the 'legume technology has not worked.' In contrast, of the 72 surveyed systems characterized by Flores and Janssen (Chapter 12, this volume), about 50% were described as having been 'well adopted' and another 35% as having had moderate adoption; additionally, most systems were described as having either increasing or stable adoption. There was seldom any quantifiable information available on adoption, however.

Key factors impacting the adoption and performance of the GMCC systems in this volume are discussed below; Table 1 summarizes factors affecting adoption as described in the case studies.

FOOD SECURITY CONSIDERATIONS

Not surprisingly, many previous studies have emphasized that the extent to which GMCC technologies can meet and improve family food security has a strong impact on their adoption (Enyong et al. 1999; Pound et al. 1999; Totongnon et al. 2000). In most of the systems presented in this volume, the GMCC has been linked to a staple crop (most commonly maize) in farming situations that are principally subsistence oriented, with surplus production sold. A notable exception to this is Santa Catarina in Brazil (Chapter 1), a state accounting for 13% of national agricultural production, although it constitutes just 1% of the land area; GMCCs in this state are cultivated with diverse cash and subsistence crops. In the subsistence situations, positive impact on food security is typically through the main crop yield. In addition, GMCCs themselves can be eaten, either by people or animals, thereby affecting food security.

There are other impacts on food security that may occur with GMCC systems, such as competition with a food crop in intercropping situations, replacement of a food crop by a GMCC and the lower diversity of edible crops in GMCC systems. These topics also are discussed below.

GMCCs Commonly Enhance the Succeeding Main Crop Yield

Most case studies report improved main crop yields after the cultivation of GMCCs (Table 2). Yield improvements of 30% or higher are commonly reported. In Santa Catarina, Brazil (Chapter 1), while yield improvements of up to 70% have been reported with GMCCs and conservation tillage, commonly, 30% higher yields are achieved.

Such increases in main crop yield are encouraging but may not be sufficient in all situations to induce adoption. Drechsel et al. (1996) considered that yield impacts of 74 and 46% in the first and second seasons were not enough in Rwanda to compensate for the reduced main crop yield and labour investments. Similarly, Morse and McNamara (2003), reporting from Kogi State in Nigeria, found a much lower adoption of leguminous cover crops than improved varieties despite the higher yield benefits of the former. An evident exception to high yield impacts are the many systems involving grain legumes with their low biomass and often low yield benefits; as pointed out for Malawi (Chapter 6), due to their edibility, lower beneficial impact on main crop yield is acceptable for them.

Table 2. Impact of green manure/cover crop (GMCC) systems on food security as reported in the case studies.

Chapter	GMCC use as food/feed	Yield impact of GMCC and its productivity if edible
1. GMCCs-Cons. Till., Brazil	Some	- Improvement of 30% over fallow control common; up to 75% observed - For various crops: maize, 11-61% (negative impacts reported after winter grains); beans, 3-85%; onion, 15-30%
2. *Tapado*, Costa Rica	No	- Average bean yield 0.49 t ha^{-1} lower than national average (0.68 t ha^{-1}) - In Acosta, most bean yields are 0.40-0.50 t ha^{-1} (range 0.40-0.90); in Coto Brus, most yields are 0.40-0.45 t ha^{-1} (range 0.25-0.70).
3. Maize-*Mucuna*, Honduras	No	- Average maize yield improvement almost 100% - Yield ranges vary by community: 2.2-3.7 (vs. 1.9 with no *Mucuna*), 4.2-4.9 (2.5), 2.7-3.7 (2.0), 1.8-3.9 (1.4), and 1.6-3.0 t ha^{-1}.
4. Maize-*Mucuna*, Mexico	No	- For intercropped system: winter maize, +50%; summer maize, no clear impact - For rotational system: winter maize, +50-120%
5. Improved fallows, Zambia	Some	- Maize yields after 2 years of *Sesbania* range 3-5 t ha^{-1}, of pigeon pea or *Tephrosia*, 2-3 t ha^{-1} vs. 3-5 t ha^{-1} for fertilized maize and < 1 t ha^{-1} for unfertilized maize.
6. Best-bet legumes, Malawi	Most	- Maize yield 8-78% higher after groundnut rotation, 13-19% higher after soybean rotation - Intercrops of pigeon pea and *Tephrosia* with maize: higher caloric output, no reduction of maize yield - *Mucuna* rotation: seed yield 1770-2130 kg ha^{-1}, positive yield impact on following maize but needs careful processing for food
7. *Marejea*, Tanzania	No	- Farmers always report higher maize yields after *marejea*; inorganic fertilizers acquired further increased maize production. - In the mission farm, maize yields have been 4.5-6.0 t ha^{-1} with *marejea* and occasional fertilizer and farmyard manure since 1960s.

Continued.

Table 2. (Continued)

Chapter	GMCC use as food/feed	Yield impact of GMCC and its productivity if edible
8. Forages, Kenya	Yes	- Higher total dry matter production when legumes intercropped in Napier grass: 16.4 t ha^{-1} (Napier only, no inorganic fertilizer) vs. 20.0-28.4 t ha^{-1} (with/without organic fertilizer) - Milk yields higher with legumes (6.1 vs. 7.8 kg d^{-1} with *Leucaena*; 88% farmers reported higher milk yield with *Clitoria*).
9. GMCCs, Uganda	Some	- Sole-cropped GMCCs increased maize yield in on-farm trials by 0-240%. - After sole-cropped *Crotalaria*: maize yields were 3990 kg ha^{-1} (vs. 2820 weedy fallow) for first season and 2630 kg ha^{-1} (2150) for second season; bean yields were 560 kg ha^{-1} (400) for first season and 740 kg ha^{-1} (660) for second season. - Competitive impacts by intercropped GMCCs on main crop yield reported (61-87%)
10. Maize-*Mucuna*, Benin	No	- After intercropped *Mucuna*, 30-70% main season and 30-50% minor season maize yield improvements reported - After sole-cropped *Mucuna*, 94-113% main season and 31% minor season yield improvements reported
11. Rice-*Sesbania*, Thailand	No	- On-station: rice yields over 6 yr improved by 18% - On-farm: rice yield impact varies by year, community and field; about 50% of fields had higher yield than the fertilized control.

In two of the cases in this volume, GMCC systems do not result in improved yield. In Costa Rica, *fríjol tapado* (Chapter 2)—a subsistence-oriented system where 60% of the beans produced are consumed at home—results in lower yields than the conventional bean production systems and cumulative yields are even less because of low cropping intensity (2-3 years production is followed by 1-4 years of fallow). Low capital requirements and risk, as well as low labour demand allowing farming families to earn additional income from other on- and off-farm activities, contribute to the continued practice of the system. Similarly, no consistent and clear positive impacts were found for the summer maize yield in the maize-*Mucuna* system in southern Veracruz (Chapter 4). Finally, great variability in on-farm performance of GMCC systems, discussed below in more detail, necessarily means that on some fields, low, no, or negative yield impacts are found.

Competition with and Replacement of Food Crops May Reduce Food Security

Negative impacts on food security are reported for several systems presented (Table 2). In intercropping situations, GMCCs may reduce yield of the main crop through competition or through partial replacement (lower plant density of main crop); this factor, however, is relatively little discussed in the case studies. In Uganda (Chapter 9), yields of maize and bean when intercropped with *Crotalaria ochroleuca* G. Don, reached only 60 and 85% of the sole-cropped yields and in Malawi (Chapter 6), pigeon pea (*Cajanus cajan* [L.] Millsp.), when planted in simultaneous intercropping arrangement with maize in a low-fertility site, reduced maize yield. In southern Veracruz, Mexico (Chapter 4), no negative (or positive) impacts from the intercropped *Mucuna* on the summer maize yield were found.

In many systems, the GMCC occupies land for a whole cropping season, thereby fully replacing the main crop. Commonly and understandably, the GMCC is assigned the less favourable of the cropping seasons, such as in Santa Catarina, Brazil (Chapter 1) where winter cultivation of GMCCs is common, in southern Veracruz (Chapter 4) and in southern Benin (Chapter 10), where *Mucuna* typically grows in the minor, second season. Also, in northern Honduras (Chapter 3), *Mucuna* occupies the fields for the season with the lower average maize yield, which in the region is the first, rainier season (this, however, may be the more beneficial season for *Mucuna* growth). In such situations, the yield loss due to replacement is lower.

If the GMCC is inedible, competitive effects from intercropping and replacement of a food crop for an entire season, even during the less beneficial season, can result in lower food production, threaten the short-term family food security and therefore accentuate a conflict between the immediate need to guarantee food supply for today and the objective to build soil fertility in the long term (Kumwenda et al. 1997). In these case studies (as more widely in many regions), the most common inedible GMCC species has been *Mucuna*, which is typically not consumed as a food crop in most areas where it has been introduced in recent decades.

In the case of the two countries with grave food insecurity and strong land pressure, Malawi (Chapter 6) and Benin (Chapter 10), allocating land to *Mucuna* is seen as problematic. In Malawi, because of intense land pressure and the paramount need for ensuring household food security, *Mucuna* rotations are judged a feasible option only if *Mucuna* were to be managed as a grain legume; for that to happen, effective ways to process its seed for human consumption would have to be found. The high *Mucuna* seed yields obtained under Malawian conditions (averaging 1770 and 2130 kg ha^{-1} in two trials) make it a particularly attractive food crop. Similarly, in southern Benin, the negative impacts of an inedible GMCC are stark: taking land out of food production during the minor rainy season, *Mucuna* does not provide food needed to carry the family through the following dry season and the subsequent hungry period. Planting GMCCs on only a portion of the land may be a solution but only for those with sufficient land. In the region, efforts to introduce *Mucuna* for soil fertility improvement therefore rightly included efforts to process it for human consumption. Unlike in Malawi and Benin, high adoption in Honduras (Chapter 3) has been possible despite the fact that *Mucuna* is inedible; in Veracruz, Mexico (Chapter 4), however, this lack of direct utility (as a food, feed, or cash crop) does affect adoption.

A number of authors in this volume emphasize that edible GMCCs are more desirable as they directly contribute to food security. Efforts described in Zambia (Chapter 5), Kenya (Chapter 8) and Uganda (Chapter 9) also include GMCCs that are readily edible, by either human beings or animals.[1] High farmer interest in continuing to use traditional legumes or testing new edible legumes has been noted (Schultz et al. 2003). Many GMCC species utilized are legumes with high protein and relatively high fat content. However, caution has been urged regarding the alleged overestimation of the potential contribution of introduced legumes to human nutrition due to, for example, poor grain quality leading to a need for excessive processing (Weber 1996).

In contrast, in the two systems that farmers have spontaneously adopted—*fríjol tapado* (Chapter 2) and maize-*Mucuna* in northern Honduras (Chapter 3)—the mulched species are inedible and similarly, in Santa Catarina, Brazil (Chapter 1), many of the cultivated GMCCs are inedible. Edibility is therefore clearly not a precondition for a viable GMCC system. What seemingly sets apart the Malawi (Chapter 6) and Benin (Chapter 10) experiences discussed above is the very high extent of food insecurity and land pressure in these two countries. In areas with either greater land and/or better food availability, edibility may not be as important.

Finally, another way GMCCs compete with food crops, although rarely mentioned in the case studies, is by decreasing diversity of food crops in a field. In southern Veracruz, Mexico (Chapter 4), farmers noted as disadvantages the lower occurrence of minor crops in the maize-*Mucuna* fields and the fact that beans cannot be planted with *Mucuna*. Several other case studies, such as Tanzania (Chapter 7), Uganda (Chapter 9) and Benin (Chapter 10), note the high diversity of food crops in farmer fields. For example, in southern Benin, a survey indicated that 72% of fields had two and 22% had three crop mixtures.

ECONOMIC AND POLICY CONSIDERATIONS

Considerations of profitability and various factors contributing to it—such as returns to labour and land use—have been an important component in many studies on GMCC technologies, with most authors emphasizing the need for clear profitability gains and direct economic benefits (Enyong et al. 1999; Snapp et al. 2002b; Carsky et al. 2003). Such considerations are often tied with macro-economic policies, for example, as they affect prices for inputs and products, off-farm employment and labour costs. Among the case studies in this volume, thorough economic reviews have been conducted only for a few. Much of the information is therefore either observational or qualitative. Some of the economic estimates that exist take into account only certain of the costs and benefits from the introduced technology, a common limitation in studies on such systems (Cairns and Garrity 1999). In general, the lack of long-term and on-farm data, in addition to such factors as quickly changing macro-economic context, prevent realistic economic conclusions over the long term.

Central economic and policy issues considered in the studies are profitability, total amount of labour used, returns to labour and land and impact on input use and on risk. They are discussed below.

Most Systems are Judged Profitable but Not Immediately So

Most of the systems for which economic studies have been conducted have been found to be profitable, at least after an initial period. Table 3 summarizes information available on the economic performance of these systems.

Table 3. Impact of green manure/cover crop (GMCC) systems on smallholder economy, particularly on land and labour use, as reported in the case studies.

Chapter	Impact on smallholder economy
1. GMCCs-Cons. Till., Brazil	- Reduced labour requirement is cited by 87-100% of farmers as reason for practicing conservation tillage. - Labour reduction from conservation tillage is 55-59% in animal traction, 10-51% in mechanized systems. - Crop rotation-zero tillage-GMCC system can have up to 100% higher economic returns than conventional systems. - For maize and soybean, returns to zero tillage were 65% higher for mechanized and 90% for animal traction systems. - Direct-seeding maize reduces costs by 9% for animal traction systems and 12% for mechanized systems and increases returns by 16% for both systems. - Special considerations: 1. Herbicide cost may be higher, 2. First years may be unfavourable, 3. Special machinery may be a bottleneck for poorer farmers.
2. *Tapado*, Costa Rica	- In *tapado*, 43% of cost is from labour (27% in dibble-stick system) and 57% from material purchases (73%); total labour use is lower (41 vs. 61 h ha^{-1}). - *Tapado* is more profitable than dibble-stick (₡55777 vs. ₡1535 ha^{-1}) and less risky (53% vs. 93% probability of losing money). - *Tapado* has more variable earnings but their range is more favourable; return to assets is also higher for *tapado*. - Marginal rate of return for *tapado* is 531% vs. 96% for an intensified dibble-stick system. - Special considerations: System minimizes financial loss by allowing other activities during off-times and by providing for food security.
3. Maize-*Mucuna*, Honduras	- Returns to land and labour are higher than for the bush-fallow system but only after 1 year for labour and 2 years for land. The system has 52% higher net profits than winter slash-and-mulch system; summer maize is slightly unprofitable. - Profitability of bean production is 3-4 times higher; for high-yielding maize-*Mucuna* sites, profitability is the same. - Dairying is as profitable, less risky and labour consuming but requires more land; chilli pepper production is more profitable but needs capital and is risky. - Maize-*Mucuna* production has higher return per unit of capital than a more profitable, mechanized fertilizer-based system.
4. Maize-*Mucuna*, Mexico	- Maize-*Mucuna* system (with other resource-conserving technologies) is unprofitable for the first few years but later may become profitable. - Researcher estimates indicate higher labour demand (by 9-29 d ha^{-1}) but farmer evaluation lower labour demand because of speedier slashing and weeding.

Continued.

Table 3. (Continued)

Chapter	Impact on smallholder economy
5. Improved fallows, Zambia	- Over 5 y, a 2-y fallow, 3-y maize system, labour requirements are 11% (unfertilized) to 32% (fertilized) less and maize yields are 40% (fertilized) to 87% (unfertilized) more. - Returns per ha are: Fertilized maize>improved fallows>unfertilized maize; to labour: improved fallows (US$1.5 0 d^{-1})>fertilized maize (US$1.45 d^{-1})>unfertilized maize (US$0.65 d^{-1}). - Improved fallows are less risky than inorganic fertilizer use at times of drought.
6. Best-bet legumes, Malawi	- Grain legume rotations bring higher economic benefits than continuous maize (groundnut by 96-157%; soybean by 44-80%; groundnut somewhat less if labour cost is included; fluctuating soybean prices endanger its profitability). - Special considerations: Labour requirements increase as grain legumes need to be returned to field and incorporated.
7. *Marejea*, Tanzania	- System clearly profitable, in good part because of fertilizer exchange scheme and *marejea* seed markets. - With inclusion of costs for *marejea* seed and fertilizer, as well as yield benefit, 2-year profit for maize-*marejea* was Tsh. 67 000 vs. 28000. - No data on labour requirements; land restricted cultivation for farmers with small plots.
8. Forages, Kenya	- Dairying households have higher income but dairying requires more capital; moreover, land and labour are preferably allocated to food crops. - A system including maize-cowpea-*Leucaena* and Napier-*Leucaena*-slurry with dairy cattle in zero grazing can fetch a profit of US$3229 ha^{-1} y^{-1}.
9. GMCCs, Uganda	- Insufficient labour (17 of 22 farmers) but not land (1 farmer) is the most common reason for discontinuing GMCC cultivation. - Labour requirement for tillage, weeding and planting may be less after GMCCs. - Land productivity is not increased with alley cropping.
10. Maize-*Mucuna*, Benin	- Positive returns from second year onwards; from first year if *Mucuna* can be sold. - Over 8-year period, cost-benefit of 1.24 for maize-*Mucuna* system, 0.62 for maize only; 3.56 if *Mucuna* sold. - Special considerations: Land is a limiting factor.
11. Rice-*Sesbania*, Thailand	- Benefit-cost ratio with data from farmer surveys is 1.62 (including costs: additional ploughing and labour and *Sesbania* seed; benefits: reduced N use and increased rice yield).

Consistently in the Brazilian studies, minimum tillage and/or GMCCs have been found to be more profitable than the conventional system but often the first year is unfavourable because of a number of factors (e.g. insufficient farmer experience, delay in soil fertility impact and lack and cost of suitable machinery). Similarly, higher profitability is found for the maize-*Mucuna* system in northern Honduras (Chapter 3), improved fallows in Zambia (Chapter 5) and grain legume rotations in Malawi (Chapter 6), among others. An exception to the cited good profitability of these systems is the smallholder maize system with (or without) *Mucuna* in southern Veracruz, Mexico (Chapter 4), which has been found to have very low profitability and negative returns if labour is valued at local wages; low valuation of family labour and subsistence orientation are considered reasons for the continued cultivation of maize in the area.

Immediate, Direct Economic Benefits are Not Always Essential

GMCCs commonly improve soil fertility and help with weed suppression. Many GMCC researchers stress that these benefits are not enough to induce adoption and instead, direct and immediate benefits are needed from the system, for example, through eating or sale of GMCC seeds or by-products (Hudson 1991; Becker et al. 1995; Weber 1996; Kaya et al. 2000; Oyewole et al. 2000; Vanlauwe et al. 2000; Carsky et al. 2001). Indeed, in the widespread adoption of *Mucuna* in the early 20th century southern USA, its utilization as a feed was at least equally important as its utilization for soil fertility improvement and weed suppression (Buckles 1995; Eilittä and Sollenberger 2001). For two of the systems, *fríjol tapado* (Chapter 2) and maize-*Mucuna* in northern Honduras (Chapter 3) the indirect benefits through weed suppression and yield improvement (or stability and lower risk, in the latter case) clearly have been enough. In contrast, most other case studies and the surveys conducted by Flores and Janssen (Chapter 12, this volume) either emphasize the need for such products, or it is still unclear whether sustained farmer acceptance may occur without such benefits.

Labour-Saving Attributes are Key for Adoptability

Labour is a resource known to have a major influence on the performance and adoption of GMCC systems and, more widely, new agricultural technologies (Becker et al. 1995; Thomas and Sumberg 1995; Drechsel et al. 1996; Becker and Johnson 1998). Smallholder farmers invariably rely on family labour for their agricultural production and few own draught animals and even fewer own farm machinery. The labour situation is made worse by HIV/AIDS (Human Immunodeficiency Virus /

Acquired Immunodeficiency Syndrome), especially in sub-Saharan Africa, and by rural-to-urban migration. While case studies from both Zambia (Chapter 5) and Uganda (Chapter 9) mention the impact of HIV/AIDS on the agricultural labour force, detailed information is lacking on the epidemic's impact on the productivity of GMCC systems.

GMCC technology generally uses labour input to improve land productivity and it is therefore not surprising that several cases have considered both the amount of labour needed and the returns to labour. The labour-saving attributes of some of the GMCC systems studied have made them economical and attractive to farmers. In Santa Catarina, Brazil (Chapter 1), the labour-saving attribute of the combined GMCC-reduced tillage has been the single most important reason for the wide adoption. Interestingly, without reduced tillage, GMCCs are seen to increase labour use but with tillage, labour use is seen to decrease. Similarly, in northern Honduras (Chapter 3), labour considerations are important and in the *fríjol tapado* system in Costa Rica (Chapter 2), the absolute amount of labour spent is far less than in the conventional system (41 vs. 61 h ha^{-1}). Improved fallows in Zambia (Chapter 5) require 11% less labour than unfertilized maize and 32% less labour than fertilized maize.

In some of the cases in this volume, higher labour requirement is cited as a bottleneck for GMCC cultivation. At the Kenyan coast (Chapter 8), shortage of labour for dairying has been a key factor contributing to the decline in adoption of planted forages and in Uganda (Chapter 9), 77% of farmers attributed labour constraints as a primary reason for abandoning of GMCC technologies. Similarly, in Thailand (Chapter 11), 25% of farmers noted that incorporating *Sesbania* in the soil made land preparation for rice transplanting less convenient and demanded more labour. For *marejea* cultivation in Tanzania (Chapter 7), labour consumption has not been described in a detailed way, although farmers in the area considered it laborious to cultivate. In southern Benin (Chapter 10), lower *Mucuna* adoption by women was partly explained by their greater labour constraints. Finally, if more effective contribution of grain legumes to soil fertility improvement is sought, additional labour needs to be expended in residue management. Taking grain legume residues home for threshing is a common practice in many locations, including in Malawi (Chapter 6) and northern Thailand (Chapter 11) but for a beneficial impact on soil, they would need to be returned to the fields.

High labour requirements are known to have caused low interest in and adoption of GMCC technologies in a number of other efforts:

- Introducing *Mucuna* to Nigeria (IIA 1936),
- In Rwanda, where perceived benefits from additional labour were considered risky (Drechsel et al. 1996),
- With labour expended on GMCC establishment and incorporation in lowland rice systems of Asia (Becker et al. 1995),

- With low adoption of organic technologies in Malawi (Snapp et al. 2002b),
- With hedgerows in the Philippines (Fujisaka 1994) and
- With several other GMCC systems (Fujisaka 1991; Weber 1996; Enyong et al. 1999; Anthofer 2000; Oyewole et al. 2000; Muhr et al. 2001; Steinmaier 2001).

Interestingly, Steinmeier (2001) notes that in Zambia, farmers assessed the labour implications more critically for the newly introduced green manure-improved variety technology than for their indigenous cropping systems, indicating that a new technology may need to clearly outperform the existing one in labour considerations. In contrast, Morse and McNamara (2003) found that in the Kogi State of Nigeria, higher labour requirement of the leguminous cover crop technology was a relatively less important factor influencing adoption than seed availability.

While consideration of total labour hours is important, the specific distribution and characteristics of labour use can be equally so (Weber 1996). In northern Honduras (Chapter 3) and in Veracruz, Mexico (Chapter 4), a particularly important labour advantage of the *Mucuna* system was the speedier slashing, the most arduous task. Additionally, less time was spent weeding after *Mucuna*, another labour bottleneck (in Honduras, reduction was from 25 to over 50%). In both locations, labour requirements were at times increased when *Mucuna* had to be slashed or its presence complicated maize harvesting. Similarly, in Uganda (Chapter 9), tillage and weeding requirements could be reduced after GMCCs and maize planting was facilitated by the holes left from their uprooting. In Santa Catarina, Brazil (Chapter 1), an interesting and important corollary benefit regarding labour use is mentioned: in GMCC-reduced tillage systems, improved moisture conditions allow planting during a wider timeframe, reducing labour peaks and allowing farmers to become more self-sufficient in labour. In *fríjol tapado* bean cultivation, labour use is concentrated at slashing/seeding and harvesting times, allowing the farmer greater flexibility at remaining times to work in other on- or off-farm activities. In contrast, a negative characteristic, constituting a disincentive for *fríjol tapado* cultivation, is the fact that labour is arduous, partly because of the steep slopes on which the system is practiced.

Finally, returns to labour in several cases have been found to be higher in GMCC systems. With improved fallows in Zambia (Chapter 5), they were found highest (US$1.50 per day) for the improved fallow systems, followed by continuous fertilized maize (1.45) and continuous unfertilized maize (0.65). (Returns to land were highest for fertilized maize, then for improved fallow system and lowest for unfertilized maize.) In northern Honduras (Chapter 3), the maize-*Mucuna* system is judged to have higher returns to labour but only after the first 2 years. Interestingly, while the *fríjol tapado* system is extensive in land use and has poor returns to land, its returns to labour and assets are relatively high.

In contrast, in Veracruz, Mexico (Chapter 4), negative returns were reported for the maize-*Mucuna* system when labour was valued at its true market price. Other studies have emphasized that greater labour returns to GMCC systems would foster future adoption (Becker and Johnson 1998).

Several Systems are Considered Less Risky but Little Information is Available

Several of the GMCC systems are regarded as less risky than the conventional systems. This is particularly the case for *fríjol tapado* (Chapter 2) whose rationale is described as a way to eliminate risk related to family food requirements. To meet family food security, the system relies mainly on family labour. Total labour use is low, however, and is concentrated at slashing/planting and harvesting times, allowing labour use for other enterprises in intervening times. In the system, 57% of costs come from material purchases, in comparison to 73% in the dibble-stick system that has been promoted by the government extension. In northern Honduras (Chapter 3), risk of the maize-*Mucuna* system was assessed not only lower than that of the bush-fallow system but also that of bean and chilli pepper cultivation and dairying. Risk decreased as farmers cultivated *Mucuna* for a longer time. Similarly, in Brazil (Chapter 1) and in Zambia (Chapter 5), GMCC systems are considered less risky during drought but in Brazil, risk is considered higher with freeze and in very wet conditions.

In contrast, in coastal Kenya (Chapter 8) dairy enterprise is risky, with or without GMCCs. Elsewhere, GMCCs have often been considered risky, as with lowland rice (Becker et al. 1995), in conditions of unreliable soil moisture and water supply (Meelu et al. 1994) and with high variability in biomass and residual impact, as in Rwanda, where 'renunciation of a crop yield to improve future yields poses an unacceptable risk precisely to those farms whose need to regenerate soil fertility is most urgent' (Drechsel et al. 1996). In contrast, a Food and Agriculture Organization (FAO) review of soil conservation projects concluded that a technology should not include any increased risk (Hudson 1991).

Extremely Small Farm Size and High Cropping Intensity Limit Adoption

Commonly, in a continuum of land-use intensity, GMCCs are seen to have potential in medium-intensity situations, that is, when land is abundant, traditional fallowing is practiced and when land is very scarce, the cropping system has no room for a GMCC (Franzel 1999) and technologies would need to be carefully integrated in the existing system (Weber 1996). These case studies lend some evidence to these hypotheses.

In southern Benin (Chapter 10), small farm size and consequent need for continuous cropping were considered obstacles to adoption and the higher number of male adopters of *Mucuna* was partially explained by their larger farm size. In northern Thailand (Chapter 11), those not interested in continuing with *Sesbania* planting indicated limited land as one of the reasons and in Tanzania (Chapter 7), farmers with small plots of land were unwilling to adopt *marejea*. Similarly, in the highlands of Rwanda, the farmers' unwillingness to fallow fields regularly partly explained low farmer interest in green manures and improved fallows (Drechsel et al. 1996). In contrast, in western Kenya's densely populated areas with small farm size, a combination of depleted land and availability of off-farm income offered incentives to fallowing in that half of farmers were reported to be periodically fallowing 10-50% of land (Swinkels et al. 1997).

In Costa Rica (Chapter 2), increasing land pressure is impacting the performance and future prospects of the *fríjol tapado* system in a different way: inadequate fallows now threaten its future sustainability. Interestingly, in the Eastern Province of Zambia (Chapter 5), land is not a constraint (only 35% of arable land is utilized) but continuous cultivation, introduced when inorganic fertilizer became available to many in the 1970s and 1980s, has continued despite the sharp curtailing of fertilizer use, resulting in soil fertility decline and perceived need for improved fallows.

Security of Land Tenure Favours Adoption

Secure land tenure has commonly and not surprisingly been associated with GMCC adoption, since many benefits of the GMCC systems do not become evident for several years of practice and farmers may not be interested in investing in a GMCC system unless they are assured of land title (Tripp et al. 1993; Fujisaka 1994; Enyong et al. 1999; Pannell 1999; Tarawali et al. 1999; Anderson et al. 2001; Flores and Janssen, Chapter 12 this volume). In Veracruz, Mexico (Chapter 4) and in southern Benin (Chapter 10), farmers with secure land tenure had greater interest in the maize-*Mucuna* system; in Benin, surveyed farmers ranked mineral fertilizers higher than the *Mucuna* technology on rented land. In Benin, insecure farm size particularly affects adoption by women who typically have to borrow land from their husbands. Also, in northern Honduras (Chapter 3), subsistence workers accessing land through rental agreements were least likely to adopt *Mucuna*. Similarly, in Costa Rica (Chapter 2), high rental prices and lack of land security were seen to act as disincentives for the continued practice of *fríjol tapado*.

Rental land is not necessarily seen as an obstacle to GMCC adoption, however. Although 25% of rural households in the project area in northern Thailand (Chapter 11) rent land, this is not expected to negatively influence *Sesbania* adoption because rents are fixed and rental periods are relatively long and renewable. Interestingly, in Zambia (Chapter 5), village leaders can withdraw land without crops from farmers and planting of improved fallow species has become a way to ensure continued access to land.

Little Information Available on Farmer Resource Access and Gender

The case studies contain varying amounts of information on the impact of farmer characteristics on farmer interest and adoption of GMCC systems. Recent studies on agroforestry technologies have emphasized the importance of household, resource-related and risk factors on adoption (Bannister and Nair 2003; Pattanayak et al. 2003). Some such factors, for example land tenure, are covered in other sections. Two issues remain, each with only scanty information available in the cases: farmer resource base and farmer gender. The cases span a great range in these two characteristics, from the relatively better-off farmers of southern Brazil (Chapter 1) to Malawi (Chapter 6) and Benin (Chapter 10), where food insecurity is common. The role of women varies equally, from the many African countries where women have a strong role in farming, to the systems in Central America and Mexico where men carry out most of the cropping activities. In Zambia (Chapter 5), the profile of the improved fallow testers is diverse. The better-off farmers are more likely to try out the improved fallow technology but the poorer farmers are more likely to continue utilizing it; both women and men appear to adopt the technology in similar proportions. In southern Benin (Chapter 10), the farmers with greatest land constraints found it difficult to leave the land fallow for the second season, with no direct, edible product. Such a characteristic was often more common for women, who also faced greater labour constraints than men, further limiting their adoption of the maize-*Mucuna* technology. In Kenya (Chapter 8), the dairy enterprise to which the legumes were integrated requires capital, constituting a disincentive to adoption for some. In Veracruz, Mexico (Chapter 4), farmers relying on maize production on poor soils, who had no other means to improve soil fertility, were seen as the target group for the *Mucuna* technology.

Impact of Macro-Economic Context is Variable

The macro-economic context can offer some of the greatest disincentives and/or strongest incentives for beneficial adoption and impact of GMCC systems. Across continents and case studies, removal of fertilizer subsidies and consequent increase in fertilizer prices (typically with stable or decreasing grain prices) has meant that, in most of the locations, inorganic fertilizer use by smallholder farmers is currently extremely low or absent. This has made GMCC systems far more attractive than they were 30 years ago. For example, in Thailand (Chapter 11), the prices of commonly used fertilizers have increased over 40% since 1997. Not surprisingly, it is the more market-oriented systems in this volume that report greater fertilizer use. Interestingly, in southern Brazil (Chapter 1), high amounts of fertilizer often continue to be used—partly a reflection of the number of non-legumes used as GMCCs and partly the higher economic status and the market-orientation of the farmers. Similarly, in northern Thailand inorganic fertilizer use continues in the rice-*Sesbania* system although some farmers reported halving it. Fertilizer use is also reported in northern Honduras (Chapter 3) and occasionally in eastern Zambia (Chapter 5) and a number of other locations. Interestingly, as mentioned above, in the Ruvuma region of Tanzania (Chapter 7), *marejea* cultivation provided an avenue for acquiring inorganic fertilizers.

In contrast, other factors in the macro-economic environment have had negative impacts on GMCC adoption and profitability. In most cases, the GMCCs in these systems are linked with subsistence crops that currently fetch low prices, such as maize, beans and rice. For example, in Veracruz, Mexico (Chapter 4), low maize prices (coupled with low baseline yields) meant that even though *Mucuna* caused quite high relative winter maize yield increases, additional cash earned from maize sales would be small. Capitalizing on yield improvement from the GMCCs may therefore mean that they need to be linked with a higher-value crop, such as proposed in Honduras (Chapter 3), where *Mucuna* is linked with more profitable second-season maize; or in coastal Kenya (Chapter 8), where good milk prices enable profits from improved forages; or with high-quality rice in Thailand (Chapter 11).

Additional negative impacts of the macro-economic context are evident in Costa Rica (Chapter 2) where in the past decade, the trade of basic grains has been liberalized, the number of institutions assisting small-scale producers has been reduced and price restrictions have been abolished. The resulting decreased production of primary food crops has also affected the practice of the *fríjol tapado* system. Moreover, the greater emphasis on export-oriented crops has meant greater land pressure, shorter fallows and the retreat of small-scale farmers to more fragile lands; all factors that can destabilize this fallow system. In Honduras (Chapter 3), rapid growth of milk and beef production has caused poorer farmers to sell their land to ranchers for establishment of permanent pastures.

NATURAL RESOURCE CONSIDERATIONS

Biomass Production and N Supply are Generally Good but Variable

The cases indicate that good productivity of the GMCCs, a necessary precondition for positive impact, seems obtainable in most locations. Table 4 highlights the average biomass amounts in the systems described and information on soil quality, pests and diseases and weeds. These average biomass amounts are sufficient to typically supply 50-100 kg of N per hectare in addition to considerable amounts of other nutrients and mirror the findings of many GMCC studies and reviews (e.g. Becker et al. 1995). Grain legumes are the exception, typically supplying far less biomass and nitrogen.

Data on biological nitrogen fixation are scanty in the case studies. In southern Benin (Chapter 10), an average amount of 167 kg ha^{-1} of N derived from atmospheric nitrogen was found in fields with a range of *Mucuna* biomass but about one-fifth of the fields had no nodulation. In Malawi (Chapter 6), over 60% of pigeon pea N was found to come from N_2 fixation.

Despite good biomass production, certain production problems are taking place. Particularly concerning is the large variability in GMCC productivity and its poor performance in some circumstances—a factor which has been reported to generally contribute to high risk and low adoption or disadoption of GMCC systems (Drechsel et al.1996; Becker et al. 1995).

Table 4. Impact of green manure/cover crops (GMCCs) on soil properties, weed incidence and pests and diseases, as reported in the case studies.

Chapter	GMCC biomass	Impact on soil chemical and physical properties	Impact on weed incidence	Relation with pests and diseases
1. GMCCs-Cons. Till., Brazil	- No details reported	- Various studies on soil properties conducted; not reported	- Some conservation tillage systems can increase susceptibility to certain weeds. - Contrasting opinions regarding herbicide use in conservation tillage	- Some minimum tillage systems increase incidence of rats and certain soil pests.
2. *Tapado*, Costa Rica	- Live biomass ranges 5.27-9.75 t DM ha^{-1} in Acosta and 2.47-17.46 t DM ha^{-1} in Coto Brus. - Dead biomass ranges 1.45-1.80 t DM ha^{-1} in Acosta and 2.75-11.87 t DM ha^{-1} in Coto Brus.	- Good physical properties - Low bulk density (<1 Mg m^{-3}) - Hydraulic conductivity >20 cm h^{-1} - Good organic matter in plots up to 80 y - P quantity exported large relative to stocks	- No weeding conducted - Weed growth inversely related to the mulch thickness - Low weed pressure: weeds germinate less and remain in seedling stage	- Fewer pests than in other bean systems, except slugs - Low fungi incidence, perhaps because residue acts as a mechanical barrier - Monocropped fallow stands are more susceptible to pests (*Tithonia* and *Mucuna*). - Seed quality a problem
3. Maize-*Mucuna*, Honduras	- Most fields: 10-12 t DM ha^{-1}. - Total biomass in 5 communities: 6.1-15.9, 7.0-16.3, 8.8-13.9, 8.6-14.5, 8.6-16.2 t DM ha^{-1}	- SOM increase of 40-50% after 10 y - Water infiltration doubled in 15 years - Marked increase in total porosity - Stable pH - Exch. bases increase through the profile - Avg. forms of most nutrients stable	- Gradual elimination of most weeds, particularly broad-leaved - Exception: *Rottboellia* grows in places without *Mucuna*, threatening the system.	- Few serious pests or soil-borne pathogens

Continued.

Table 4. (Continued)

Chapter	GMCC biomass	Impact on soil chemical and physical properties	Impact on weed incidence	Relation with pests and diseases
4. Maize-*Mucuna*, Mexico	- Relay-cropped, slashed before winter maize: 2.8-3.2 t DM ha^{-1} - Relay-cropped, no winter maize: 3.7-5.1 t DM ha^{-1} - Rotational: 7.3 t DM ha^{-1}	- No data	- Farmers report fewer weeds.	- Adopting farmers consider that *Mucuna* helps control pests and diseases. - Farmers mention increased incidence of snakes and rats as a disadvantage of *Mucuna.*
5. Improved fallows, Zambia	- Not reported	- Not reported	- Lower weed pressure (including of *Striga*) reported as a benefit of the system.	- *Sesbania*: Insect or fungal attacks in the nursery; several pests and diseases, particularly Mesoplatys beetles - Pigeon pea: Insect attacks on seeds and flowers; poor seed quality often a problem - *Tephrosia*: fewer pest problems - In areas with pest problems, farmers mix fallow species to reduce the risk and impact.

Continued.

Table 4. (Continued)

Chapter	GMCC biomass	Impact on soil chemical and physical properties	Impact on weed incidence	Relation with pests and diseases
6. Best-bet legumes, Malawi	- *Mucuna* in rotation with maize: 5.7-6.7 t DM ha^{-1} - *Tephrosia* undersown in maize: 2.1-6.6 t DM ha^{-1} (early); 0.4-2.4 t DM ha^{-1} (late)	- Average residual N left for maize: Groundnut: 5.3-12.0 kg ha^{-1} Soybean: 5.9-8.5 kg ha^{-1} 2-yr pigeon pea: 24-107 kg ha^{-1} *Mucuna* rotation: 100 kg ha^{-1}	- Preliminary evidence of reduced incidence of *Striga* after *Mucuna*	- Grain-legume rotations commonly reduce pest and disease pressure. - Leguminous technologies may also increase incidence, emphasizing the need for diverse germplasm
7. *Marejea*, Tanzania	- No data	- No data	- Positive impact on weed infestation reported - Eradication of couch grass reported	- None reported
8. Forages, Kenya	- Intercropped with Napier grass, *Leucaena* yields 4.9-6.1 t DM ha^{-1} and *Clitoria* yields 4.8 t DM ha^{-1} - On-farm biomasses highly variable	- *Clitoria* increases soil N - *Leucaena* decreases soil N and P but improved soil moisture - Field observations indicate erosion control.	- Not reported	- Earlier, negative impact of psyllid on *Leucaena* production; natural enemies have presumably developed.
9. GMCCs, Uganda	- No data reported - Intercropping reduces biomass by 40-70%	- Short-term fallow (20% of annual crop land) + intercropping (50% of banana area) reduces N requirement to 30 kg ha^{-1} + makes balance positive under banana.	- Reduced weed pressure reported	- *Tephrosia* results in higher nematode pressure but controls mole rats.

Continued.

Table 4. (Continued)

Chapter	GMCC biomass	Impact on soil chemical and physical properties	Impact on weed incidence	Relation with pests and diseases
10. Maize-*Mucuna*, Benin	- Average biomass of 2.2-3.3 at 12 weeks after planting (WAP) and 4.3-5.3 t DM ha^{-1} at 20 WAP	- Farmer-managed fields at 12 WAP: average atmospheric N contribution 167 kg ha^{-1} but 7 of 34 fields had no nodulation.	- Reduction in *Imperata* an important reason for farmer interest - 1-year *Mucuna* fallow reported to reduce *Imperata* shoot density to almost one-tenth; confirmed in several studies - *Mucuna* accessions vary in *Imperata* suppression	- None reported
11. Rice-*Sesbania*, Thailand	- On-station biomasses: 2.54-3.68 t DM ha^{-1} - On-farm, highly variable (0.9-8.4 t DM ha^{-1}), with averages of 4.70, 1.79, and 3.04 t DM ha^{-1} in three districts	- No data - Some farmers able to reduce inorganic fertilizer consumption by half	- Researcher and farmer observations confirm lower weed pressure in the rice crop after *Sesbania*, with the 55-60 day *Sesbania* crop suppressing all weeds.	- Blister beetles feed on *Sesbania* flowers and damage seed production. - Problems with rodents

A number of factors are reported in the case studies to contribute to such variability:

- Climatic factors: In southern Benin (Chapter 10) and northern Thailand (Chapter 11), dry spells have caused poor GMCC performance. In southern Veracruz, Mexico (Chapter 4), *Mucuna* growth over the winter season is limited by the dry and often windy conditions. This is not surprising, given that the temporal GMCC niche is often one where main crop production is low (i.e., drier second or minor season, or the pre-planting period with some rainfall).
- Productivity in very degraded soils: In Malawi (Chapter 6) and Benin (Chapter 10), *Mucuna* growth has been poor in badly degraded soils. For example, in southern Benin, biomass production in soils from two degraded farmer fields was reduced by 58-74% when N was absent from complete fertilizer treatment and 28-66% when P was absent. Similarly in Veracruz, Mexico (Chapter 4), soil fertility has impacted *Mucuna* productivity. Although not surprising, this finding is disconcerting given the role of GMCCs in the system and the fact that degraded soils often are considered the most suitable green manure niche (Schulz et al. 2001). Productivity in degraded soils – the 'real fallow land' (Drechsel et al. 1996) - has also been reported to be a problem in Rwanda (Drechsel et al. 1996) and in Malawi (Snapp et al. 1998).
- Farmer management: In Veracruz, Mexico (Chapter 4) and Benin (Chapter 11), late planting dates (and in Veracruz, low planting densities) have been mentioned as factors contributing to low biomass production. Late planting dates have also been observed in northern Thailand (Chapter 11) where, due to the short growing period available for *Sesbania*, delayed planting is particularly harmful. Farmers practicing *fríjol tapado* (Chapter 2) on rented lands are said to favour shorter fallows and have little interest in investing in measures to improve the sustainability of land use.
- Persistence of GMCCs and mulch: In some locations persistence of GMCCs or mulch is poor. In southern Veracruz, Mexico (Chapter 4), significant reduction in the mulch layer can occur during the dry season due to windy conditions and livestock grazing; in addition, many farmers continue the practice of burning prior to the planting of the following maize crop. In southern Benin (Chapter 10), bush fires may decimate mulch. In eastern Zambia (Chapter 5) and in northern Malawi (Chapter 6), livestock form a threat to the GMCC species that survive the dry period. With *fríjol tapado* (Chapter 2), landowners' livestock may graze on bean residues on rented land. As discussed above, farmers can also remove the GMCC biomass, as in the case of grain legumes that are brought home for threshing. In contrast, in northern Honduras (Chapter 3), the semi-permanent dead mulch

component is very large (10.8-12.4 t ha^{-1}, or 45-69% of the total biomass) and, in southern Benin (Chapter 10), new growth as a response to intermittent dry-season rains is able to offset any losses. Fires and livestock are a common problem with mulch persistence (Weber 1996; Anthofer 2000; Oyewole et al. 2000). Improved integration in GMCCs with livestock systems has been commonly urged (Schulz et al. 2001) but is a challenging and still largely unresolved issue.

It is unclear how much GMCC productivity in general is constrained by preferential allocation of land and labour to non-GMCCs. In coastal Kenya (Chapter 8), farmers give low priority to fodder systems both in terms of their land and labour use; this factor was seen to reduce the performance and adoption potential of the improved forage systems. In Veracruz, Mexico (Chapter 4), suboptimal farmer management of *Mucuna* and its allocation to areas with poorest soil fertility may indicate similar prioritization. Given the fact that many GMCCs are inedible, this factor may constrain the adoption of GMCCs more widely also (e.g. Oyewole et al. 2000; Muhr et al. 2001). Of the cases in this volume, few studies have been conducted on farmer management; many such conclusions rely on field observations only.

Impact on Soil Characteristics Is Generally Positive When Measured

For a few of the systems, a great deal of information is available on the impacts on soil, particularly on soil chemical characteristics (Table 4). Research on soil issues in southern Brazil (Chapter 1) has been wide-ranging and has covered physical and chemical characteristics and soil biology; typically, positive impacts on soil have been reported. Two other case studies from the Latin American region, *fríjol tapado* (Chapter 2) and maize-*Mucuna* in northern Honduras (Chapter 3), also have been relatively well studied in regard to soil parameters and report positive impacts. For the other systems examined, reported data on impacts on soils is more limited. Few cases report detailed nutrient dynamic studies but a study in Malawi (Chapter 6), indicating very long N immobilization periods even when green pigeon pea leaves were combined with maize residues, is disconcerting. Many other agronomic studies have emphasized the positive soil impacts from GMCCs but Drechsel et al. (1996), in describing poor performance and low adoption of improved fallows and green manures in Rwanda, pointed to a limited amount of potentially recyclable nutrients because of high acidity and very low subsoil nutrient reserves of the local soils; in addition, rapid mineralization of N and K led to high leaching losses.

Similarly, data on positive impacts on soil physical and biological characteristics are available for only a few cases. In southern Brazil (Chapter 1), erosion rates with conservation tillage and GMCCs are estimated at 5% of those in conventional systems. In Veracruz, Mexico (Chapter 4), farmers that had adopted *Mucuna* identified reduction in erosion and moisture conservation as advantages of the *Mucuna* system. In coastal Kenya (Chapter 8), field observations suggest that tree species reduce erosion if planted across slopes and in northern Honduras (Chapter 3), use of a permanent mulch layer was considered to reduce erosion but it was believed that the thick mulch layer may contribute to localized landslides. In Uganda (Chapter 9), planting of *Mucuna* and *Lablab purpureus* (L.) Sweet were considered best for soil erosion.

Reduced Weed Pressure is Typically Reported

Most cases, such as those from Honduras (Chapter 3), Veracruz, Mexico (Chapter 4), Zambia (Chapter 5), Malawi (Chapter 6), Tanzania (Chapter 7), Uganda (Chapter 9) and Thailand (Chapter 11) report decreased weed pressure after GMCC cultivation. In the *fríjol tapado* system, no weeding is conducted at all and weed growth has been shown to be inversely related to the mulch thickness. Low weed pressure originates both from lower weed germination under the mulch layer and from the fact that non-bean species remain in seedling stages until bean pod formation. In southern Benin (Chapter 10), apparent reduction in *Imperata cylindrica* (L.) Raeusch. infestation was an important reason for early farmer interest and testing; such effect has later been confirmed in a number of studies.

An exception to generally favourable impacts on weeds has been *Rottboellia cochinchinensis* (Lour.) Clayton in northern Honduras, which has been able to utilize the weed-free spaces left in areas where *Mucuna* does not germinate and has become a threat to the system itself. And, interestingly, in Santa Catarina, Brazil (Chapter 1), debate continues on the relationship between reduced tillage with GMCCs and weed pressure, and the consequent effect on the need for herbicides. Several factors seem to determine whether weed pressure is greater or lower with reduced tillage, especially the farmer's crop husbandry skills. For now, at field level, weed pressure and consequent herbicide use are greater in many farms with reduced tillage.

Reported Problems with Pests and Disease Are Few

Pests seem to be a relatively minor problem in the systems described in this volume. In *fríjol tapado* (Chapter 2), a lower incidence of fungi but a higher incidence of slugs is reported and in Honduras (Chapter 3), few serious pests or soil-borne pathogens are evident. In Zambia (Chapter 5), several pests of the improved fallow species are mentioned except for *Tephrosia vogelii* Hook. f.. In at least southern Brazil (Chapter 1), Mexico (Chapter 4) and Thailand (Chapter 11), problems with rats are mentioned and in Veracruz, snakes as well. Despite the seeming resilience of GMCCs, one should not overlook the potential of disease and pest outbreaks. As discussed later, particular attention should be paid to having sufficiently wide genetic base in the GMCCs introduced to an area.

Information on Regional and Long-Term Impacts is Typically Lacking

Most cases lack data on long-term impacts and on impacts beyond the field and farm. Long-term data are available from Honduras (Chapter 3) where fields planted with *Mucuna* for up to 15 years have been studied and in the *fríjol tapado* system (Chapter 2) where one case of documented use of up to 80 years without detrimental effects on soil has been reported. Given the relatively short timespan of most of the efforts described in this volume, the lack of long-term data is not surprising.

Similarly, few regional impacts are discussed or described. Monitoring water discharge in the Chapecó micro-watershed in western Santa Catarina, Brazil (Chapter 1) over a 10-year period (1986-96) when GMCCs and conservation tillage increased greatly, showed increased discharge rates and greater water availability along with lower sediment load and improved water quality. The maize-*Mucuna* system is accredited for the stabilization of frontier agriculture in the northern Honduras region (Chapter 3). For the maize-*Mucuna* system in southern Benin (Chapter 11), a study found positive economic returns even at the regional level from second year onwards but noted that declining yield trends necessitated additional inputs with time. Within the province of Mono, *Mucuna* use was estimated to annually save 6.5 million kg of N, or US$1.85 million. Few studies are available on the impact of these systems on erosion rates; they are described above.

DIFFUSION FACTORS

Diffusion of GMCC technologies poses greater challenges than diffusion of many other agricultural technologies. GMCC systems are typically more complex and require more knowledge and management skills (Franzel et al. 2001; Morse and McNamara 2003). Extension approaches developed for simple technologies are not adequate for complex GMCC systems (Tripp et al. 1993; Drechsel et al. 1996; Snapp et al. 2002b).

Incentives May Harm Long-Term Adoption

To diffuse a number of the systems described in the case studies, incentives were used to facilitate GMCC introduction, including distribution of free seed, purchases of seed, credit, inputs and other goods. A common action—and often only a *de facto*, not an intended, incentive—has been the creation of artificial markets for GMCC seeds. This is understandable because, at the start of diffusion efforts, lack of seed is typically a major obstacle to overcome. The problem arises because farmers come to view the GMCCs as a cash crop whose markets fail—when the eventual and inevitable discontinuation of the artificial seed markets takes place. Table 5 summarizes information on the diffusion methods used, while Table 3 includes data, where available, on the impact of incentives on profitability.

Few studies have considered the impact of such incentives but it is clear that (as often also hoped) in many cases short-term profitability was improved. In the Ruvuma region of Tanzania (Chapter 7) for example, *marejea* seed could be sold or exchanged for inorganic fertilizers (100 kg of *marejea* for 6 bags of sulphate of ammonium). This created a great deal of enthusiasm and induced *marejea* cultivation at large scale, often as a sole crop. In southern Benin (Chapter 10), where large-scale extension efforts generated *Mucuna* seed markets, positive returns from *Mucuna* cultivation were found from the first year onwards if *Mucuna* seed could be sold; in the absence of seed markets, the first year was unprofitable. The cost-benefit ratio in Benin over an 8-year period was almost triple if *Mucuna* could be sold.

In other case studies also, incentives were used but few details are provided. Free GMCC seed was distributed in southern Veracruz, Mexico (Chapter 4), and only at the beginning of the extension efforts, in eastern Zambia (Chapter 5).[2] Seed was purchased from farmers in Veracruz at times at very attractive prices and occasionally in eastern Zambia. No data on their impact on economic profitability or adoption are available.

In several cases, these incentives led to greater short-term adoption but often they may have negatively impacted the potential of long-term adoption. This is most evident in the cultivation of *marejea* in the Ruvuma region of Tanzania (Chapter 7) where, when the seed purchases stopped, cultivation of the legume virtually came to a halt. Negative effects from seed purchases seem also evident with *Mucuna* in Veracruz, Mexico (Chapter 4) and in Benin (Chapter 10). Negative impacts of incentives on adoption have also been noted with, for example, agroforestry technologies in Panama (Fischer and Vasseur 2002) and elsewhere in Central America (Hellin and Schrader 2003). Carsky et al. (2003) advocate emphasis on true benefits, instead of incentives and a disfavourable view of incentives has become commonplace (Hudson 1991).

Table 5. Characterization of the diffusion efforts with green manure/cover crop (GMCC) systems in the case studies.

Chapter	Milestones and activities	Methods	Partners other than farmers	Approach
1. GMCCs-Cons. Till., Brazil	- 1978-90: State's integrated programme for soil and water conservation - 1983-: Increasing research - Mid-1980s-: Tobacco company promotion - 1987-: Federal government's micro-watershed management programme - 1991-99: State's Micro-watershed Project	- Farmer training courses - Work through agricultural cooperatives - Research and extension by tobacco companies - Municipally-sponsored field days - Rural extension (state, municipal governments, private enterprises and agricultural cooperatives) - Diverse initiatives of agro-industrial consortia	- Government research and extension - Cooperatives - Municipalities - Agro-industrial enterprises - International donor	- Large-scale, diverse partners, diverse media
2. *Tapado*, Costa Rica	- Pre-Columbian, based on spontaneous farmer diffusion - 1980s and 1990s: Research to understand and improve the system; little promotion of research results, limited diffusion efforts	- Spontaneous farmer adoption - On-farm research - Some diffusion work focusing on post-harvest processing, bean collection and distribution and organic bean production	- National and international universities - Local non-governmental organizations (NGOs) - International research institute	- Spontaneous farmer-to-farmer diffusion
3. Maize-*Mucuna,* Honduras	- 1970s: Introduced in the region - 1980s: Rapid spontaneous adoption, research to understand the system and determine adoption - 1990s: Rapid disadoption; adoption research	- Spontaneous farmer adoption - On-farm experiments and observational studies - Surveys	- International universities and centres - Local NGOs	- Spontaneous farmer-to-farmer diffusion

Continued.

Table 5. (Continued)

Chapter	Milestones and activities	Methods	Partners other than farmers	Approach
4. Maize-*Mucuna*, Mexico	- 1991-92: Preliminary on-farm research - 1992-93: One NGO initiated extension - 1994: NGO collaboration with state in extension - 1995: Two NGOs' diffusion efforts - 1996-99: One NGO focuses on on-farm research; another continues diffusion and research - 1998-99: Both NGOs collaborate with government in seed production; work of one evolves to cooperative	- Farmer extension workers - Seed distribution - Farmer meetings and field visits - Community-to-community visits	- Local NGOs - International centre - State and national government - Farmer extension workers	- Participatory, farmer-to-farmer diffusion
5. Improved fallows, Zambia	- 1987-: On-station research - 1992-: On-farm trials - 1996-: Large-scale dissemination efforts	- On-station and on-farm trials - Farmer-to-farmer training and farmer field days - Demonstration fields and on-station field days - Farmer field schools - Radio, theatre, songs, video - Distribution of leaflets, brochures, etc.	- Farmers' clubs and cooperatives, women's clubs - Government ministry - International research centre, NGOs and university - Churches, church associations - Volunteer groups	- Participatory, emphasis on giving options to farmers
6. Best-bet legumes, Malawi	- 1990s: Research efforts - 1995-: Coordination of research-diffusion	- Maize Productivity Taskforce - On-farm and on-station research	- Local research and extension - Int'l research centres	- Nationwide research/ demonstration

Continued.

Table 5. (Continued)

Chapter	Milestones and activities	Methods	Partners other than farmers	Approach
7. *Marejea*, Tanzania	- 1940s: Introduced to area - 1970s-early 1990s: Large-scale promotion - 1994: Seed exchange discontinued; *marejea* one of many technologies in organic production	- Village-level promoter - Seed distribution and borrowing - Seed purchases and exchange for fertilizer - Radio - Involvement of politicians and mission	- Catholic missionaries - Village promoters - Politicians - Few researchers	- Emphasis on seed production, exchanging with fertilizer
8. Forages, Kenya	- 1970s: Screening - 1988: Research and development	- Research and extension - On-farm trials and demonstrations - Farmer training and meeting	- National research and extension organization - International research centre	- Emphasis on interaction with farmers
9. GMCCs, Uganda	- 1992-: On-farm efforts, on-station research	- Meetings - On-station and on-farm trials - Seed exchange, printed materials, agricultural shows, extension, mini-kits, seed sales through stockists	- National research and extension organization - International research centre	- Participatory, interdisciplinary, systems-oriented
10. Maize-*Mucuna*, Benin	- 1986-89: On-farm research - 1990-92: Pre-extension tests by government extension - 1992-96: Large-scale extension by government extension with international NGO collaboration; research efforts continued	- Demonstration trials - Demonstrations - Credit - Seed purchase and distribution - Brochure - On-station and on-farm research	- National research and extension organization - International NGO - International research centre	- Initial efforts: more participatory
11. Rice-*Sesbania*, Thailand	- 1993-: On-station research - 1997-: On-farm research	- On-station and on-farm research - Work with farmer groups or individual farmers	- National university	- Participatory - Focus on the intensification of the rice system

Credit is just another incentive and often may have similar consequences, as in southern Benin (Chapter 10), where a package of improved maize variety, inorganic fertilizer and *Mucuna* was available for farmers with attractive credit but when discontinued also led to declined interest in the *Mucuna* system. Interestingly, while in southern Brazil (Chapter 1), incentives for soil conservation practices have been provided, they are not considered an important factor in terms of dissemination of the system.

Importance of Efforts that are Long-Term, Concerted, Flexible and Diverse

Flexibility and a problem-solving approach in diffusion efforts on organic technologies have been urged to ensure that technology development is driven by smallholders' needs (Kumwenda et al. 1997; Snapp et al. 2002a). In addition, long-term commitment and pursuit of a diversity of options are important. An FAO review of soil conservation projects emphasized the importance of duration. Of those projects that were considered successful, 75% lasted 5 years or more (Hudson 1991). In contrast, much of the work on GMCCs in the 1980s and 1990s took place in the context of small projects because, for many projects, working with various partners over large areas would have required unavailable human and financial resources (Snapp et al. 2002a).

The long-term, focused commitment of many institutions that offer diverse options for farmers seems to have been particularly successful in two cases. In Santa Catarina, Brazil (Chapter 1), the high adoption rates of GMCCs and conservation tillage can be largely attributed to the involvement of many players in the development and promotion of the technologies. A multi-institutional approach, including the private sector, farmer organizations and extension and research organizations, enabled the development of the systems and the equipment for draught animals and tractors that they necessitated. These efforts were consolidated and given a greater focus by the Micro-watershed Basin Management Project of Santa Catarina State. A multitude of species and systems, corresponding to the diverse agro-ecology, cropping systems and farmer needs, were screened, researched and introduced to the farmers. What also enabled such efforts were policy-related factors that have been cited as particularly important in adoption of such technologies, namely support to public sector research, to extension services and for getting inputs to the farms (Tripp et al. 1993). Similarly, in Zambia (Chapter 5), a regional consortium with diverse organizations, from NGOs to government extension agencies, research organizations, church organizations and farmers, is successfully promoting the improved fallow systems. The consortium members used an array of extension approaches, including farmer-to-farmer extension, farmer field schools, rural radio, community theatre and training videos. In southern Benin (Chapter 10), initial efforts focused on several options to farmers, including fertilizer, alley cropping and pigeon pea but because of the farmer interest in *Mucuna*, efforts on the others decreased in importance.

In contrast, several other cases describe efforts that are shorter in duration and rather than focusing on multiple options, center on a single species.

Availability of Seed Constrains Adoption of GMCC Systems

Availability of appropriate legume seed commonly has been cited as a constraint to GMCC adoption, as in lowland rice systems (Becker et al. 1995), in Nigeria (Oyewole et al. 2000; Muhr et al. 2001; Morse and McNamara 2003), as well as generally (Meelu et al. 1994). This has also been the case for most of the case studies, for example, those from Brazil (Chapter 1), Zambia (Chapter 5), Malawi (Chapter 6), Kenya (Chapter 8), Uganda (Chapter 9) and Thailand (Chapter 11) but interestingly, the demand for GMCC seeds in the *Mucuna* system in Honduras (Chapter 3) is low and for the *fríjol tapado* system, nil, because of the importance of natural reseeding. Insufficiency of seed can be due to a number of factors. Most commonly, it is a constraint occurring particularly at the beginning of diffusion efforts, when the number of farmers is expanding but seed production and supply methods have not been formalized. Seeds of most GMCC species are not available commercially because existing demand is usually not sufficient for seed companies to invest in their production.

Insufficiency of seed can also be caused by a number of system-, location- or species-specific factors. In some GMCC systems, legume biomass is incorporated at flowering, and such systems do not allow for seed harvesting. In certain areas, seed production may be poor because of environmental factors or pests and diseases. Species-related problems may include seed shattering and dormancy, which are common in legumes (Weber 1996).

Surprisingly, after almost two decades of efforts, seed still continues to be a major bottleneck to greater cultivation of GMCCs in Santa Catarina, Brazil (Chapter 1). However, there are positive signs, including the commercial interest to produce GMCC seed, specialization of some farmers in seed production and the contracts provided by some agricultural commodity chains for GMCC seed production.

In many of the cases described, as in Mexico (Chapter 4), Zambia (Chapter 5), Tanzania (Chapter 7) and Benin (Chapter 10), farmers were initially provided with seed, often after buying the seed from the region's other farmers (later, in Zambia, seed was given on a loan basis). Soon, farmers in several locations were developing their own methods of seed production, indicating that with further experimentation, location-specific solutions may be found. This has been the case in Zambia (Chapter 5), where *Gliricidia sepium* (Jacq.) Walp. trees are planted along borders and in the improved fallow plots; when coppiced, *Gliricidia* also produces seed in the improved fallow. In Thailand (Chapter 11), a farmer group utilized banks of irrigation canals and upper slopes to produce *Sesbania* seed.

Base of Germplasm Options is Narrow in Most Cases

Most of the case studies in this volume describe efforts centring on one or a few GMCC species. Wide adoption of a single species in one or two cropping systems should perhaps not be expected. Nor, as pointed out with improved fallows in the case from Zambia (Chapter 5), should maintenance of a narrow genetic base be considered desirable because of its potential negative consequences in the even of pest and disease occurrences. In contrast, in Santa Catarina, Brazil (Chapter 1), a large number of GMCC species, both legume and non-legume, conventional and non-conventional, have been screened, introduced to farmers and adopted.

Further progress in species selection and germplasm management in GMCC technologies is likely to have high payoffs (Becker et al. 1995). Indeed, detailed assessment of site-specific constraints—which is commonly advocated before the initiation of any on-farm research/development project—is only valuable if a large array of diverse technologies is available for diffusion efforts (Weber 1996). Few available studies have evaluated a large number of GMCC species and in general, only a small proportion of the legume collections currently held in forage gene banks—many of which could presumably be effective GMCCs—have been tested (Thomas and Sumberg 1995). An exception to this approach is Becker and Johnson (1998), who tested 50 accessions representing 39 species in four rice production environments in Côte d'Ivoire. Promising approaches include also those by Wheeler et al. (1999), Qi et al. (1999) and Keatinge et al. (1999), who used modelling of physiological responses to identify cover crops for suitable environments.

Participatory Methods are Considered Promising

The past decade has seen increasing emphasis on participatory approaches to technology diffusion and the case studies reflect this with most mentioning reliance on participatory methods (as evident in Table 5). In the absence of detailed studies comparing various extension methods and the particular difficulties of assessing such context-sensitive methods as participatory research (Martin and Sherington 1997), it is hard to draw conclusions on the impact of such methods on adoption. Many authors comment on their positive experiences with participatory research and extension methods, including the value of constant feedback from farmers. Researchers in Uganda (Chapter 9) and in northern Thailand (Chapter 11), for example, report positive experiences with participatory research methods with farmer groups. In southern Benin (Chapter 10), initial research efforts were participatory and undertaken with constant farmer feedback; however, the later large-scale diffusion efforts have been criticized as top-down. In Veracruz, Mexico (Chapter 4), experiences with the diffusion efforts through farmer extension agents were mainly positive but difficulties were experienced because of, for example, political and religious divisions within communities. Similarly, Morse and McNamara (2003) found that direct involvement with on-farm trials was a factor linked with adoption. Improved integration of farmer perspectives into the research process is likely to lead to widening researchers' views on the impacts and management of new technologies (Kristjanson et al. 2002).

Farmer Experimentation Gives Valuable Feedback

It can be expected that a range in farmer management and in consequent performance of the GMCC system would typically be found and, as has been discussed earlier, this variability also characterizes the GMCC systems in this volume. Part of this comes from farmer experimentation and part from factors such as delays with planting, which may at times originate from farmers giving low priority to GMCCs (discussed above). The first is desirable, as it ensures the adaptation of GMCC technologies to the farmer's varied circumstances (Tripp et al. 1993) and these modifications are sources of potential improvements and give insights into farmers' perceptions (Carsky et al. 2003). Such modifications also affirm that farmers evaluate components and management options rather than cropping patterns (Fujisaka 1991). The second may be necessary and also offers lessons for those involved in research and diffusion. It may also impact performance so negatively that no sustained adoption occurs, as in Rwanda, where alley cropping was modified into cultivation of shrubs in abandoned fields and where wide alleys were tested due to land scarcity (Drechsel et al. 1996). Such factors and their impact on productivity have hardly been studied at the farm level and most of the information relies on field observations.

In many of the cases described in this volume, farmers modified components of GMCC systems to suit their circumstances:

- Brazil (Chapter 1): Farmers and farmer organizations have been active partners in the diffusion process and modified and designed GMCC systems in addition to inventing machinery suitable for such systems.
- Honduras (Chapter 3): The management of the maize-*Mucuna* system is considered remarkably uniform, with some variation in timing of

Mucuna slashing and maize planting, choice and timing of weed control, density of maize planting and by the extent to which *Mucuna* self-reseeding is complemented by manual reseeding.
- Veracruz, Mexico (Chapter 4): Farmers have experimented with *Mucuna* management, for example, by slashing it prior to maturity or using herbicides to control aggressive self-reseeding, by allowing it to regenerate mainly through self-reseeding and by utilizing seed plots.
- Zambia (Chapter 5): Some farmers did not raise their tree seedlings in plastic pots or bags in nurseries as recommended but transplanted bare root seedlings. Others direct-seeded in the field. This probably saved on input costs and labour but was likely to slow establishment of the improved fallows. Farmers also experimented with pest control by cropping multiple, instead of single, improved fallow species and introduced the practice of intercropping food crops in the initial year.
- Kenya (Chapter 8): Farmers established *Leucaena leucocephala* (Lam.) De Wit and *Gliricidia* hedgerows along farm boundaries and plot borders instead of in the alleys as recommended. They also planted legume trees with coconut (*Cocos nucifera* L.) and cashew (*Anacardium occidentale* L.) and utilized wider spacings and variable cutting heights for Napier grass (*Pennisetum purpureum* Schumach.).
- Uganda (Chapter 9): Farmer experimentation introduced many new options for GMCC integration, such as intercropping with food crops and cultivation of widely spaced *Tephrosia* for mole rat control.
- Benin (Chapter 10): Some farmers were reported to rely on self-reseeding of *Mucuna*, slashing it during the subsequent growing season to control its growth.

Other researchers (e.g. Kanmegne and Degrande 2002) have reported similar modifications.

CONCLUSIONS

A principal motivation behind the documentation of these case studies was to better understand the on-farm performance of GMCC systems and factors that seem to foster and limit their performance. Put simply, those who initiated these case studies and their analyses wanted to find out whether these systems are capable of 'working' and seem to 'work'—can they and do they improve food security, agricultural productivity and profitability, and sustainability of farming? And, in situations of variable performance, what factors would seem to be particularly important in making systems 'work' or not?

Simplifying again, the answer for most of the cases documented here is 'yes'; the systems are capable of resulting in higher main crop yields, better profitability and improved soil conditions. Many studies confirm that they can 'work'. They, or some of them, also seem to work in certain conditions of smallholder farmers, as evidenced by the sustained adoption of some. Clearly, farmers in southern Brazil, northern Honduras and Costa Rica have found better profitability and sustainability of their farming through the practice of such a system.

The often low and decreasing adoption of the systems in this volume would seem to contradict the assertion that GMCC systems do 'work' at farm level. In fact, as evidenced by these case studies and other experiences, many research-development efforts on GMCC systems have not led to their widespread adoption. For a variety of reasons, the hoped-for farm-level impact has not been realized at larger scale and in a multitude of locations.

Below, we summarize the main lessons regarding success and failure of GMCC systems. While derived from the case studies, they also integrate some of the experiences of the editors of this volume. These main lessons are separated into three groups, those that are technology- and farming systems-related, those that are related to the way the diffusion efforts are conducted and those that are market- and policy-related.

Technology and Farming System Factors

Three factors are particularly important in favouring adoption. First, the GMCCs should be labour saving and preferably reduce labour inputs. The success of the southern Brazil experience with GMCCs and conservation tillage (Chapter 1) primarily originates from the labour-saving attributes of conservation tillage; experiences with *fríjol tapado* and with maize-*Mucuna* in northern Honduras (Chapter 3) also emphasize the importance of labour considerations. In contrast, in Malawi (Chapter 6), Uganda (Chapter 9), and Thailand (Chapter 11), incorporation of GMCC biomass is considered arduous and a factor limiting the potential of adoption and impact of GMCCs. Second, as evident in the cases of Malawi (Chapter 6) and Benin (Chapter 10), in areas of food insecurity and/or land pressure, only GMCCs that are edible by human beings or livestock or that are readily marketable should be introduced. Where these conditions do not exist, non-edible GMCCs seem successful, such as in Costa Rica (Chapter 2) or northern Honduras (Chapter 3). Third, whether or not livestock are integrated in the technology design and diffusion process, they are likely to be a part of its success or failure; for improved orientation in technology design and diffusion, livestock should be considered an opportunity, not an obstacle, for GMCC introduction. GMCC species with high seed yield, such as *Mucuna*, can fulfil multiple functions in soil improvement (if foliage is left as cover) and ruminant animal feed (if seed or pods are fed to ruminants). And, finally, multiple and preferably direct benefits are important, with many authors of this volume and others having emphasized that GMCCs should provide benefits beyond soil fertility.

Diffusion Factors

Future diffusion efforts on GMCCs should have a long-term commitment. Some projects in this volume attempted to achieve their goals in a few years' timeframe, often unsuccessfully. In contrast, in southern Brazil (Chapter 1), efforts have now lasted 25 years and are still ongoing. These efforts have also involved diverse organizations and institutions that have contributed to technology development. Importantly, as bottlenecks developed, there was time to try solving them. Long-term commitment is needed for a number of reasons, for example, because of the knowledge-intensive nature of GMCC technologies, along with their farm-level complexity and the often poor availability of germplasm and insufficient information on it. Field-level experiences and farmer feedback generate ideas for improvements and time is needed to solve problems in, for example, equipment or other technology needed. Moreover, if diverse actors are engaged, it helps ensure that the multitude of efforts needed is done and a large numbers of farmers reached. Working with extension services is imperative but may not be enough to reach farmers in the most effective way. Finally, many cases in this volume indicate negative impact of incentives on long-term adoption.

Market and Policy Factors

In the market and policy sphere, three factors are particularly influential. Almost without exception, the GMCCs in these case studies are tied with subsistence crops that currently have low prices. Yield improvement through GMCC integration is made far more attractive if crops or products that fetch higher prices in the market are used, such as milk in coastal Kenya (Chapter 8) or high-quality rice in Thailand (Chapter 11). Second, the southern Brazil (Chapter 1) case study demonstrates the importance of the commitment of policy-makers for ensuring that actors at different levels work towards similar goals. Conflicting policy messages, such as are evident for intensification and extensification options in southern Veracruz, Mexico (Chapter 4) are often more common. Finally, most studies of the GMCC systems in this volume point to the importance of secure land tenure in fostering GMCC adoption and impact.

NOTES

1. In other situations where edibility by livestock constitutes a threat to persistence of the GMCCs, authors have pointed out the advantage of an inedible GMCC, as discussed for *Tephrosia* in Malawi.

2. In Veracruz, Mexico, additional incentives were offered during 1 year only when the NGO collaborated with state government.

REFERENCES

Anderson S, Gündel S, Pound B, Triomphe B. Cover crops in smallholder agriculture: Lessons from Latin America. ITDG Publishing, London. 136 p.

Anthofer J (2000) Farmers' experiences with *Mucuna* cover crop systems in Ghana. *In*: Carsky R J, Etèka A C, Keatinge J D H, Manyong V M (Eds.) Cover crops for natural resource management in West Africa. Proceedings of a workshop organized by the International Institute of Tropical Agriculture (IITA) and the Centre for Cover Crops Information and Seed Exchange in Africa (CIEPCA), 26-29 October 1999. CIEPCA, IITA and the Cornell International Institute for Food, Agriculture and Development (CIIFAD), Ibadan, Nigeria. p. 120-130.

Bannister M E, Nair P K R (2003) Agroforestry adoption in Haiti: The importance of household and farm characteristics. Agrofor. Syst. 57:149-157.

Becker M, Johnson DE (1998) Legumes as dry season fallow in upland rice-based systems of West Africa. Biol. Fertil. Soils 27:358-367.

Becker M, Ladha J K, Ali M (1995) Green manure technology: Potential, usage, and limitations. A case study for lowland rice. Plant Soil 174:181-194.

Boonman, J G (1999) Green-and organic manures in mixed-farming systems of East Africa. Report of Boma Consult, The Hague, the Netherlands. 55 p.

Buckles, D (1995) Velvetbean: A 'new' plant with a history. Econ. Bot. 49:13-25.

Cairns M, Garrity D P (1999) Improving shifting cultivation in Southeast Asia by building on indigenous fallow management strategies. Agrofor. Syst. 47:37-48.

Carsky R J, Becker M, Hauser S (2001) *Mucuna* cover crop fallow systems: Potential and limitations. *In:* Tian G, Ishida F, Keatinge J D H (Eds.) Sustaining soil fertility in West Africa. Soil Science Society of America (SSSA) Special Publication No. 58. SSSA and American Society of Agronomy, Madison, Wisconsin. p. 111-135.

Carsky R J, Douthwaite B, Manyong V M, Sanginga N, Schulz S, Vanlauwe B, Diels J, Keatinge J D H (2003) Amélioration de la gestion des sols par l'introduction de légumineuses dans les systèmes céréaliers des savanes Africaines. Cah. Agric. 12:227-233.

Chettri G B, Ghimiray M, Floyed C N (2003) Effects of farmyard manure, fertilizers and green manuring in rice-wheat systems in Bhutan: Results from a long-term experiment. Exp. Agric. 39:129-144.

Drechsel P, Steiner K G, Hagedorn F (1996) A review on the potential of improved fallows and green manures in Rwanda. Agrofor. Syst. 33:109-136.

Eilittä M, Sollenberger L E (2001) The many uses of *Mucuna*: Velvetbean in the southern United States in the early 20th century. *In*: Flores B, Eilittä M, Myhrman R, Carew L B, Carsky R J (Eds.) Food and feed from *Mucuna*: Current uses and the way forward. Centro Internacional de Información sobre Cultivos de Cobertura (CIDICCO), Center for Cover Crops Information and Seed Exchange in Africa (CIEPCA) and World Hunger Research Center, Judson College, Tegucigalpa, Honduras. p. 73-100.

Enyong L A, Debrah S K, Bationo A (1999) Farmers' perceptions and attitudes towards introduced soil-fertility enhancing technologies in western Africa. Nutrient Cycl. Agroecosyst. 53:177-187.

Fischer A, Vasseur L (2002) Smallholder perceptions of agroforestry projects in Panama. Agrofor. Syst. 54:103-113.

Franzel S (1999) Socioeconomic factors affecting the adoption potential of improved tree fallows in Africa. Agrofor. Syst. 47:305-321.

Franzel S, Coe R, Cooper P, Place F, Scherr S J (2001) Assessing the adoption of agroforestry practices in sub-Saharan Africa. Agric. Syst. 69:37-62.

Fujisaka S (1991) Improving productivity of an upland rice and maize system: Farmer cropping choices or researcher cropping pattern trapezoids. Exp. Agric. 27:253-261.

Fujisaka S (1994) Learning from six reasons why farmers do not adopt innovations intended to improve sustainability of upland agriculture. Agric. Syst. 46:409-425.

Hellin J, Schrader K (2003) The case against direct incentives and the search for alternative approaches to better land management in Central America. Agric. Ecosyst. Environ. 99(1-3):61-81.

Hudson N W (1991). A study of the reasons for success or failure of soil conservation projects. FAO Soils Bulletin 64. Food and Agriculture Organization of the United Nations (FAO), Italy, Rome.

IIA – International Institute of Agriculture (1936) Use of leguminous plants in tropical countries as green manure, as cover and as shade. IIA, Rome, Italy.

Kanmegne J, Degrande A (2002) From alley cropping to rotational fallow: Farmers' involvement in the development of fallow management techniques in the humid forest zone of Cameroon. Agrofor. Syst. 54:115-120.

Kaya B, Hildebrand P E, Nair P K R (2000) Modelling changes in farming systems with the adoption of improved fallows in southern Mali. Agric. Syst. 66(1):51-68.

Keatinge J D H, Qi A, Wheeler T R, Musitwa F, Franks N A P, Kuribuza D, Ellis R H, Summerfield R J (1999) Selecting potential annual-sown legumes for cover crops, trash lines, mulching material, green manure or short-fallow replacement in low-input systems in the east African Highlands: The example of S.W. Uganda. Mountain Res. Dev. 19:345-53.

Kristjanson P, Place F, Franzel S, Thornton P K (2002) Assessing research impact on poverty: The importance of farmers' perspectives. Agric. Syst. 72:73-92.

Kumwenda J D T, Waddington S R, Snapp S S, Jones R B, Blackie M J (1997) Soil fertility management in southern Africa. *In*: Byerlee D, Eicher, C (Eds.) Africa's emerging maize revolution. Lynne Rienner Publishers, Boulder, Colorado, USA. p. 157-172.

Martin A, Sherington J (1997) Participatory research methods–implementation, effectiveness and institutional context. Agric. Syst. 55(2):195-216.

Meelu O P, Yadvinder-Sing, Bijaya-Sing (1994) Green manuring for soil productivity improvement. World Soil Resources Report 76. Food and Agricultural Organization of the United Nations (FAO), Rome, Italy.

Morse S, McNamara N. (2003) Factors affecting the adoption of leguminous cover crops in Nigeria and a comparison with the adoption of new crop varieties. Exp. Agric. 39:81-97.

Muhr L, Tarawali S A, Peters M, Schultze-Kraft R (2001) Acceptability of forage legumes for improved fallows – first experiences of agro-pastoralists in sub-humid southwest Nigeria. Exp. Agric. 2001: 495-507.

Oyewole B, Carsky R J, Schulz S (2000) On-farm testing of *Mucuna* and cowpea double cropping with maize in the Guinea savanna of Nigeria. *In*: Carsky R J, Etèka A C, Keatinge J D H, Manyong V M (Eds.) Cover crops for natural resource management in West Africa. Proceedings of a workshop organized by the International Institute of Tropical Agriculture, (IITA) and the Centre for Cover Crops Information and Seed Exchange in Africa (CIEPCA), 26-29 October 1999. CIEPCA, IITA and the Cornell International Institute for Food, Agriculture and Development (CIIFAD), Ibadan, Nigeria. p. 138-147.

Pannell D J (1999) Social and economic challenges in the development of complex farming systems. Agrofor. Syst. 45:393-409.

Pattanayak S K, Evan Mercer D, Sills E, Yang J C (2003) Taking stock of agroforestry adoption studies Agrofor. Syst. 57(3): 173-186.

Pound B, Anderson S, Gündel S (1999) Species for niches: When and for whom are cover crops appropriate? Mountain Res. Dev. 19(4): 307-312.

Qi A, Wheeler T R, Keatinge J D H, Ellis R H, Summerfield R J, Craufurd P Q (1999) Modelling the effects of temperature on the rates of seedling emergence and leaf appearance in legume cover crops. Exp. Agric. 35:327-44.

Schulz S, Carsky R J, Tarawali S A (2001) Herbaceous legumes: The panacea for West African soil fertility problems? *In*: Sustaining soil fertility in West Africa. SSSA Special Publication no. 58. Soil Science Society of America and American Society of Agronomy. p. 175-195.

Schulz S, Honlonkou A N, Carsky R J, Manyong V M, Oyewole B D (2003) Alternatives to *Mucuna* for soil fertility management in southern Bénin: Farmer perception and use of traditional and exotic grain legumes. Exp. Agric. 39:267-278.

Snapp S S, Magongooya P L, Waddington S (1998) Organic matter technologies for integrated nutrient management in smallholder cropping systems of southern Africa. Agric. Ecosyst. Environ. 71:185-200.

Snapp S S, Kanyama-Phiri G, Kamanga B, Gilbert R, Wellard K (2002a) Farmer and researcher partnerships in Malawi: Developing soil fertility technologies for the near-term and far-term. Exp. Agric. 38:411-431.

Snapp S S, Rohrbach D D, Simtowe F, Freeman H A (2002b) Sustainable soil management options for Malawi: Can smallholder farmers grow more legumes? Agric. Ecosyst. Environ. 91:159-174.

Steinmaier N (2001) Potential of pasture legumes in low-external-input and sustainable agriculture (LEISA). 2. Farmer adaptation of starter technology by farmer research groups in Luapula Province, Zambia. Exp. Agric. 37:309-317.

Sumberg J (2002) The logic of fodder legumes in Africa. Food Policy 27(3):285-300.

Swinkels R A, Franzel S, Shepherd K D, Ohlsson E, Ndufa J K (1997) The economics of short rotation improved fallows: Evidence from areas of high population density in Western Kenya. Agric. Syst. 55(1): 99-121.

Tarawali G, Manyong V M, Carsky R J, Vissoh P V, Osei-Bonsu P, Galiba M (1999) Adoption of improved fallows in West Africa: Lessons from *Mucuna* and *Stylo* case studies. Agrofor. Syst. 47(1-3):93-122.

Thomas D, Sumberg J E (1995) A review of the evaluation and use of tropical forage legumes in sub-Saharan Africa. Agric. Ecosyst. Environ. 54:151-163.

Totongnon J K, Lame B, Zannou P (2000) Quelques performances et contraintes d'adoption du pois d'Angole (*Cajanus cajan*) et du pois mascate (*Mucuna pruriens* variété *utilis*) en milieu paysan dans le Départament de l'Atlantique, République du Bénin. *In*: Carsky R J, Etèka A C, Keatinge J D H, Manyong V M (Eds.) Cover crops for natural resource management in West Africa. Proceedings of a workshop organized by the International Institute of Tropical Agriculture (IITA) and the Centre for Cover Crops Information and Seed Exchange in Africa (CIEPCA), 26-29 October 1999. CIEPCA, IITA and the Cornell International Institute for Food, Agriculture and Development (CIIFAD), Ibadan, Nigeria. p. 109-119.

Tripp R, Buckles D, van Nieuwkoop M, Harrington L (1993) Land classification, land economics, and technical change: Awkward issues in farmer adoption of land-conserving technologies. Presented at the Seminario/Taller Internacional para la definición de una metodología de evaluación de tierras para una agricultura sostenible en México, 10-13 August 1993, El Batán, Mexico.

Vanlauwe B, Aihou K, Iwuafor E N O, Houngnandan P, Diels J, Manyong V M, Sanginga N (2000) Recent developments in soil fertility management of maize-based systems: The role of legumes in N and P nutrition of maize in the moist savanna zone of West Africa. *In*: Carsky R J, Etèka A C, Keatinge J D H, Manyong V M (Eds.) Cover crops for natural resource management in West Africa. Proceedings of a workshop organized by the International Institute of Tropical Agriculture (IITA) and the Centre for Cover Crops Information and Seed Exchange in Africa (CIEPCA), 26-29 October 1999. CIEPCA, IITA and the Cornell International Institute for Food, Agriculture and Development (CIIFAD), Ibadan, Nigeria. p. 3-13.

Weber G (1996) Legume-based technologies for African savannas: Challenges for research and development. Biol. Agric. Hortic. 13:309-333.

Wheeler T R, Qi A, Keatinge J D H, Ellis R H, Summerfield R J (1999) Selecting legume cover crops for hillside environments in Bolivia. Mountain Res. Dev. 19(4): 318-324.

Future Perspectives

The Editors

In the coming years, many of the forces that are now causing soil degradation in the developing countries—population growth, increasing land pressure, shortening fallows and poor access to fertilizers—will certainly continue and are likely to intensify. Green manure/cover crop (GMCC) systems are one possible option for sustainable cropping in the future.

The success—when measured either in terms of impact or adoption—of many of the GMCC systems described in this volume has been mixed and in some cases it has been quite poor. Future contribution of GMCC systems on the productivity and sustainability of smallholder farming can be improved by learning from these experiences. The following seven recommendations for directions in research and diffusion efforts on GMCC systems are derived from the case studies.

EMPHASIZE DIVERSITY OF OPTIONS

Much of the diffusion work in the GMCC systems in the 1980s and 1990s took place within the context of projects working in a limited geographical area, such as a few communities or a district. Such efforts have a number of shortcomings that inhibit large-scale impact. They typically focus on just one, or at most a few, GMCCs and main crops. Farmers, on the other hand, typically cultivate a number of main crops both within a farm and a region. Many such localized projects also are not able to conduct adequate screening of GMCC germplasm, nor are they able to carry out the adaptive systems-focused research necessary to develop locally appropriate GMCC systems. They often also have short-term and insecure funding, another obstacle to lasting and widespread impact. These features have characterized several of the efforts that have been presented in this volume.

Lasting and wider positive impacts of GMCC systems are more likely to be felt if efforts are taking place at a larger scale and if they focus on diverse technologies. Such efforts need not be centrally directed but can consist of

M. Eilittä et al. (eds.), Green Manure/Cover Crop Systems of Smallholder Farmers, 349-353.

small projects working under a common umbrella. The ability to offer farmers a menu of GMCC options over a long time-frame has characterized the successful efforts in the state of Santa Catarina, and more widely in southern Brazil, which have led to wide adoption, improved profitability of farming and reduced erosion. Diversity should be sought in both the main crop (i.e. subsistence/cash orientation, annual/perennial nature and conventional/ organic systems) as well as in the GMCC (i.e. various species and varieties of both legumes and non-legumes). Particular attention should be paid to systems with higher profitability. As the world prices of major grains are likely to continue to be low, the profitability of such systems can be improved by linking them with high-value crops.

EXPLOIT GMCC GERMPLASM DIVERSITY

Available GMCC germplasm needs to be better exploited in future efforts. In southern Brazil, screening work at large scale preceded the diffusion efforts on GMCCs. An example of these efforts is the German Technical Cooperation-Instituto Agronômico do Paraná (GTZ-IAPAR) project in Paraná, which alone screened more than 100 plant species in 10 different regions of the state between 1977 and 1984; at least 10-15 of these species are now widely used in the southern states of Brazil. The project consciously sought species from other locations, even those that seemed unlikely at the time—such as oilseed radish (*Raphanus sativus* [L.] var. *oleiferus*) from Germany—and kept a wide definition of cover crops by bearing in mind that any species, either legume or non-legume, given abundant biomass and adequate soil cover can act as a GMCC. (Despite such efforts, quality and in some cases availability of germplasm remains a problem in the region.)

Most efforts to date, including many of those presented in this volume, have focused on only a few GMCC species and/or accessions within species. This can bring about a number of problems, including utilization of species or accessions that are not well adapted to the location, that have an increased risk of pest/disease outbreak and that may offer too limited and narrow choices for farmers. Disappointingly little information is available on diverse accessions of even the most popular GMCC species and of that which is available, little has been documented in a systematic fashion. Currently active networks in GMCC work, such as the International Centre for Information on Cover Crops (CIDICCO, the Spanish acronym), the Legume Research Network Project (LRNP) and the Soil Fertility Management and Policy Network for Maize-based Farming Systems in Southern Africa (Soil Fertility Network), conduct some GMCC screening and multiplication but their resources for this work are limited. A similar situation regarding GMCC screening and multiplication characterizes those Consultative Group on International

Agricultural Research (CGIAR) centres that conduct research on relevant themes (e.g. International Institute for Tropical Agriculture [IITA], International Livestock Research Institute [ILRI], International Center for Tropical Agriculture [CIAT, the Spanish acronym] and International Plant Genetic Resources Institute [IPGRI]).

We therefore propose improved coordination of GMCC screening and multiplication. This could be achieved regionally but ensuring that collaboration across regions would take place. An existing centre would ideally be charged with coordination. Such a centre could collaborate in screening efforts with partner institutions located in diverse agro-ecologies and should be responsible for documentation of the research results. When appropriate, it could multiply or coordinate multiplication of GMCC seed and keep track of GMCC seed markets on a global scale.

INCORPORATE MINIMUM OR NO TILLAGE

Many of the experiences in this volume have focused on inserting GMCCs in the local farming systems, which often have included tillage. A more varied and positive effect of GMCCs can be expected when reduced or no-tillage technology is applied together with GMCCs, particularly in short- or long-term fallow systems. In warm climates, tillage results in the mineralization of organic matter and consequent release of CO_2 in the atmosphere; in contrast, in no-tillage systems, soil is transformed into a carbon sink. Utilization of cover leads to improved moisture conservation and reduced erosion that has both farm-level and wider impacts. Importantly, reduced tillage decreases both labour use and cost of production. A breakthrough in the no-tillage technology in Brazil was achieved by the combination of GMCCs with no tillage and has resulted in the adoption of no tillage on more than 17 million hectares in 2002 (Derpsch and Benites 2003).

The minimum or no-tillage option should be included in research efforts on GMCC systems and its impacts on both soil productivity and labour use should be monitored over several years.

MINIMIZE INCENTIVES

Most commonly, incentives are created in the form of seed purchases—an understandable action at the beginning of diffusion efforts since, commonly, no GMCC seed is available through conventional seed markets. However, several of the case studies demonstrate the pitfall of such incentives: farmer 'adoption' is easy to achieve as long as the incentives last but when they are withdrawn, adoption quickly dwindles. Another case of incentives, little discussed in the case studies, is that of subsidies for production that achieves societal goals, such as reduction of erosion or decrease in CO_2 emissions or in inorganic nitrogen use.

Simple recommendations for the use of incentives and subsides are difficult to give but the experiences in this volume and others affirm that the context of the diffusion efforts should be as realistic as possible, that is, similar to the conditions that will exist when the project is no longer operating. In most cases, therefore, use of incentives is not recommended. An exception to this may be when the society is willing to pay long term for societal services, often of environmental type.

INVESTIGATE OPTIONS FOR DEGRADED SOILS

Relatively little attention has focused on GMCC species and systems that may help recover extremely degraded soils, although certainly their share of agricultural soils will further increase in the future. Such soils are characterized by low organic matter content (less than 1%, often as low as 0.2%) and poor biological activity, among others. Choice of GMCC species alone may go a long way towards the rehabilitation of such soils. A number of cases in this volume also mention the poor performance of *Mucuna pruriens* (L.) DC. in degraded soils. In such locations, alternative species should be tested, such as pigeon pea (*Cajanus cajan* [L.] Millsp.) or in some cases *Canavalia ensiformis* (L.) DC., in appropriate cropping sequences and tillage and fertilization regimes. An example of a successful rehabilitation of extremely degraded soils is from the departments of Paraguarí, Central, Cordillera and Guairá in Paraguay, where about 90% of soils are extremely degraded. To ensure that management changes with the changing soil fertility, the project recommended a cropping sequence that maximizes biomass production and thereby gives the farmer greater options with increasing fertility.

IMPROVE UPON MULTIPLE USES

Many of the popular GMCC species in the 1980s and 1990s either lacked or had limited uses as food and feed. These case studies, as well as many other experiences, have emphasized the importance of such uses, particularly in areas of extreme land scarcity and poverty. A well-known—in the minds of some, notorious—species is *Mucuna*, whose excellence as a GMCC generated a great deal of enthusiasm among research and development organizations in the 1990s. Diffusion efforts on *Mucuna* typically found initial farmer enthusiasm but no sustained adoption; most of those involved cited its perceived unsuitability as a food and feed and consequent lack of markets as the major reason for this. Coordinated research on *Mucuna*'s food and feed uses initiated in 2000 (Flores et al. 2002; Eilittä et al. 2003) but the food and/or feed potential of

a number of other similarly productive GMCCs, such as *Canavalia*, await the needed resources and research attention.

CONSIDER SOCIO-ECONOMIC IMPLICATIONS

A factor hindering improved understanding of many of the GMCC systems in this volume is the lack of socio-economic and adoption studies conducted on them. This seems to be particularly true with the earlier experiences, as the 1990s saw an increasing interest in these themes. The issue of trade-offs among the various uses becomes particularly relevant as more attention is given to the multiple uses of GMCCs.

CONCLUSION

GMCC systems are a promising option for the sustainable and productive management of soils in smallholder farms in tropical and subtropical regions. Field-level experiences in this volume include ones that have been clearly successful or indicate signs of having future promise, others that have been mixed and ones that have not led to improvement at farm level. It is hoped that lessons from them will help improve future research and diffusion efforts.

REFERENCES

Derpsch R, Benites J R (2003) Situation of conservation agriculture in the world. Paper presented at the II World Congress on Conservation Agriculture, 'Producing in harmony with nature,' 11-15 August 2003, Iguacu Falls, Paraná, Brazil. 10 p.

Eilittä M, Muinga R, Mureithi J, Sandoval C, Szabo N (Eds.) (2003) Increasing *Mucuna*'s potential as a food and feed crop. Proceedings of an international workshop, 23-26 September 2002, Mombasa, Kenya. Trop. Subtrop. Agroecosyst. 1:2-3. (Special issue) www.uady.mx/sitios/veterina/servicios/journal.

Flores M, Eilittä M, Myhrman R, Carew L, Carsky R (Eds.) (2002) *Mucuna* as a food and feed: Current uses and the way forward. Proceedings of a workshop, 26-29 April 2000, Tegucigalpa, Honduras. International Cover Crops Clearinghouse (CIDICCO), Center for Cover Crops Information and Seed Exchange in Africa (CIEPCA) and Judson College, Tegucigalpa, Honduras. 413 p. www.cidicco.hn.

Acronyms and Abbreviations Used

ACRONYMS

AAU	Action Aid Uganda
ACARESC	Associação de Crédito e Assitência Técnica e Extensão Rural de Santa Catarina, Brazil
ACCS	Asociación Costarricense de la Ciencia del Suelo, San José, Costa Rica
ACIAR	Australian Centre for International Agricultural Research, Australia
AFC	Agricultural Finance Corporation, Kenya
AFRENA	Agroforestry Research Network for Africa, Kenya
AFRNET	African Feed Resources Network, of ILCA, Ethiopia
ARDN	Adaptive Research and Dissemination Network for Agroforestry, Zimbabwe
ARI	Agricultural Research Institute
ARS	Agricultural Research Service, USDA, Washington D.C., USA
AS	Assessoria e Serviços, PTA, Rio de Janeiro, Brazil
ASP	Artisanal Seed Production, Tanzania
ASPRO	Agro Silvo Pastoral Project, Tanzania
BAAC	Bank of Agriculture and Agricultural Cooperatives, Thailand
CADESCA-CEE	Comité de Acción de Apoyo al Desarrollo Económico y Social de Centroamérica-Comunidad Económico Europea, Costa Rica
CARDER	Centre d'Action Régionale pour le Développement, Benin
CARS	Coast Agricultural Research Station of KARI, Kenya.
CATI	Coordenadoria de Assistência Técnica Integral, Brazil
CATIE	Centro Agronómico Tropical de Investigación y Enseñanza, Costa Rica
CEDECO	Corporacíon Educativa para el Desarrollo Costarricense, Costa Rica
CEPA-SC	Instituto de Planejamento e Economia Agrícola de Santa Catarina, Brazil
CGIAR	Consultative Group on International Agricultural Research, Washington D.C., USA
CIAT	Centro Internacional de Agricultura Tropical, Colombia

CIDA	Canadian International Development Agency, Canada
CIDICCO	Centro Internacional de Información sobre Cultivos de Cobertura, Honduras
CIEPCA	Centre d'Information et d'Echanges sur les Plantes de Couverture en Afrique, Benin
CIID	Centro Internacional de Investigaciones para el Desarrollo, Costa Rica
CIIFAD	Cornell International Institute for Food, Agriculture and Development of Cornell University, NY, USA
CIMMYT	Centro Internacional de Mejoramiento de Maíz y Trigo, Mexico
CIRAD	Centre de Coopération Internationale en Recherche Agronomique pour le Développement, France
CLUSA	Co-operative League of the USA, Washington D.C., USA
CNP	Consejo Nacional de la Producción, Costa Rica
CPAC	Centro de Pesquisa Agropecuária do Cerrados, of EMBRAPA, Brazil
CREDESA	Centre Régional pour le Développement et la Santé, Benin
CSO	Central Statistics Office, Zambia
CURLA	Centro Universitario Regional del Litoral Atlántico, Honduras
DACO	District Agricultural Coordinator, Zambia
DEAG	Dirección de Extensión Agraria, Paraguay
DECOTUX	Desarrollo Comunitario de Los Tuxtlas, Mexico
DGC	Directoria de Geociências, Brazil
DIA	Dirección de Investigación Agrícola, Paraguay
DLD	Department of Land Development, Thailand
DOA	Department of Agriculture, Thailand
DOAE	Department of Agricultural Extension, Thailand
DOL	Department of Livestock, Thailand
DSE	German Foundation for International Development, Germany
ECF	East Coast Fever
EMATER	Empresa de Assistência Técnica e Extansao Rural, Brazil
EMBRAPA	Empresa Brasileira de Pesquisa Agropecuária, Brazil
EPAGRI	Empresa de Pesquisa Agropecuária e difusão de tecnologia de Santa Catarina, Brazil
ETH	Eidgenössische Technische Hochschule, Switzerland
FAO	Food and Agriculture Organization of the United Nations, Italy

FEWS	Famine Early Warning System
GAP	Good Agricultural Practice Program of DOA, Thailand
Gorta	Freedom from Hunger Council of Ireland
GSB	Government Savings Bank, Thailand
GTZ	Deustsche Gesellschaft fur Technische Zusammen-arbeit, Germany
HIV-AIDS	Human Immunodeficiency Virus - Acquired Imm-unodeficiency Syndrome
HPI	Heifer Project International, AR, USA
IAEG	Impact Assessment and Evaluation Group, of the CGIAR, Washington D.C., USA
IAPAR	Instituto Agronômico do Parana, Brazil
IARC	International Agricultural Research Centre
IBGE	Instituto Brasileiro de Geografía e Estatísticas, Brazil
ICABE	International Conference on Agriculture and Biological Environment
ICIS	International Crop Information System
ICRAF	International Centre for Research in Agroforestry, now World Agroforestry Centre, Kenya
ICRISAT	International Crops Research Institute for the Semi-Arid Tropics, India
IDEAS	Instituto para el Desarrollo y Acción Social, Costa Rica
IDRC	International Development Research Centre, Canada
IFAD	International Fund for Agricultural Development, Italy
IFDC	International Fertilizer Development Council, AL, USA
IFPRI	International Food Policy Research Institute, Washington D.C., USA
IIA	International Institute of Agriculture, Italy
IICA	Instituto Interamericano de Cooperación para la Agricultura, Costa Rica
IITA	International Institute of Tropical Agriculture, Nigeria
ILCA	International Livestock Centre for Africa, Ethiopia, now ILRI, Kenya
ILDIS	International Legume Database and Information Service
ILRI	International Livestock Research Institute, Nairobi
INSURF	International Network on Soil Fertility and Sustainable Rice Farming, Thailand
IPGRI	International Plant Genetic Resources Institute, Italy

IRRI	International Rice Research Institute, the Philippines
ISNAR	International Service for National Agricultural Research, the Netherlands
ISRIC	International Soil Reference and Information Centre, Wageningen, the Netherlands
KARI	Kenya Agricultural Research Institute, Kenya
KIHATA	Kilimo Hai Tanzania, a national organization on organic farming, Tanzania
LRNP	Legume Research Network Project, of KARI, Kenya
LVC	Lake Victoria Crescent, Uganda
LWF	Lutheran World Federation
MAFE	Malawi Agroforestry Extension project, Malawi
MAFF	Ministry of Agriculture, Food and Fisheries, Zambia
MAG	Ministerio de Agricultura y Ganadería, Costa Rica and Paraguay
MCC	Multiple Cropping Centre, Chiang Mai University, Thailand
MERCOSUR	Mercado Común del Sur, the regional trading organization, South America
MISEREOR	Development agency of the German Catholic Church
MOA	Ministry of Agriculture, Kenya
MOAC	Ministry of Agriculture and Cooperatives, Thailand
MOAI	Ministry of Agriculture and Irrigation, Malawi
MoALDM	Ministry of Agriculture, Livestock Development, and Marketing, Kenya
MOIST	Management of Organic Inputs in Soils of the Tropics, Cornell University, NY, USA
MoLD	Ministry of Livestock Development and Marketing, now the Ministry of Agriculture and Rural Development, Kenya
MPTF	Maize Productivity Task Force, Malawi
MWK	Malawi Kwacha currency
NAFTA	North American Free Trade Agreement
NARO	National Agricultural Research Organization, Uganda
NDDP	National Dairy Development Project, Kenya
NRG	Natural Resources Group of CIMMYT, Mexico
NRI	Natural Resources Institute, UK
OAE	Office of Agricultural Economics, Thailand
OCEPAR	Organizacão das Cooperativas do Estado do Paraná, Brazil
OTS	Organization for Tropical Studies, Costa Rica

PCCMCA	Programa Cooperativo Centroamericano para el Mejoramiento de Cultivos Alimenticios, Costa Rica
PDBL	Proyecto de Desarrollo del Bosque Latifoliado, Honduras
PDEBE	Projet de Développement de l'Elevage dans le Borgou Est, Benin
PRIAG	Programa Régional de Reforzamiento a la Investigación Agronómica sobre los Granos en Centro América
PROCAMPO	El Programa de Apoyos Directos al Campo, Mexico
PROSCARP	Promotion of Soil Conservation and Rural Production project of NRI in Malawi
PSSM	Production System and Soil Management Research Group, CIAT, Colombia
PTA	Projetos de Tecnologias Alternatives, Brazil
RAMR	Recherche Appliquée en Milieu Réel, Benin
RCMP	Resource and Crop Management Program of IITA, Nigeria
RED	A network of organizations working in sustainable agriculture, Mexico
RRC	Regional Research Centre, Kenya
SAA-SC	Secretaria de Agricultura e Abastecimento de Santa Catarina, Brazil
SADC	Southern Africa Development Community
SCAFE	Soil Conservation and Agroforestry Extension Project, Zambia
SEMANARP	Secretaria del Medio Ambiente, Recursos Naturales, Agua y Pesca, Mexico
SG2000	Sasakawa-Global 2000
SIDA	Swedish International Development Agency, Sweden
SSSA	Soil Science Society of America
TROPSCORE	Tropical Soil Cover and Organic Resources Exchange, Cornell University, NY, USA
TSBF	Tropical Soils Biology and Fertility Program, Kenya
UCR	Universidad de Costa Rica, Costa Rica
UFSC	Universidad Federal de Santa Catarina, Brazil
UFSM	Universidade Federal de Santa Maria, Brazil
UMADEP	Uluguru Mountain Agricultural Development Project, Tanzania
UNAH	Universidad Nacional Autónoma de Honduras, Honduras
UNAM	Universidad Nacional Autónoma de México, Mexico

UNDP	United Nations Development Programme, NY, USA
UNEP	United Nations Environment Programme, Switzerland
UNESCO	United Nations Educational, Scientific, and Cultural Organization, France
USAID	United States Agency for International Development, Washington D.C., USA
USDA	United States Department of Agriculture
VAST	Vascular Tropicos database of the Missouri Botanical Gardens, MO, USA
WAFSRN	West African Farming Systems Research Network
WRI	World Resources Institute, Washington D.C., USA
ZARD	Zambia Agricultural Research Division

ABBREVIATIONS

AEZ	agroecological zone
C&D	characterization and diagnosis
CEC	cation exchange capacity
CT	conventional tillage
DAP	days after planting
DM	dry matter
ECEC	effective cation exchange capacity
ETP	evapotranspiration
FARMERS	recently developed by farmers
FOL	foliage
FPR	farmer participatory research
GIS	geographic information systems
GMCC	green manure/cover crop
HQ	headquarters
HYV	high yielding variety
INM	integrated nutrient management
LEISA	low-external-input and sustainable agriculture
LER	land equivalent ratio
N.d.	no data, not reported in surveys
NF	natural fallow
NGO	non-governmental organization
NOTH	nothing
NPV	net present value
OTH	other
P	precipitation
PLER	partial land equivalent ratio
REL	relay cropping

RES	researcher developed
RES+RES	researcher developed with follow-up
ROT	rotation
RP	rock phosphate
SE	seed
SIC	simultaneous intercropping
SOM	soil organic matter
T	temperature
TRAD	traditional system
TRAD+RES	traditional with research follow-up
WAP	weeks after planting
ZT	zero tillage